OPTOELECTRONICS
Infrared-Visible-Ultraviolet
Devices and Applications

SECOND EDITION

OPTICAL SCIENCE AND ENGINEERING

Founding Editor
Brian J. Thompson
University of Rochester
Rochester, New York

1. Electron and Ion Microscopy and Microanalysis: Principles and Applications, *Lawrence E. Murr*
2. Acousto-Optic Signal Processing: Theory and Implementation, *edited by Norman J. Berg and John N. Lee*
3. Electro-Optic and Acousto-Optic Scanning and Deflection, *Milton Gottlieb, Clive L. M. Ireland, and John Martin Ley*
4. Single-Mode Fiber Optics: Principles and Applications, *Luc B. Jeunhomme*
5. Pulse Code Formats for Fiber Optical Data Communication: Basic Principles and Applications, *David J. Morris*
6. Optical Materials: An Introduction to Selection and Application, *Solomon Musikant*
7. Infrared Methods for Gaseous Measurements: Theory and Practice, *edited by Joda Wormhoudt*
8. Laser Beam Scanning: Opto-Mechanical Devices, Systems, and Data Storage Optics, *edited by Gerald F. Marshall*
9. Opto-Mechanical Systems Design, *Paul R. Yoder, Jr.*
10. Optical Fiber Splices and Connectors: Theory and Methods, *Calvin M. Miller with Stephen C. Mettler and Ian A. White*
11. Laser Spectroscopy and Its Applications, *edited by Leon J. Radziemski, Richard W. Solarz, and Jeffrey A. Paisner*
12. Infrared Optoelectronics: Devices and Applications, *William Nunley and J. Scott Bechtel*
13. Integrated Optical Circuits and Components: Design and Applications, *edited by Lynn D. Hutcheson*
14. Handbook of Molecular Lasers, *edited by Peter K. Cheo*
15. Handbook of Optical Fibers and Cables, *Hiroshi Murata*
16. Acousto-Optics, *Adrian Korpel*
17. Procedures in Applied Optics, *John Strong*
18. Handbook of Solid-State Lasers, *edited by Peter K. Cheo*
19. Optical Computing: Digital and Symbolic, *edited by Raymond Arrathoon*
20. Laser Applications in Physical Chemistry, *edited by D. K. Evans*
21. Laser-Induced Plasmas and Applications, *edited by Leon J. Radziemski and David A. Cremers*
22. Infrared Technology Fundamentals, *Irving J. Spiro and Monroe Schlessinger*

23. Single-Mode Fiber Optics: Principles and Applications, Second Edition, Revised and Expanded, *Luc B. Jeunhomme*
24. Image Analysis Applications, *edited by Rangachar Kasturi and Mohan M. Trivedi*
25. Photoconductivity: Art, Science, and Technology, *N. V. Joshi*
26. Principles of Optical Circuit Engineering, *Mark A. Mentzer*
27. Lens Design, *Milton Laikin*
28. Optical Components, Systems, and Measurement Techniques, *Rajpal S. Sirohi and M. P. Kothiyal*
29. Electron and Ion Microscopy and Microanalysis: Principles and Applications, Second Edition, Revised and Expanded, *Lawrence E. Murr*
30. Handbook of Infrared Optical Materials, *edited by Paul Klocek*
31. Optical Scanning, *edited by Gerald F. Marshall*
32. Polymers for Lightwave and Integrated Optics: Technology and Applications, *edited by Lawrence A. Hornak*
33. Electro-Optical Displays, *edited by Mohammad A. Karim*
34. Mathematical Morphology in Image Processing, *edited by Edward R. Dougherty*
35. Opto-Mechanical Systems Design: Second Edition, Revised and Expanded, *Paul R. Yoder, Jr.*
36. Polarized Light: Fundamentals and Applications, *Edward Collett*
37. Rare Earth Doped Fiber Lasers and Amplifiers, *edited by Michel J. F. Digonnet*
38. Speckle Metrology, *edited by Rajpal S. Sirohi*
39. Organic Photoreceptors for Imaging Systems, *Paul M. Borsenberger and David S. Weiss*
40. Photonic Switching and Interconnects, *edited by Abdellatif Marrakchi*
41. Design and Fabrication of Acousto-Optic Devices, *edited by Akis P. Goutzoulis and Dennis R. Pape*
42. Digital Image Processing Methods, *edited by Edward R. Dougherty*
43. Visual Science and Engineering: Models and Applications, *edited by D. H. Kelly*
44. Handbook of Lens Design, *Daniel Malacara and Zacarias Malacara*
45. Photonic Devices and Systems, *edited by Robert G. Hunsberger*
46. Infrared Technology Fundamentals: Second Edition, Revised and Expanded, *edited by Monroe Schlessinger*
47. Spatial Light Modulator Technology: Materials, Devices, and Applications, *edited by Uzi Efron*
48. Lens Design: Second Edition, Revised and Expanded, *Milton Laikin*
49. Thin Films for Optical Systems, *edited by Francoise R. Flory*
50. Tunable Laser Applications, *edited by F. J. Duarte*
51. Acousto-Optic Signal Processing: Theory and Implementation, Second Edition, *edited by Norman J. Berg and John M. Pellegrino*
52. Handbook of Nonlinear Optics, *Richard L. Sutherland*
53. Handbook of Optical Fibers and Cables: Second Edition, *Hiroshi Murata*

54. Optical Storage and Retrieval: Memory, Neural Networks, and Fractals, *edited by Francis T. S. Yu and Suganda Jutamulia*
55. Devices for Optoelectronics, *Wallace B. Leigh*
56. Practical Design and Production of Optical Thin Films, *Ronald R. Willey*
57. Acousto-Optics: Second Edition, *Adrian Korpel*
58. Diffraction Gratings and Applications, *Erwin G. Loewen and Evgeny Popov*
59. Organic Photoreceptors for Xerography, *Paul M. Borsenberger and David S. Weiss*
60. Characterization Techniques and Tabulations for Organic Nonlinear Optical Materials, *edited by Mark G. Kuzyk and Carl W. Dirk*
61. Interferogram Analysis for Optical Testing, *Daniel Malacara, Manuel Servin, and Zacarias Malacara*
62. Computational Modeling of Vision: The Role of Combination, *William R. Uttal, Ramakrishna Kakarala, Spiram Dayanand, Thomas Shepherd, Jagadeesh Kalki, Charles F. Lunskis, Jr., and Ning Liu*
63. Microoptics Technology: Fabrication and Applications of Lens Arrays and Devices, *Nicholas Borrelli*
64. Visual Information Representation, Communication, and Image Processing, *edited by Chang Wen Chen and Ya-Qin Zhang*
65. Optical Methods of Measurement, *Rajpal S. Sirohi and F. S. Chau*
66. Integrated Optical Circuits and Components: Design and Applications, *edited by Edmond J. Murphy*
67. Adaptive Optics Engineering Handbook, *edited by Robert K. Tyson*
68. Entropy and Information Optics, *Francis T. S. Yu*
69. Computational Methods for Electromagnetic and Optical Systems, *John M. Jarem and Partha P. Banerjee*
70. Laser Beam Shaping, *Fred M. Dickey and Scott C. Holswade*
71. Rare-Earth-Doped Fiber Lasers and Amplifiers: Second Edition, Revised and Expanded, *edited by Michel J. F. Digonnet*
72. Lens Design: Third Edition, Revised and Expanded, *Milton Laikin*
73. Handbook of Optical Engineering, *edited by Daniel Malacara and Brian J. Thompson*
74. Handbook of Imaging Materials: Second Edition, Revised and Expanded, *edited by Arthur S. Diamond and David S. Weiss*
75. Handbook of Image Quality: Characterization and Prediction, *Brian W. Keelan*
76. Fiber Optic Sensors, *edited by Francis T. S. Yu and Shizhuo Yin*
77. Optical Switching/Networking and Computing for Multimedia Systems, *edited by Mohsen Guizani and Abdella Battou*
78. Image Recognition and Classification: Algorithms, Systems, and Applications, *edited by Bahram Javidi*
79. Practical Design and Production of Optical Thin Films: Second Edition, Revised and Expanded, *Ronald R. Willey*
80. Ultrafast Lasers: Technology and Applications, *edited by Martin E. Fermann, Almantas Galvanauskas, and Gregg Sucha*
81. Light Propagation in Periodic Media: Differential Theory and Design, *Michel Nevière and Evgeny Popov*

82. Handbook of Nonlinear Optics, Second Edition, Revised and Expanded, *Richard L. Sutherland*
83. Polarized Light: Second Edition, Revised and Expanded, *Dennis Goldstein*
84. Optical Remote Sensing: Science and Technology, *Walter Egan*
85. Handbook of Optical Design: Second Edition, *Daniel Malacara and Zacarias Malacara*
86. Nonlinear Optics: Theory, Numerical Modeling, and Applications, *Partha P. Banerjee*
87. Semiconductor and Metal Nanocrystals: Synthesis and Electronic and Optical Properties, *edited by Victor I. Klimov*
88. High-Performance Backbone Network Technology, *edited by Naoaki Yamanaka*
89. Semiconductor Laser Fundamentals, *Toshiaki Suhara*
90. Handbook of Optical and Laser Scanning, *edited by Gerald F. Marshall*
91. Organic Light-Emitting Diodes: Principles, Characteristics, and Processes, *Jan Kalinowski*
92. Micro-Optomechatronics, *Hiroshi Hosaka, Yoshitada Katagiri, Terunao Hirota, and Kiyoshi Itao*
93. Microoptics Technology: Second Edition, *Nicholas F. Borrelli*
94. Organic Electroluminescence, *edited by Zakya Kafafi*
95. Engineering Thin Films and Nanostructures with Ion Beams, *Emile Knystautas*
96. Interferogram Analysis for Optical Testing, Second Edition, *Daniel Malacara, Manuel Sercin, and Zacarias Malacara*
97. Laser Remote Sensing, *edited by Takashi Fujii and Tetsuo Fukuchi*
98. Passive Micro-Optical Alignment Methods, *edited by Robert A. Boudreau and Sharon M. Boudreau*
99. Organic Photovoltaics: Mechanism, Materials, and Devices, *edited by Sam-Shajing Sun and Niyazi Serdar Saracftci*
100. Handbook of Optical Interconnects, *edited by Shigeru Kawai*
101. GMPLS Technologies: Broadband Backbone Networks and Systems, *Naoaki Yamanaka, Kohei Shiomoto, and Eiji Oki*
102. Laser Beam Shaping Applications, *edited by Fred M. Dickey, Scott C. Holswade and David L. Shealy*
103. Electromagnetic Theory and Applications for Photonic Crystals, *Kiyotoshi Yasumoto*
104. Physics of Optoelectronics, *Michael A. Parker*
105. Opto-Mechanical Systems Design: Third Edition, *Paul R. Yoder, Jr.*
106. Color Desktop Printer Technology, *edited by Mitchell Rosen and Noboru Ohta*
107. Laser Safety Management, *Ken Barat*
108. Optics in Magnetic Multilayers and Nanostructures, *Štefan Višňovský*
109. Optical Inspection of Microsystems, *edited by Wolfgang Osten*
110. Applied Microphotonics, *edited by Wes R. Jamroz, Roman Kruzelecky, and Emile I. Haddad*
111. Organic Light-Emitting Materials and Devices, *edited by Zhigang Li and Hong Meng*
112. Silicon Nanoelectronics, *edited by Shunri Oda and David Ferry*

113. Image Sensors and Signal Processor for Digital Still Cameras, *Junichi Nakamura*
114. Encyclopedic Handbook of Integrated Circuits, *edited by Kenichi Iga and Yasuo Kokubun*
115. Quantum Communications and Cryptography, *edited by Alexander V. Sergienko*
116. Optical Code Division Multiple Access: Fundamentals and Applications, *edited by Paul R. Prucnal*
117. Polymer Fiber Optics: Materials, Physics, and Applications, *Mark G. Kuzyk*
118. Smart Biosensor Technology, *edited by George K. Knopf and Amarjeet S. Bassi*
119. Solid-State Lasers and Applications, *edited by Alphan Sennaroglu*
120. Optical Waveguides: From Theory to Applied Technologies, *edited by Maria L. Calvo and Vasudevan Lakshiminarayanan*
121. Gas Lasers, *edited by Masamori Endo and Robert F. Walker*
122. Lens Design, Fourth Edition, *Milton Laikin*
123. Photonics: Principles and Practices, *Abdul Al-Azzawi*
124. Microwave Photonics, *edited by Chi H. Lee*
125. Physical Properties and Data of Optical Materials, *Moriaki Wakaki, Keiei Kudo, and Takehisa Shibuya*
126. Microlithography: Science and Technology, Second Edition, *edited by Kazuaki Suzuki and Bruce W. Smith*
127. Coarse Wavelength Division Multiplexing: Technologies and Applications, *edited by Hans Joerg Thiele and Marcus Nebeling*
128. Organic Field-Effect Transistors, *Zhenan Bao and Jason Locklin*
129. Smart CMOS Image Sensors and Applications, *Jun Ohta*
130. Photonic Signal Processing: Techniques and Applications, *Le Nguyen Binh*
131. Terahertz Spectroscopy: Principles and Applications, *edited by Susan L. Dexheimer*
132. Fiber Optic Sensors, Second Edition, *edited by Shizhuo Yin, Paul B. Ruffin, and Francis T. S. Yu*
133. Introduction to Organic Electronic and Optoelectronic Materials and Devices, *edited by Sam-Shajing Sun and Larry R. Dalton*
134. Introduction to Nonimaging Optics, *Julio Chaves*
135. The Nature of Light: What Is a Photon?, *edited by Chandrasekhar Roychoudhuri, A. F. Kracklauer, and Katherine Creath*
136. Optical and Photonic MEMS Devices: Design, Fabrication and Control, *edited by Ai-Qun Liu*
137. Tunable Laser Applications, Second Edition, *edited by F. J. Duarte*
138. Biochemical Applications of Nonlinear Optical Spectroscopy, *edited by Vladislav Yakovlev*
139. Dynamic Laser Speckle and Applications, *edited by Hector J. Rabal and Roberto A. Braga Jr.*
140. Slow Light: Science and Applications, *edited by Jacob B. Khurgin and Rodney S. Tucker*
141. Laser Safety: Tools and Training, *edited by Ken Barat*
142. Near-Earth Laser Communications, *edited by Hamid Hemmati*

143. Polarimetric Radar Imaging: From Basics to Applications, *Jong-Sen Lee and Eric Pottier*
144. Photoacoustic Imaging and Spectroscopy, *edited by Lihong V. Wang*
145. Optoelectronics: Infrared-Visible-Ultraviolet Devices and Applications, Second Edition, *edited by Dave Birtalan and William Nunley*

OPTOELECTRONICS

Infrared-Visible-Ultraviolet Devices and Applications

SECOND EDITION

Edited by
DAVE BIRTALAN
WILLIAM NUNLEY

CRC Press is an imprint of the
Taylor & Francis Group, an **Informa** business

CRC Press
Taylor & Francis Group
6000 Broken Sound Parkway NW, Suite 300
Boca Raton, FL 33487-2742

© 2009 by Taylor & Francis Group, LLC
CRC Press is an imprint of Taylor & Francis Group, an Informa business

No claim to original U.S. Government works
Printed in the United States of America on acid-free paper
10 9 8 7 6 5 4 3 2 1

International Standard Book Number-13: 978-1-4200-6780-4 (Hardcover)

This book contains information obtained from authentic and highly regarded sources. Reasonable efforts have been made to publish reliable data and information, but the author and publisher cannot assume responsibility for the validity of all materials or the consequences of their use. The authors and publishers have attempted to trace the copyright holders of all material reproduced in this publication and apologize to copyright holders if permission to publish in this form has not been obtained. If any copyright material has not been acknowledged please write and let us know so we may rectify in any future reprint.

Except as permitted under U.S. Copyright Law, no part of this book may be reprinted, reproduced, transmitted, or utilized in any form by any electronic, mechanical, or other means, now known or hereafter invented, including photocopying, microfilming, and recording, or in any information storage or retrieval system, without written permission from the publishers.

For permission to photocopy or use material electronically from this work, please access www.copyright.com (http://www.copyright.com/) or contact the Copyright Clearance Center, Inc. (CCC), 222 Rosewood Drive, Danvers, MA 01923, 978-750-8400. CCC is a not-for-profit organization that provides licenses and registration for a variety of users. For organizations that have been granted a photocopy license by the CCC, a separate system of payment has been arranged.

Trademark Notice: Product or corporate names may be trademarks or registered trademarks, and are used only for identification and explanation without intent to infringe.

Visit the Taylor & Francis Web site at
http://www.taylorandfrancis.com

and the CRC Press Web site at
http://www.crcpress.com

Contents

Preface xix
The Authors xxi

Part 1   The Source (Infrared-Emitting Diodes)

Chapter 1
Basic Theory 3

1.1 Introduction 3
1.2 P-N Junction Injection Electroluminescence 3
1.3 Relative Efficiency 5

Chapter 2
Light-Emitting Diode Fabrication 7

2.1 Introduction: Epitaxial Growth 7
2.2 Liquid-Phase Epitaxy 7
    2.2.1 Gallium Arsenide 9
    2.2.2 Gallium Aluminum Arsenide 10
2.3 Metal Organic Chemical Vapor Deposition 13
2.4 Hydride Vapor-Phase Epitaxy 14
2.5 Molecular Beam Epitaxy 15

Chapter 3
IRLED Packaging 17

3.1 Introduction 17
3.2 Techniques for Improving Photon Emission Efficiency 19
3.3 Packaging the IRLED 21
3.4 Characterization of the Packaged IRLED 25
    3.4.1 Package Side Emission 28
    3.4.2 Package Height 29
    3.4.3 Lens Quality 30
    3.4.4 Chip Centering 31
    3.4.5 Heat Dissipation 32
    3.4.6 Reliability 32
    3.4.7 Storage Temperature Range 34
    3.4.8 Operating Temperature 34
    3.4.9 Thermal Shock 34
    3.4.10 Solvent Resistance 34

		3.4.11 Hermeticity	35
		3.4.12 Mechanical Shock and Vibration	35
		3.4.13 Optical Consideration	35
3.5		Understanding Thermal Impedance	39
		3.5.1 Thermal Impedance Calculations	39
3.6		Understanding the Measurement of Radiant Energy	46
		3.6.1 General Discussion	46
		3.6.2 Parameter Definitions and Measurement Techniques	51
3.7		Reliability	52
3.8		Conclusion	57

Part 2   *The Receiver (Silicon Photosensor)*

Chapter 4
The Photodiode ... 61

4.1	Basic Theory	61
	4.1.2 Photoelectric Effect	61
4.2	Optimization	64
4.3	Characterization	66

Chapter 5
The Phototransistor and Photodarlington ... 75

5.1	Basic Theory	75
	5.1.1 Phototransistor	77
	5.1.2 Photodarlington	78
	5.1.3 R_{BE} Phototransistor	79
5.2	Characterization	80
	5.2.1 Switching Characteristics	82
	5.2.2 Package Lens Effects	97

Chapter 6
The Photointegrated Circuit ... 99

6.1	Basic Theory	99
6.2	IC Processing Steps	101
	6.2.1 Starting Wafer	101
	6.2.2 Buried Layer	101
	6.2.3 Epi	102
	6.2.4 Isolation	102
	6.2.5 Deep N^+	102
	6.2.6 P^+	103
	6.2.7 N^+	104
	6.2.8 Resistors	104

Contents   xiii

      6.2.9   Contact 105
      6.2.10 Metal 105
      6.2.11 Passivation 106
      6.2.12 Backside Processing 106
6.3   Other IC Devices 107
      6.3.1   Photodiodes 108
      6.3.2   Lateral PNP 109
      6.3.3   Vertical PNP 110
      6.3.4   Capacitor 110
      6.3.5   Summary 110
6.4   Characterization 110

Chapter 7
Special-Function Photointegrated Circuits 115

7.1   Basic Theory 115
7.2   Triac Driver Photosensors 115
7.3   Synchronous Driver Detector (SDD) 118
7.4   Color Sensor 120

Part 3   *The Coupled Emitter (IRLED) Photosensor Pair*

Chapter 8
The Transmissive Optical Switch 125

8.1   Slotted (Transmissive) Optical Switch 125
8.2   Encoder Wheel Design 131
8.3   Performance Characteristics 131
8.4   Flag Switches 136
8.5   Slotted Switch Summary 136

Chapter 9
The Reflective Optical Switch 141

9.1   Electrical Considerations 141
9.2   Mechanical Considerations 151

Part 4   *The Optical Isolator and Solid-State Relay*

Chapter 10
Electrical Considerations 155

10.1 Background 155
10.2 Function 156

10.3	Different Types	160
10.4	Introduction to Solid-State Relays (SSR)	167
	10.4.1 Applications	168
10.5	Theory of Operation	168
10.6	DC Input–DC Output SSR Operation	169
10.7	DC Input–AC Output SSR Operation	170
10.8	AC Input–AC Output SSR Operation	171
10.9	Temperature Considerations	172
10.10	Overvoltage and Overcurrent Surge Protection	173
10.11	Zero Voltage Crossover	173
10.12	Conclusion	175

Chapter 11
Mechanical and Thermal Considerations 177

11.1	Mechanical Considerations	177
	11.1.1 Discrete Components	177
	11.1.2 Surface Mount Chip Carriers (SMCCs)	177
	11.1.3 SMCC Advantages	178
	11.1.4 The Emitter and Photosensor Assemblies	179
	11.1.5 Optoisolators	179
11.2	Managing Junction Temperature of High-Power Light-Emitting Diodes	181
	11.2.1 Reduced MTTF and Accelerated Degradation	182
	11.2.2 Generating Heat	182
	11.2.3 Removing Heat	182
	11.2.4 Thermal Equilibrium	183
	11.2.5 Analogy to Electrical Circuits	183
	11.2.6 Determining Junction Temperature from Forward Voltage	184
	11.2.7 Passive Thermal Management	184
	11.2.8 Active Thermal Management	185

Part 5 Open Air and Fiber-Optic Communication

Chapter 12
Fiber-Optic Communication 191

12.1	Basics	191
12.2	Advantages	195
	12.2.1 Less Expensive	195
	12.2.2 Small Diameter and Light Weight	196
	12.2.3 Higher Capacity	196

Contents   xv

      12.2.4   Nonconductivity 196
      12.2.5   Low Power 196
      12.2.6   Less Signal Loss 196
      12.2.7   Digital Signals 196
      12.2.8   Security 196

Chapter 13
Wireless Communication 197

13.1   Basic Theory 197
13.2   Background 200

Part 6   *IR Applications*

Chapter 14
Pulse Operation 205

14.1   The IRLED 205
14.2   The Photosensor 209

Chapter 15
Driving a Light-Emitting Device 211

15.1   Introduction 211
15.2   Variables 211
15.3   Linear Operation 213

Chapter 16
Interfacing to the Photosensor 215

16.1   The Photodiode 215
16.2   The Phototransistor and Photodarlington 219
         16.2.1   Basic Interface Circuits for CMOS 221
         16.2.2   Basic Interface Circuits for TTL 222
16.3   The Photointegrated Circuit 225

Chapter 17
Computer Peripheral and Business Equipment Applications 227

17.1   Copiers and Printers 227
17.2   Keyboards and Mice 227
17.3   Touch Screens 227
17.4   Data Interface 227
17.5   Check/Card Reader 229
17.6   Optical Couplers/Isolators 229

Chapter 18
Industrial Applications 231

18.1 Safety-Related Optical Sensors 231
18.2 Hazardous Fluid Sensing 232
18.3 Security and Surveillance Systems 235
18.4 Mechanical Aids, Robotics 236
18.5 Miscellaneous Optical Sensors 238
18.6 Optical Couplers/Isolators 238

Chapter 19
Automotive Applications 241

19.1 Existing Applications 241

Chapter 20
Military Applications 245

20.1 Military Applications for Optical Sensors 245

Chapter 21
Consumer Applications 249

21.1 TV and Game Controls 249
21.2 CD and DVD Discs 249
21.3 Dollar Bill Changers 250
21.4 Coin Changers 250
21.5 Smoke Detectors 251
21.6 Slot Machines 253
21.7 Camera Applications 254
21.8 Optical Golf Games 254
21.9 Household Appliance Controls 254

Chapter 22
Medical Applications 255

22.1 Introduction 255
22.2 Pill-Counting Systems 255
22.3 Electrical Isolation Systems 255
22.4 Infusion Pump Application 255
22.5 Hemodialysis Equipment Application 257
22.6 Fluid and Bubble Sensing for Medical Applications 257
22.7 Intravenous Drop Monitor 260
22.8 Pulse Rate Detection 260

Chapter 23
Telecommunications 263

23.1 Telecommunications Overview 263

Contents

Part 7 Visible and Ultraviolet Technologies

Chapter 24
Visible-Light-Emitting Diodes (VLEDs) 269

24.1 Introduction 269
24.2 What Is an LED? 269
    24.2.1 Chip (Die) 269
    24.2.2 Packaging 271
    24.2.3 Phosphor Coatings 272
    24.2.4 Secondary Optics 273
24.3 Generating Light 273
24.4 Optical, Thermal, and Electrical Measurements 274
    24.4.1 Measurement of Parameters 274
24.5 Laboratory Testing 279
24.6 Reliability 283
    24.6.1 Lifetime and Operation of LEDs 283
    24.6.2 Factors Affecting LED Lifetime 284
24.7 Applications 287

Chapter 25
Ultraviolet Electromagnetic Radiation 289

25.1 Overview 289
25.2 The UV Spectrum 289
25.3 Fluorescent Lamp 291
25.4 Types of UV Lamps 292
25.5 UV Lamp Aging 294
25.6 RoHS and WEEE Concerns for UV Lamps 295
25.7 Failure Modes in UV Gas Discharge Lamps 295
25.8 Overview of UV Applications 296
25.9 Germicidal Effects of UV Light 297
25.10 Ultraviolet LED 303
25.11 UV Packaging Technology 305
25.12 Currency and Document Validation Applications 309
25.13 Industrial UV Curing Applications 310
25.14 Air–Water–Surface Disinfection 312
25.15 Medical and Forensic Applications 313

Bibliography 315

Glossary 317

Index 327

Preface

The second edition updates all the comprehensive information and illustrations regarding infrared semiconductor optoelectronics provided in the first edition. The same easy-to-read format developed in the first edition was again utilized, and this provides engineers a primary introductory reference guide for LED light sources and silicon photosensors. The updates include epitaxial growth techniques, IC process flows, packaging, new devices, applications and solid-state relays. We have also added two entirely new chapters for both the visible and ultraviolet portion of electromagnetic radiation spectrum, which design engineers will find very valuable.

Since the first edition was published in 1987, both the significance of optoelectronics in our everyday lives and the market size for such devices has grown considerably. The market size for optoelectronic devices is now measured in billions of dollars and will continue to grow even larger in the coming years. A major market driver in the future will be the conversion of traditional light bulbs to visible LED lighting to gain longer operating life, higher efficiencies, and a significant reduction of mercury released into the environment. Governing bodies around the world are legislating new lighting efficiency standards that will drive the next round of growth. The commercially emerging ultraviolet (UV) LED market is beginning to make inroads into applications monopolized by traditional low- and medium-pressure lamps.

The authors wish to thank Reyne Parks for her patience and assistance in the helping to organize the manuscript material. A special note of thanks goes to Gary Clements for all the wonderful illustrations he created for the second edition of this book. The authors also wish to express our deepest gratitude to the technical staff at Optek Technology/TT electronics—comprising Jim Plaster, Richard Saffa, Bob Procsal, Larry Overholtzer, and Walter Garcia Brooks—for their technical contributions in the second edition along with all the contributors for the first edition. It is with our sincerest wish that the information presented will provide design engineers with practical and useful guide in their engineering endeavors.

Dave Birtalan
William Nunley

The Authors

Dave Birtalan began his career at the General Electric's Semiconductor Division in 1979 where he held various sales, marketing, product management/development, and engineering positions, most notably as a laser diode engineer associated with the SDI program, before joining a joint semiconductor venture between General Electric and Mitsubishi. In addition to lasers and LEDs, the author has been involved with RF transistor, MOSFET, and power semiconductor technologies. His various optoelectronic industry leadership positions include director of Vishay Telefunken–Americas. More recently he joined Optek Technology/TT electronics as vice president of Component BU in 2005. Dave received his bachelor of science degree in electrical engineering from Pennsylvania State University in 1979 and conducted his graduate studies at Syracuse University.

William Nunley is an innovator in the semiconductor field, where he worked from 1957 until his retirement in 1993. This encompassed the primary growth period of transistors, integrated circuits, and related components including LEDs and photodiodes. He was involved in production engineering, quality control, development engineering, application engineering, technical sales, marketing, and related management positions. He is the author of numerous technical papers, application notes, and articles. A recognized authority on optoelectronics, he has conducted numerous seminars on this subject. His bachelor of science degree in electrical engineering was obtained from Vanderbilt University in 1953.

Part 1

The Source
(Infrared-Emitting Diodes)

1 Basic Theory

1.1 INTRODUCTION

This chapter is devoted to developing a basic understanding of the transmitting portion of the optical switch. The transmitting portion of the optical switch is considered the light source and typically takes the form of an infrared light-emitting diode. Other terms sometimes used by engineers to refer to these devices include the following: *emitters*, *IR emitters*, *LED*, *IRED*, or *IRLED*. Although all terms refer to essentially the same device, this book uses infrared-emitting diode and IRLED for clarity. The reader will understand how energy is transformed from current to infrared energy, the criteria used to select the materials normally used to generate this energy, how the infrared-emitting diode is made, how the emitting wavelength is controlled, and how the energy is emitted. The discussion will not delve deeply into semiconductor physics, because only a conceptual understanding of the element is required for successful use of the device in applications, which is the major goal of this reference book.

1.2 P-N JUNCTION INJECTION ELECTROLUMINESCENCE

A P-N junction can be formed in a semiconductor material by doping one region with donor atoms and an adjacent region with acceptor atoms. This produces a nonuniform distribution of "impurities" with an abrupt change from one type of doped material to another. When the donor atoms dominate, the material is known as *N-type* and there is an excess of N-type atoms or majority carriers. For N-type material, the majority carriers are negatively charged and are electrons. When the acceptor atoms dominate, the material is known as *P-type*; the majority carriers are positively charged and are called *holes*. Figure 1.1 shows pictorially the P-N junction with the excess of negatively charged electrons and the positively charged holes.

The P-N junction is characterized by an abrupt change from material containing a majority of positively charged carriers or holes to material containing an excess of negatively charged carriers or electrons. When the junction is formed, these electrons and holes will flow in opposite directions across the junction (without applied bias) until equilibrium is reached. This gives rise to a built-in potential barrier. If an external electrical bias is applied across the junction that counteracts this built-in potential, additional electrons and holes will be injected or flow across the P-N junction. These carriers will then recombine by either a *radiative* or *nonradiative* process. In the case of a radiative process, the recombination requires that the energy level of the electron drop to facilitate recombination with a hole. The electron sheds this excess energy in the form of a discrete amount of energy known as a *photon* (Figure 1.2). In a nonradiative process, excess energy is still released but takes the form of heat and is quantified as *phonons*. In both cases, this recombination occurs in close proximity to the P-N junction.

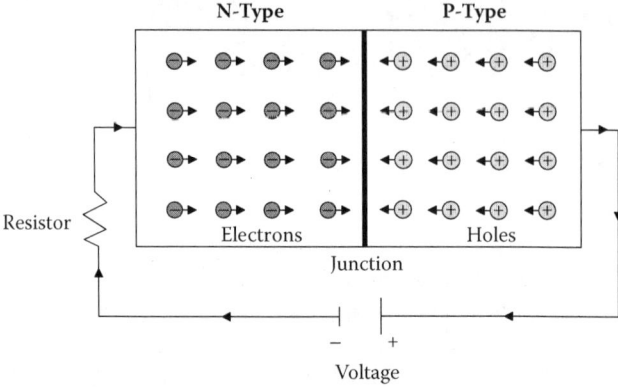

FIGURE 1.1 The P-N junction.

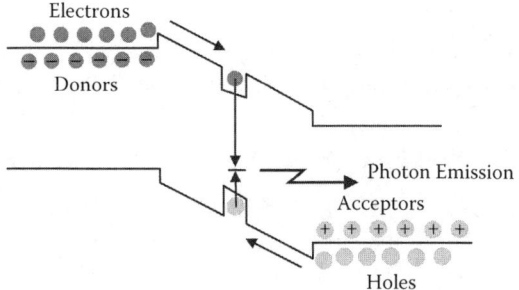

FIGURE 1.2 P-N Junction under forward bias.

The maximum possible energy of the emitted photons is determined by the band gap energy of the solid in which the P-N junction is formed. There are numerous elements and elemental compounds that have band gap energies that lie in the region from ultraviolet to infrared. Table 1.1 lists several materials utilized for commercial devices.

The band gap energy in Table 1.1 is related to emission wavelength according to this equation:

$$E_g = \frac{hc}{\lambda} = \frac{1240\,\text{eV} - \text{nm}}{\lambda\,\text{nm}}$$

where
E is the energy transition in electron volts
h = Planck's constant = 4.135×10^{-15} eV-s
c = speed of light = 2.998×10^8 m/s
λ = wavelength in nanometers (nm)

Example: GaAs; $E = 1240/930 = 1.33$ eV.

TABLE 1.1
Some of the Materials Utilized for LED Devices

Material Composition	Formula	Band Gap[a] Energy Electron Volts (eV)	Emission Wavelength Nanometers (nm)
Aluminum nitride	AlN	6.20	200
Gallium nitride	GaN	3.44	360
Gallium phosphide	GaP	2.26	550
Aluminum arsenide	AlAs	2.16	574
Gallium arsenide	GaAs	1.33	930
Indium phosphide	InP	1.35	918
Gallium aluminum nitride	GaAlN	4.96–2.97	250–460
Gallium arsenide phosphide	GaAsP	2.12–1.88	585–660
Aluminum gallium indium phosphide	AlGaInP	2.21–1.92	560–644
Gallium aluminum arsenide	GaAlAs	1.94–1.39	640–890
Indium gallium arsenide phosphide	InGaAsP	1.13–0.74	1100–1670

[a] The band gap, and therefore the wavelength, is affected by impurity concentrations and also by temperature. These are typical calculations for 300 K.

1.3 RELATIVE EFFICIENCY

The percentage of the current that results in recombinations giving rise to photons of the desired wavelength is a measure of the internal conversion efficiency of the P-N diode. A material with low internal conversion efficiency would offer little practical interest as an electroluminescent device. However, even a material with high internal conversion efficiency may not be useful if the emitted photons cannot be efficiently "emitted" from the diode structure or "coupled" to the external environment. Two major factors control the internal-to-external coupling coefficient. One factor is the opacity of the diode material. This is directly related to the reabsorption of emitted photons. The second factor is the internal reflection at the interface of the semiconductor crystal and the encapsulation material. This may cause the photon to be reflected back into the crystal and subsequently reabsorbed. This efficiency increases if there is a decrease in the crystal's probability of reabsorbing or reflecting back photons emitted at the junction. If more photons escape from the diode material, the relative output efficiency increases.

It becomes very obvious that the GaAlAs and GaAs IRLEDs are preferred transmitters because of the efficiency with which they emit infrared energy. However, to be used commercially, it is necessary to have detectors available that respond to the same wavelength of infrared. In other words, the spectral response of the detector must be a reasonable match to the emission wavelength of the material chosen for the IRLED. Silicon is the material normally utilized because its peak absorption falls in the 750 to 950 nm range. This will be discussed in more detail in Part 2, which deals with the photosensor.

The selection of GaAs and GaAlAs for use in optical switches is primarily because of improved efficiency (conversion of input power to usable output energy); however, its choice also depends on the availability of photosensors capable of detecting the wavelengths emitted.

2 Light-Emitting Diode Fabrication

2.1 INTRODUCTION: EPITAXIAL GROWTH

Epitaxy comes from the Greek words *epi,* meaning "on," "to," or "above," and *taxis,* meaning "order" or "ordering." Atoms are deposited on top of a substrate in an ordered fashion. The atoms form a crystal layer that has similar characteristics as the structure of the substrate. The purpose of epitaxy is to create a thin high-quality layer or layers of different composition such as abrupt P-N junctions or heterojunction interfaces. Different techniques for doing this have been developed, and in the following sections, a brief description will be given for a few of the more important epitaxial growth techniques.

2.2 LIQUID-PHASE EPITAXY

The epitaxial technique adds more semiconductor material to the material originally present, commonly referred to as the *substrate*. This growth or addition of semiconductor material having the same crystalline structure as the original material allows flexibility in both how the donor or acceptor atoms are introduced and what their concentration profile looks like.

One of the epitaxial techniques is described as *liquid-phase epitaxy,* which means the growth occurs from a liquid solution source located adjacent to the semiconductor material. This is accomplished by using what is commonly called a *graphite slider boat* positioned in a liquid-phase epitaxial reactor.

There are many, very different designs that have been developed over the years by various companies, but the basic principles are similar. Figure 2.1 shows a basic system. It consists of a quartz tube positioned in a furnace that is capable of precise temperature control. The chemical elements used in this process as well as the graphite boat are adversely affected by the presence of oxygen. The quartz tube and all the gas-handling equipment must be manufactured in such a way as to eliminate oxygen from the system. The gas used during the process is high-purity hydrogen. Prior to applying heat to the system, it is purged with nitrogen, or perhaps argon, to remove any air that is introduced into the system during loading of the components.

The graphite slider boat is loaded with gallium arsenide substrates, gallium, and the appropriate dopants for the material that is to be deposited. Figure 2.2 shows a cross-sectional view of a simplified portion of a graphite slider boat. The substrate is placed in a slot that is on the top surface of a long flat plate called the *slider*. It is used to move the substrate from one bin (or compartment) to another during the deposition process. Gallium, gallium arsenide, and the dopants are placed in the deposition bins.

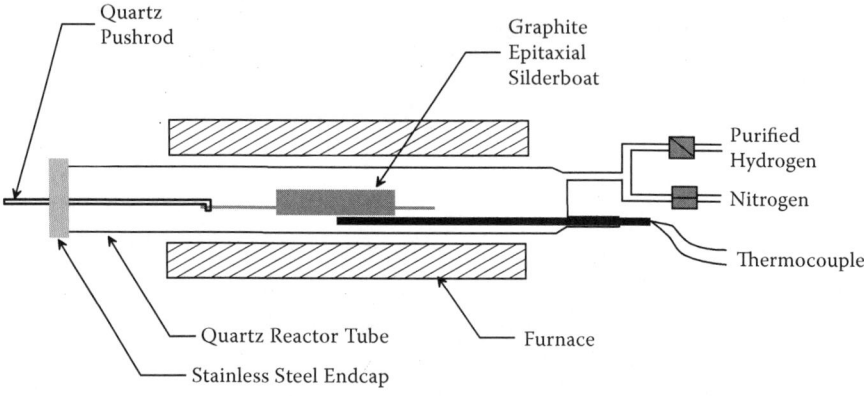

FIGURE 2.1 Liquid-phase epitaxial reactor.

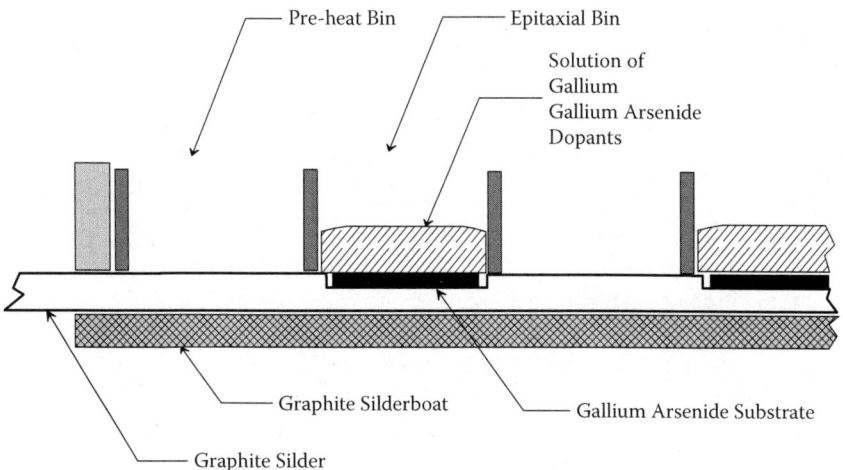

FIGURE 2.2 Liquid-phase epitaxial deposition slider boat.

The fully loaded slider boat is then placed into the quartz tube, purged with the gases, and heated to the appropriate growth temperature, which is typically between 800°C and 950°C. The system is allowed enough time for the temperature to stabilize and for all the solution materials to become mixed uniformly.

The graphite slider is then moved in such a way as to reposition the substrate under the solution. The growth process is started by lowering the temperature at the rate of a few tenths of a degree a minute. As the temperature is lowered, the solution becomes supersaturated and, in order to maintain equilibrium, the gallium arsenide and dopants start depositing on the gallium substrate. The process for growing the 880 nm and 930 nm IRLEDs discussed in this chapter is to allow the substrate and solution to cool all the way to room temperature. This results in a single epitaxial layer with a P-N junction formed during the growth process.

Light-Emitting Diode Fabrication

Other types of structures that include several layers of slightly different material can be grown with this technique by modifying the slider boat and using different solution compositions.

2.2.1 Gallium Arsenide

An early commercially successful product was developed using silicon as a dopant. Silicon has the unusual capability of doping gallium arsenide and gallium aluminum arsenide, both N-type and P-type, depending on the growth conditions.

For instance, a single-crystal GaAs substrate, which has majority donor or N-type impurity atoms, is placed in a graphite carrier and raised to a temperature in excess of 920°C. The melt, a solution containing gallium, silicon impurities, and polycrystalline GaAs, is then placed on the substrate and the epitaxial layer of GaAs starts to grow. The temperature is gradually reduced. As growth of the epitaxial layer continues, the silicon impurity behaves as a donor (N-type). At the higher initial growth temperatures, silicon replaces gallium atoms in the crystal structure, which results in a more N-type material. This N-type concentration will gradually decrease. The silicon atoms start to replace arsenic atoms instead of gallium atoms, resulting in a more P-type material. At approximately 900°C, the donor (N-type) atoms and the acceptor (P-type) atoms become equal in concentration and the junction is formed.

The temperature continues to decrease, and the concentration of the acceptor (P-type) atoms increases. Figure 2.3 shows these changes graphically.

The starting substrate, which was approximately 14 mils thick, has been increased in thickness to approximately 20 mils by the epitaxial growth. The junction is approximately 4 mils down from the top. Figure 2.2 shows a cross section of these layers, including the profile after removal of the excess starting material. Approximately 12 mils of the original starting substrate is removed by a mechanical abrading process commonly called *lapping*.

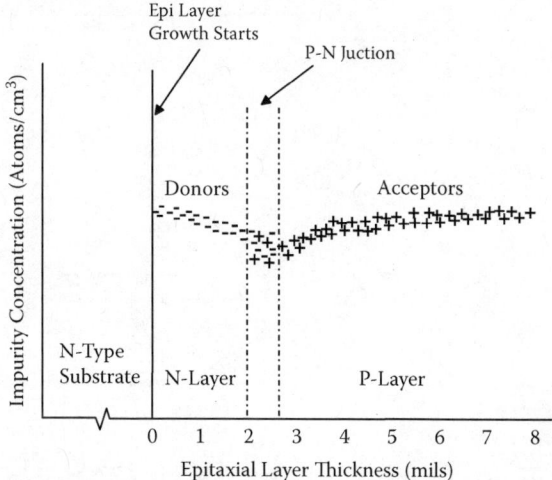

FIGURE 2.3 Impurity profile of GaAs epitaxial layers grown by liquid-phase epitaxy.

The final processing step is to add metal contacts to the P-type side and N-type side of the slice, prior to separation into individual IRLEDs. The separation process is generally performed by high-speed saws that use precision saw blades coated with diamond particles that are only a few micrometers in size.

2.2.2 Gallium Aluminum Arsenide

GaAlAs IRLEDs are made in a similar manner. The process requires more stringent controls because there is an additional variable of the aluminum added to the melt or solution source. The growth process is essentially the same with minor changes to the starting temperature. As the temperature is decreased, the epitaxial layer first grows N-type GaAlAs. The aluminum that forms in the crystalline structure has the heaviest concentration at the initial starting crystal–growth interface and decreases along a logarithmic curve. The doping level is controlled to be approximately 5% by volume at the P-N junction.

The doping level of the aluminum atoms at the junction controls the wavelength emitted. Increasing it will shorten the wavelength, and decreasing it will lengthen the wavelength (variation in emitted wavelength can change from approximately 800 to 930 nm). The 5% aluminum concentration by volume gives emission at 880 nm. This provides a close spectral match to the silicon phototransistor discussed in Part 2. Figure 2.5 shows the same type of concentration profile that was shown in the section on GaAs (Figure 2.3) but with the addition of the aluminum.

The starting substrate, which was approximately 14 mils thick, has been increased in thickness to approximately 22 mils by the epitaxial growth. The junction is 2 to 3 mils from the top surface of the IRLED chip. Figure 2.4 shows the

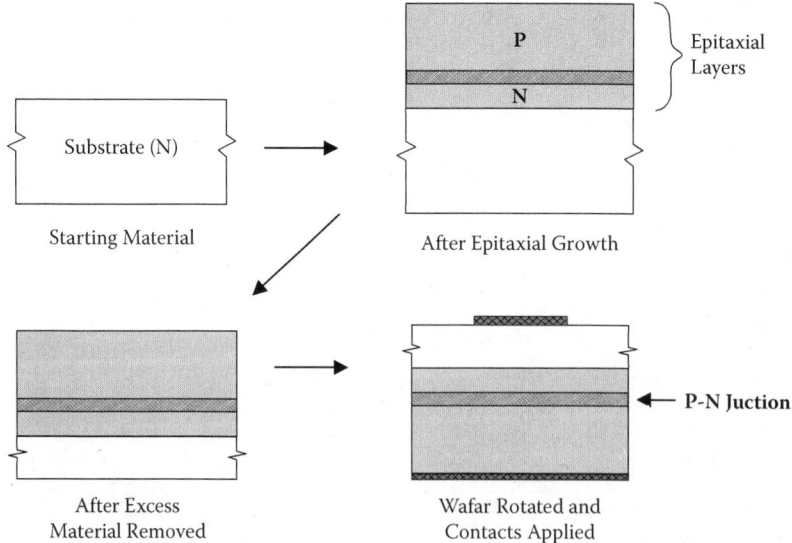

FIGURE 2.4 GaAs slice with liquid-phase-grown GaAs epitaxial layer. N-type GaAs is used as the substrate to grow additional N-type material transitioning to P-type material to form the P-N junction.

Light-Emitting Diode Fabrication

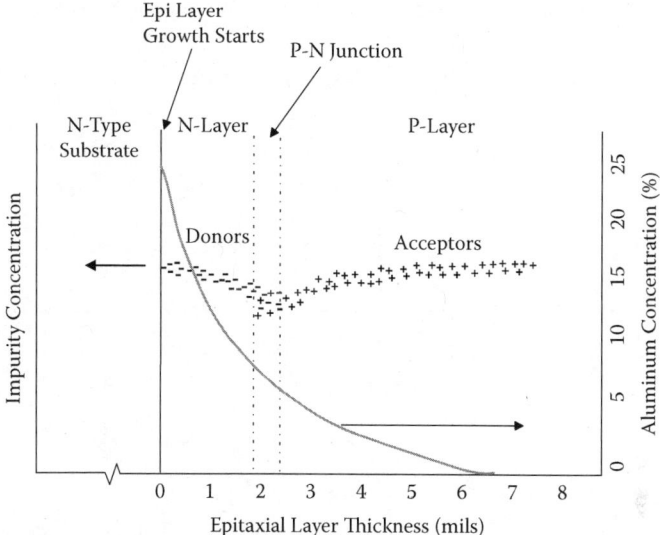

FIGURE 2.5 Impurity profile of gallium aluminum arsenide epitaxial layers doped with silicon.

profile of these layers, including the profile after removal of the substrate. The addition of the aluminum to the grown N-type region allows the use of an etchant for substrate removal. The etchant (a mixture of hydrogen peroxide and ammonium hydroxide) selectively attacks the GaAs material but not the material containing aluminum.

The three major differences between GaAs and GaAlAs are shown in Table 2.1. The epitaxial growth stage for GaAlAs diodes requires more stringent controls of the total growth phase. This mandates increased processing steps and more sophistication in the process and equipment. The excess material removal of the starting GaAs substrate is less difficult with GaAlAs, because selective etching can be used to remove the GaAs substrate. Mechanical abrasion, which is a more expensive process, is required to remove excess material on the GaAs wafers (Figure 2.6).

After removal of the excess material, metallization is applied to the complete P-type side. This side will subsequently become the mounting surface of the chip. The N-type side then becomes the topside and has only a localized contact added. This provides the necessary electrical contact but minimizes the area covered to allow the maximum amount of infrared energy to escape. Some manufacturers use the reverse process, in which the N-type side is mounted down rather than the P-type side. Figure 2.7 is an example of two of these chips, along with a cross section showing the location of the junction. The small chips are 0.011 in. long × 0.011 in. wide × 0.008 in. high, whereas the larger chips are 0.016 in. long × 0.016 in. wide × 0.008 in. high.

One of the characteristics of these chips is that a majority of the light is emitted from the sides of the chips. For this reason, most of the time these chips are mounted

TABLE 2.1
Major Differences between GaAs and GaAlAs

	Epitaxial Growth	Excess Material Removal	Doping
GaAs	Relatively simple	Mechanical thinning of substrate	Relatively simple
GaAlAs	More complex	Complete removal of substrate	More complex

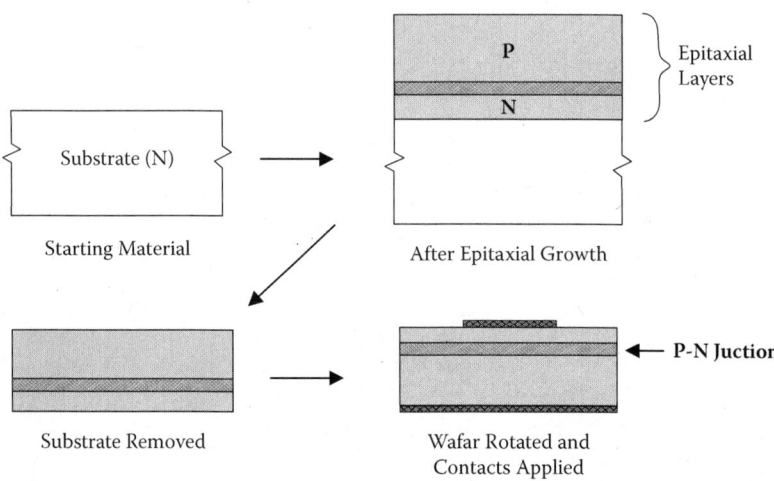

FIGURE 2.6 Major process steps of gallium aluminum arsenide wafer grown by liquid-phase epitaxy.

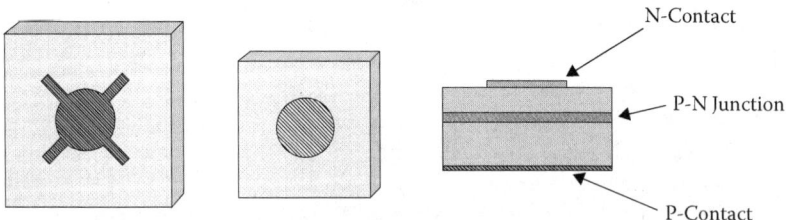

FIGURE 2.7 GaAs and GaAlAs chips. The chip size is directly related to its maximum current rating. The contact pad design and size affect current and forward voltage ratings.

on headers or lead frames that have cups with reflective, slanted sides. This allows most of the light to be emitted in the same direction. Figure 2.8 illustrates this for the gallium arsenide IRLED, and Figure 2.9 shows the structure for the gallium aluminum arsenide IRLED.

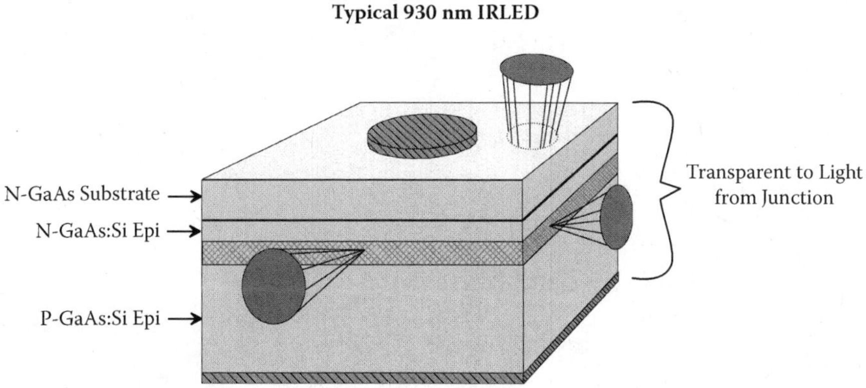

Light-Emitting Diode Fabrication

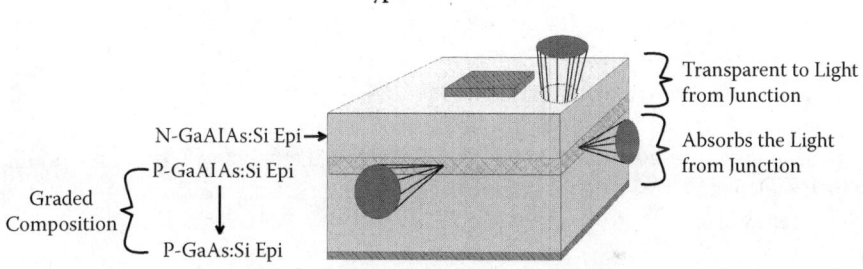

FIGURE 2.8 Gallium arsenide IRLED light emission.

FIGURE 2.9 Gallium aluminum arsenide IRLED light emission.

2.3 METAL ORGANIC CHEMICAL VAPOR DEPOSITION

MOCVD stands for Metal Organic Chemical Vapor Deposition. This is another technique for depositing thin layers of atoms onto a semiconductor wafer. Using MOCVD, you can build up many layers, each with a precisely controlled thickness, to create a material that has specific optical and electrical properties. It has become the dominant process for the manufacture of material for light-emitting diodes and laser diodes.

The process takes place in a reactor chamber that may be operated at atmospheric or reduced pressure. A mixture of gases that contain metal organic precursors flows through the chamber, where the substrates have been heated to a temperature of 600 to 800°C for most III–V materials and as high as 1200°C for gallium nitride. The reactive gases decompose and react at the heated surface of the substrate. Very thin epitaxial layers can be grown with this process, and layers of different composition are possible by varying the types of materials that are used.

The reactor, illustrated in Figure 2.10, is made of materials, such as stainless steel alloys, that do not react to the chemicals being used. The chamber must also be able to withstand high temperatures. The metal organic compounds are introduced into

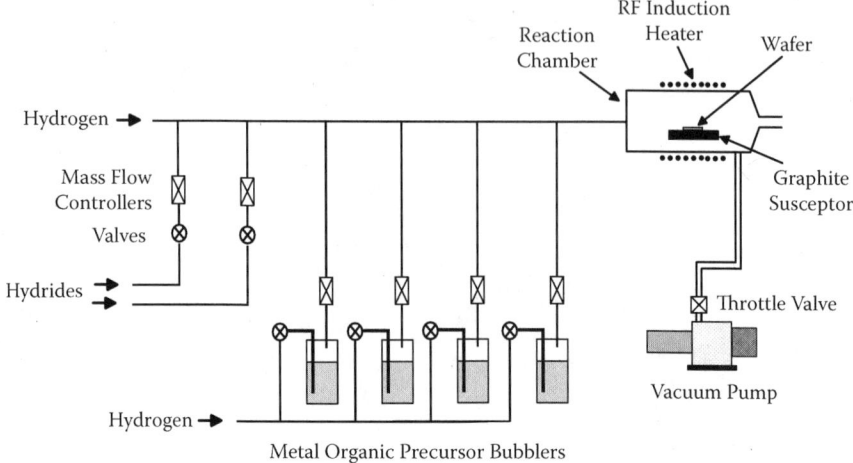

FIGURE 2.10 Schematic drawing of a metal organic vapor-phase reactor.

the system via devices known as *bubblers*. A carrier gas (usually nitrogen or hydrogen) is bubbled through the metal-organic liquid, which picks up some of the metal-organic material and transports it into the reactor. The amount that is transported depends on the gas flow rate and the temperature of the bubbler. The entire system, which includes valves, switches, and various monitors, is carefully controlled by a computer system.

2.4 HYDRIDE VAPOR-PHASE EPITAXY

HVPE technology has been a successful technique employed to produce epitaxial material for commercial applications for many years. In the late 1960s, it was used to produce gallium arsenide phosphide, which was processed into red visible LEDs. These early diodes were used as digital displays in calculators, watches, and a variety of other applications.

Recently, there has been a renewed interest in this technology for producing Group III Nitride substrates and device structures for ultraviolet- and blue-light-emitting devices. One approach to using this technique to produce epitaxial layers is illustrated in Figure 2.11.

The metals comprising the III–V layers are transported as gaseous metal halides to the reaction zone of the reactor. Accordingly, gallium, aluminum, and indium metals are used as source material. Hydrogen chloride gas flows over the heated metals, forming gallium chloride, etc. The metal halides are transported to the reaction chamber, where they combine with ammonia gas and the appropriate dopants to form layers of epitaxial material on the heated substrate.

It offers several advantages over the more commonly used MOCVD process. The equipment to produce this type of material is simpler and therefore less expensive to set up and operate. Another advantage is that the growth rate is inherently greater, and thicker layers become more economical. However, it is more difficult to produce

Light-Emitting Diode Fabrication

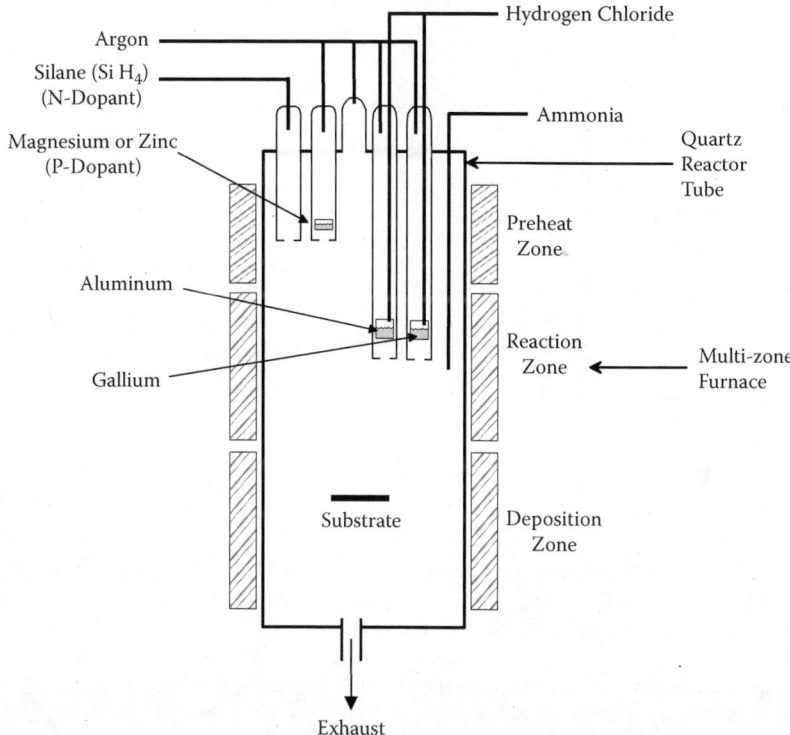

FIGURE 2.11 Hydride vapor-phase reactor.

the more complex structures with thin, multiple layers that are required for most of the high-brightness visible light-emitting diodes and lasers.

2.5 MOLECULAR BEAM EPITAXY

Molecular beam epitaxy (MBE) is another technique for depositing well-structured layers on high-purity substrates. The principle of MBE is based on the evaporation of ultrapure elements of the structure to be grown. This process takes place in a furnace under ultrahigh vacuum (where the pressure can be as low as 10^{-8} Pa) in order to create a pure, high-quality layer or layers of material. Figure 2.12 shows a sketch of a generic MBE system. One of the key aspects of MBE is its slow deposition rate (1 to 300 nM per minute), which allows the films to grow epitaxially. However, this slow growth rate requires proportionately better vacuum to achieve the low impurity rates necessary for high-quality material preparation.

One or more thermal beams of atoms or molecules react on the surface of a single-crystal wafer placed in a high-temperature furnace (several hundred degrees Centigrade). Thus, it becomes possible to stack ultrathin layers that are composed of a thickness of several atoms. The elements are evaporated or sublimated from ultrapure source material that is placed in an effusion cell. These cells are enclosures

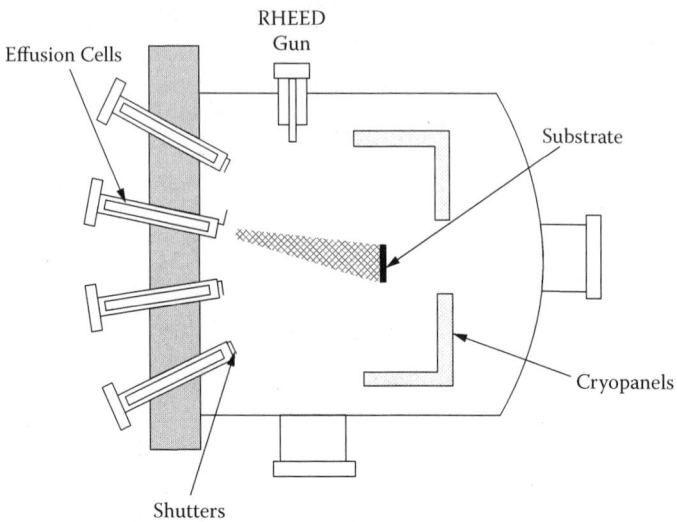

FIGURE 2.12 Basic components of a molecular beam reactor. A typical system has multiple analytic instruments and can be the most complex and expensive of all the different techniques for growing epitaxial material.

where molecular flux moves from a region with a given pressure to another region of lower pressure.

A range of structural and analytical probes can monitor film growth in situ and in real time, using surface quality analysis and grazing angle phase transitions by LEED (low-energy electron diffraction), or RHEED (reflection high-energy electron diffraction). Various spectroscopic methods are also used, including Auger electron spectroscopy, SIMS (secondary ion mass spectrometry), XPS (X-ray photoelectron spectrometry), or UPS (ultraviolet photoelectron spectrometry). The ultrahigh vacuum necessary for this process relies on several techniques using ion pumps, cryopumps, titanium sublimation pumps, along with diffusion pumps or turbomolecular pumps.

3 IRLED Packaging

3.1 INTRODUCTION

In Chapter 1, Section 1.2, the mechanism of photon formation was described, and in Section 1.3 the emission of these photons was described. A brief review will clarify this mechanism. Figure 3.1 shows the bias arrangement for forward conduction of a P-N junction.

Figure 3.2 optically models this recombination as a point source. It should be noted that the planes formed across the cross section parallel to the junction and within a few diffusion lengths of the junction will contain a very large number of these point sources emitting energy. By reviewing a single point source, the overall mechanism can be understood.

The ability of the photons that are generated to escape into the outside world is affected by the index of refraction of the various materials that are used in the construction of light-emitting devices. The light coming from the junction first must exit the chip itself. One of the controlling factors is the critical angle, which defines the angle at which there is total internal reflection (see Figure 3.3). The critical angle is determined by the formula:

$$\text{SIN } \theta_C = \frac{n_2}{n_1} \tag{3.1}$$

where n_2 is the index of refraction of air, or 1, and n_1 is the index of refraction of the chip material.

With GaAs, $n_1 = 3.6$:

$$\text{SIN } \theta_C = \frac{1}{3.6} \quad \theta_C = 16° \tag{3.2}$$

With GaAlAs, $n_1 = 3.4$:

$$\text{SIN } \theta_C = \frac{1}{3.4} \quad \theta_C = 17° \tag{3.3}$$

At angles less than the critical angle, there is partial reflection. At angles greater than the critical angle, there is total internal reflection.

Some of the photons pass vertically through the n_1/n_2 (IRLED/outside world) interface. n_1 is the normal method of referring to material; the energy is moving "from"

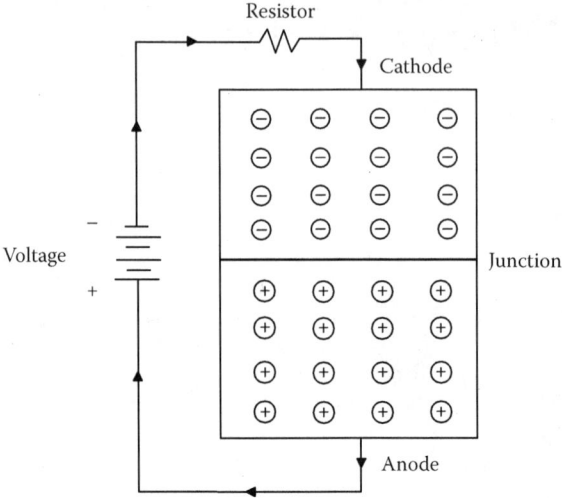

FIGURE 3.1 Bias diagram for forward conduction of a P-N junction. When the supply or battery voltage is below approximately 0.9 V, conduction does not occur. Once conduction starts, recombination will occur. The most efficient recombination occurs within a few diffusion lengths of the P-N junction in both the P-type and the N-type regions. The energy release in this recombination is a radiative process generating photons.

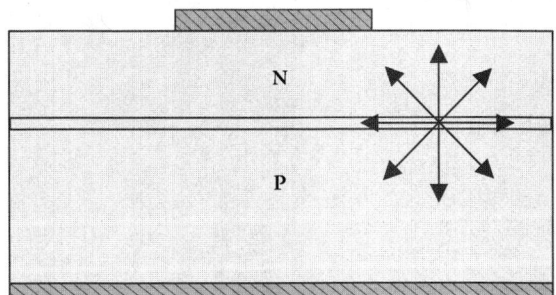

FIGURE 3.2 Energy emissions from a point source near the P-N junction. The photons created during this recombination process will radiate out in all directions from the point source.

or "in," with n_2 being the adjacent material. As the angle between the vertical and the horizontal path of the photons is increased, the photon path becomes more and more bent at this n_1–n_2 interface. At a point called the *critical angle,* the photons do not escape and are reflected back into the semiconductor material. The critical angle with GaAs or GaAlAs as the n_1 material and air as the n_2 material approximates 16°. This critical angle will change as the n_1 and n_2 materials change. The photons that escape become potentially useful energy for transfer, whereas the remaining photons are reabsorbed and lost within the IRLED (infrared light-emitting diode).

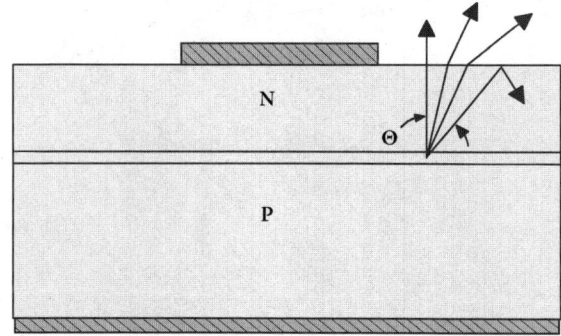

FIGURE 3.3 Energy emissions from the IRLED to the outside world.

3.2 TECHNIQUES FOR IMPROVING PHOTON EMISSION EFFICIENCY

There are many techniques that have been practiced for improving the percentage of photons that escape from the semiconductor material. Figure 3.4 shows an early technique used by an IRLED manufacturer.

Two IRLEDs are fastened together with the P-type sides common. They are then tumbled in a process similar to polishing gems. When a sphere is formed, the process is stopped and the unit is segregated into two half-spheres. The two units that result will have improved emission efficiency because the number of photons approaching the n_1–n_2 interface at less than the critical angle has been increased. Most manufacturers of IRLEDs practice another similar technique. It is simply an artificial roughening of the surface that creates more surface area. This, in turn, allows more photons to escape. This is shown in Figure 3.5.

A common technique is to mount the IRLED in a well or cavity contoured in such a fashion that the photon energy escaping the sides of the chip will be reflected toward the receiving area. A silicone compound can be placed over the IRLED chip. By virtue of refractive index matching, this increases the size of the critical angle to approximately 22°. This results in approximately 30% more photons being able to escape and become potential usable energy. Figure 3.6 shows the IRLED mounted in a well with the silicone gel compound over it.

As mentioned in Section 3.1, the maximum generation of photons occurs adjacent to the P-N junction. If we assume an IRLED chip 0.016 in. long × 0.016 in. wide × 0.008 in. high with the junction centered in the chip, the top surface has a radiating area of 256 sq. mils and the regions on the sides have a radiating area of 0.008 in. × 0.016 in. × 4, or 512 sq. mils. The longer the distance the photons have to travel toward the sides to escape, the larger the number that will be reabsorbed (the vertical distance is 0.004 in., whereas the horizontal distance is an average of 0.008 in. to 0.010 in.). The net result of this geometry is that the top radiation will be approximately 60% of the total, with the side radiation being 40% of the total. These ratios are significantly altered as the geometry of the chip changes. The addition of the well will improve the effective energy radiated by approximately 40%, by utilizing the energy radiated from the sides of the IRLED.

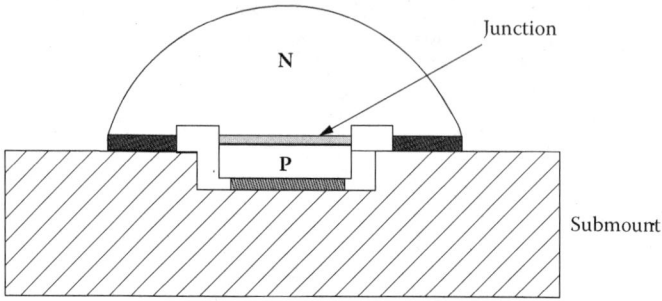

FIGURE 3.4 Contouring the n_1–n_2 interface to improve efficiency. Altering the device geometry lessens the problem created by critical angle reflections. A hemispherical shape increases the number of escaping photons by lowering the probability of photons encountering the interface at an angle in excess of the critical angle.

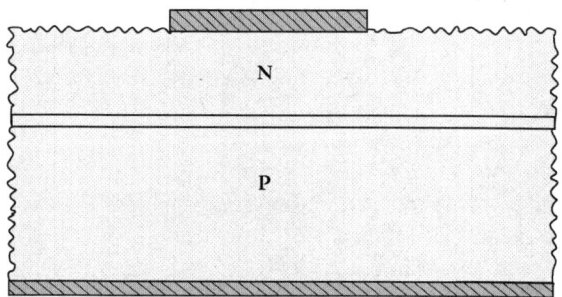

FIGURE 3.5 Increasing surface area by roughening the chip. A second version of efficiency improvement by altering chip geometry is to roughen the surface of the chip. From a cost standpoint, this is a more practical method of increasing efficiency than that shown in Figure 3.4.

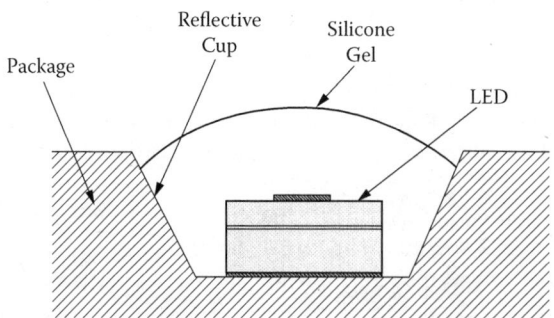

FIGURE 3.6 IRLED mounted in a well with silicone gel coverage. Usable energy emitted from the sides of the IRLED is directed upward by the reflective well, where it encounters the hemispherical interface of silicone gel and air (or plastic encapsulation material). The semiconductor–silicone gel interface has a larger critical angle, resulting in 30% greater output.

IRLED Packaging

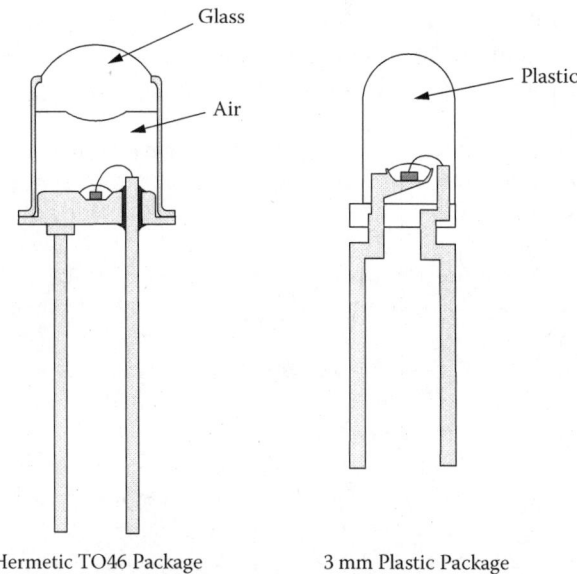

FIGURE 3.7 Package outline for hermetic and plastic case.

A significant difference in effective energy radiated also occurs as different materials are used for the package. Figure 3.7 shows the package outline for both a hermetic metal and a plastic-encapsulated unit.

The hermetic package has four interfaces that the infrared energy must traverse in order to escape the package (the chip–silicone, the silicone–air, the air–glass, the glass–air). In the plastic package there are only three interfaces (chip–silicone, silicone–plastic, and plastic–air). In actual practice, the silicone–plastic interface has little effect on transmission because the refractive indices for the silicone and the plastic are virtually the same. This reduction by two interfaces in the plastic package allows an improvement of 40 to 100% in escaping IR energy.

3.3 PACKAGING THE IRLED

Packaging of the IRLED consists of four basic operations. The chip is placed in the well, and the mounting provides both an electrical path for current flow as well as a thermal path to remove the heat from the chip. There are two techniques commonly used for this mounting:

Conductive epoxy mount: A paste containing metallic particles such as silver or gold contained in a carrier or binder (with a thinner that allows the material to be dispensed) is placed in the well. The chip is then mounted into this paste. Heat causes the paste to cure, driving off the thinner and forming a mechanical bond to both the chip and the mount area.

Alloy mount: Layers of deposited metal on the P-type side of the chip allow ohmic and mechanical contact to the chip as well as ohmic and mechanical contact to the mount area. This is different from the conductive epoxy mount in that the metal deposited on the P-type region of the chip actually penetrates into the chip and forms an alloyed region.

Each of these techniques has advantages and disadvantages. If proper controls are used, the manufacturer can select the optimal system for volume/cost trade-offs. Figure 3.8 shows an IRLED lead frame of 20 units and a detailed enlargement of the chip-mounting area.

The pictures shown are for a plastic-packaged IRLED with conductive epoxy mount. The process would be the same in principle for any of the methods of mounting discussed previously.

The mounting of the LED die is typically performed by automated equipment. This allows for very rapid and accurate placement of the die in the various packages.

The second operation is the bonding that allows contact to be made to both the N-type region of an N-side-up IRLED and the cathode lead exiting the package. This operation is also performed by high-speed automated equipment. There are two different bonding techniques used to make electrical and mechanical contact with the chip and lead frame: thermocompression and ultrasonic bonding. Thermocompression bonding is normally made by utilizing gold wire and a combination of temperature and pressure to form the electrical contact. Figure 3.9 illustrates this technique.

The procedure for performing thermocompression ball-and-wedge bonding is to apply pressure by either a weighted capillary (dispenser and guide of the gold wire) or a wedge (chisel that deforms the gold wire). The heat is supplied by heating the substrate and/or the capillary or wedge. The combination of heat and pressure forms a metallurgical bond between the bond wire and the contact pad. Ball bonding is usually the preferred method of contacting the IRLED pad, whereas wedge bonding is usually the method of contacting the header or lead frame post.

In ultrasonic bonding, the wire is deformed in a controlled fashion while it is being vibrated at ultrasonic speeds. This vibration creates a metallurgical bond between the wire and substrate. This is similar in appearance to wedge bonding but allows bonding to be performed at lower substrate temperatures.

Thermosonic bonding is a combination of thermocompression and ultrasonic bonding techniques.

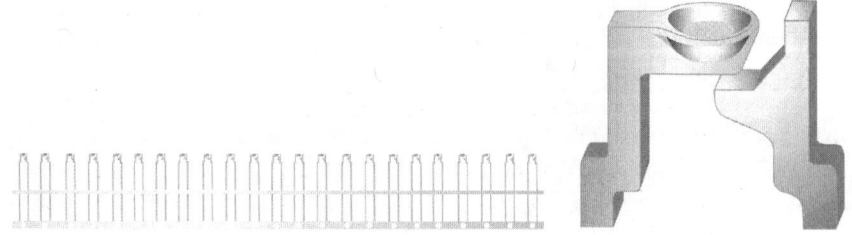

FIGURE 3.8 A 20-unit IRLED lead frame with detail enlargement of IRLED chip mounting area. Production economies are gained by processing 20 IRLEDs per lead frame. Conductive epoxy mounting is typical with this configuration.

IRLED Packaging

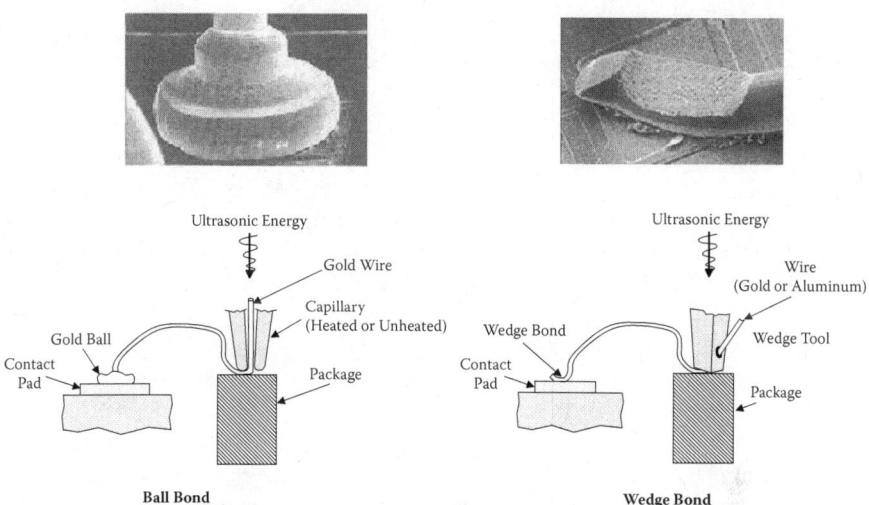

FIGURE 3.9 Thermocompression ball and wedge bonding. Heat and pressure combine to form a metallurgical bond between the contact pad and the gold wire. Ball bonding is used to make the IRLED connection, and wedge bonding is used to attach the gold wire to the lead frame post. Ultrasonic energy allows less heat to be used in making a strong bond.

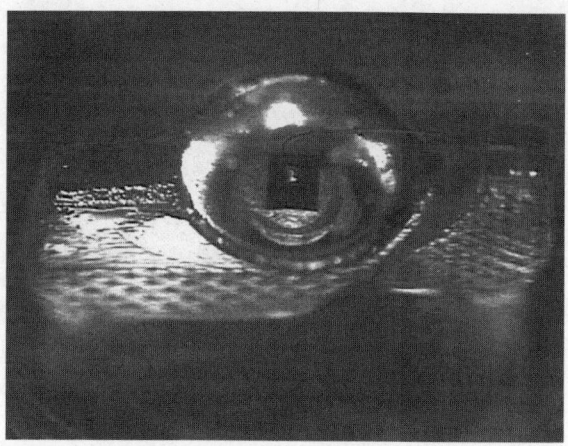

FIGURE 3.10 Detail of the IRLED mounted, bonded, and coated with the silicone gel. In addition to the refractive index matching feature of silicone gel, it also serves to make the wire bond more robust against shock, vibration, and thermal cycles.

The third operation is the addition of the silicone compound. This serves both to increase the number of photons that escape the chip and, in plastic packages, to protect the chip and the bond area from the shear stresses that the device goes through during changes in temperature. Figure 3.10 shows an IRLED chip mounted in the well, with the silicone gel around it.

The fourth operation is the final encapsulation of the device. For metal or hermetic parts, this is accomplished by either welding or soldering. The piece parts are shown in Figure 3.11.

In the solder process, the lens ring, the solder preform, and the main portion of the package are placed into a graphite boat, which is then heated to a temperature that will allow the solder to flow. The convex lower portion of the lens that will extend into the base of the package allows a self-centering feature. The extra gold bond wire that was wedge-bonded to the top of the base section of the package goes into solution and is absorbed by the solder. The device is now complete and ready for testing.

In the welded enclosure, the can containing the glass lens is placed in an electrode that presses against the extrusion at the base of the can. The bottom electrode is beneath the header base. As the electrodes are pressed together, the extrusion or lip of the can will press against the header base, creating a low-resistance contact area. Current is then passed through the electrodes and the metal package, creating heat that merges the metal can to the header base. Examination of the weld area will show smooth metal flow without extrusions. The series resistance between the upper and lower electrodes must be uniform throughout the circumference of the contact to ensure a satisfactory weld. Viewing this area will reveal a smooth and continuous surface of the metal melted from the package and header base. No voids or extruded metal should be visible. The device is now complete and ready for test.

In plastic-encapsulated devices, the final package is formed by either casting or transfer molding. The cast parts usually utilize molded plastic inserts, which are contained in a machined holder. Mold material is dispensed into each of the 20 cavities and the 20 unit lead frame is then inserted. Heat and time cures the molding compound. The plastic inserts are discarded after several uses. This system leaves a relatively uncontrolled surface at the lead egress point of the package but allows the total lens area and sides of the package to be free of mold marks.

Transfer molding is a similar process in that the lead frame is placed in a holder that has 20 machined cavities. The holder is in two parts, a top section and a bottom

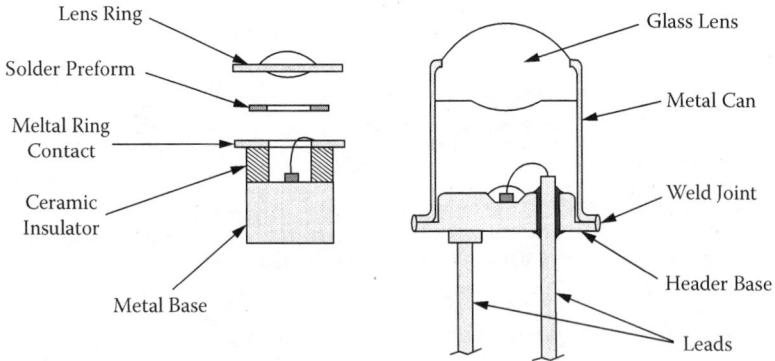

FIGURE 3.11 Sealing or final encapsulation of hermetic or metal can package. Solder process or welding may be used to create the hermetic seal completing the device. Electrical testing and inspection will take place prior to shipment.

IRLED Packaging

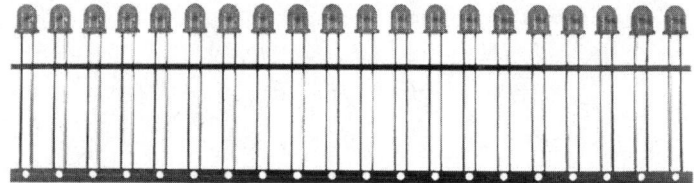

FIGURE 3.12 A 20-unit lead frame that has been cast and molded. The extra material, or flash, around the molded parts is removed in the next operation, called *deflash*. The bottom of the cast parts has a meniscus between the egressing lead frame and the edge of the base.

section, both of which enclose the unit. The plastic flows under pressure and temperature and subsequently fills the 20 cavities. The holder is removed from the press, the mold halves separated, and the encapsulated devices removed. The separation point of the two sections of the mold is called the *parting line*. Plastic is usually extruded into this interface during the molding process and must subsequently be removed. This excess material is called *flash*. The lead frames are then segregated, and the devices are ready for testing. Figure 3.12 shows the 20 unit frames after removal from the molding fixture.

3.4 CHARACTERIZATION OF THE PACKAGED IRLED

This section will cover the characterization of both GaAs and GaAlAs IRLEDs in various size chips and in different packages. The material will be discussed in both a comparative manner and an explanatory manner. The first portion of the discussion will cover the forward voltage drop of the IRLED as shown in Figure 3.13. When current is passed through the IRLED in the forward direction, very little current flows until the voltage reaches approximately 0.9 V. This voltage overcomes the built-in potential barrier discussed in Section 1.1 and is known as the *diode threshold voltage*. The slope of the V_F–I_F curve then changes to a fairly constant logarithmic slope.

Once that voltage drop is exceeded, photons begin to be formed. As the current is increased in a well-controlled IRLED, the contact resistance on both the N-type and P-type regions will give an increase in voltage drop versus current. Another factor enters into the picture with respect to the N-type regions. Because the N-type contact is smaller than the chip cross section, the resistance from the contact through the N-type region will not linearly increase with increasing current. The total junction must first be turned on and then as the current density increases, it becomes saturated. Figure 3.14 graphically shows this condition. If we assume the chip is 0.016 in. × 0.016 in., then 10 mA of current flow would correspond to 6.0 A/cm². This IRLED would be reasonably linear in the range of 10 mA to 3 A. A smaller IRLED, of dimensions 0.010 in. × 0.010 in., would be reasonably linear in the range of 4 mA to 1.2 A. This curve will change in scale with different impurity concentrations. As the impurity concentrations at the IRLED are decreased, the curve will shift to the left. Planar diffused IRLEDs have a current density pattern that shifts this curve to the left due to the lower impurity concentration or doping level.

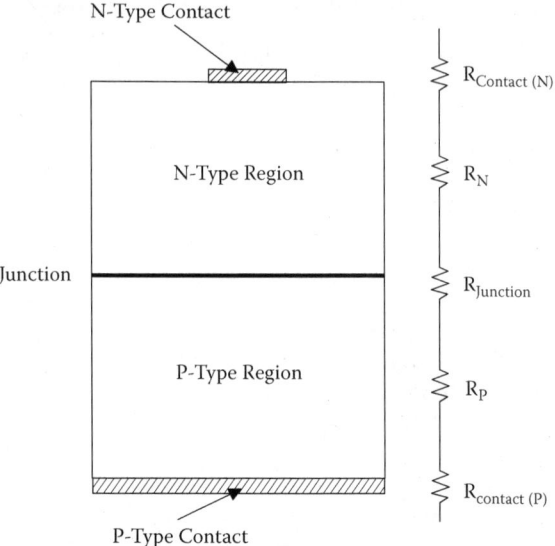

FIGURE 3.13 An IRLED cross section with conduction resistances. As conduction occurs in the IRLED, the initial voltage drop will equal the voltage required to overcome the diode threshold voltage of approximately 0.9 V. The other conduction resistances control the subsequent slope of the V_F–I_F curve.

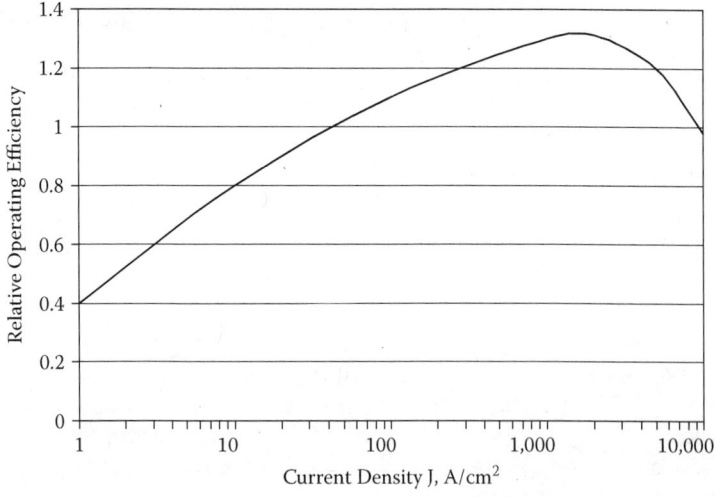

FIGURE 3.14 Normalized operating efficiency versus current density for an IRLED. The curve shown is for solution-grown epitaxial IRLEDs. As the impurity concentration changes, the shape of the curve will remain the same but will shift along the horizontal axis.

IRLED Packaging

The forward voltage drop will be reasonably linear in the range of 10 mA to 3 A on the 0.016 in. × 0.016 in. chip and from 4 mA to 1.2 A on the 0.010 in. × 0.010 in. chip. At low forward currents (<1 mA), nearly all of the current flow is utilized for surface recombination, tunneling phenomena, space charge recombination, and bulk recombination due to anomalous impurities. As the current increases, these nonideal mechanisms become saturated and become a lower and lower portion of the total current. Conversely, as the forward current approaches 2000 A/cm^2, current crowding or saturation tends to reduce the emitting efficiency as illustrated in Figure 3.14.

The net effect is to cause the IRLED to have a peak operating efficiency at a current that is a function of the geometry of the junction and the size of the electrical contact. This ignores the heating effect caused by the voltage across and the current through the IRLED. This factor will be discussed in Section 3.5.

Figure 3.15 shows the forward voltage drop versus forward current for current ranges of 0.1 mA to 100 mA over temperature. These curves are taken from GaAlAs IRLEDs measuring 0.011 in. × 0.011 in.

These curves will decrease slightly with the change to GaAs and will significantly change with chip size or change in the contact pattern to the N-type side. Note that these curves do not show the effect of heating resulting from the voltage drop across the IRLED. This factor is also discussed later in Section 3.5.

The forward voltage drop has a negative temperature coefficient. That is, the voltage drop at a fixed forward current will decrease with increasing temperature. The units are pulsed so that the device's internal temperature is held close to that of the outside ambient temperature. The plots are for a 0.016 in. × 0.016 in. GaAlAs IRLED.

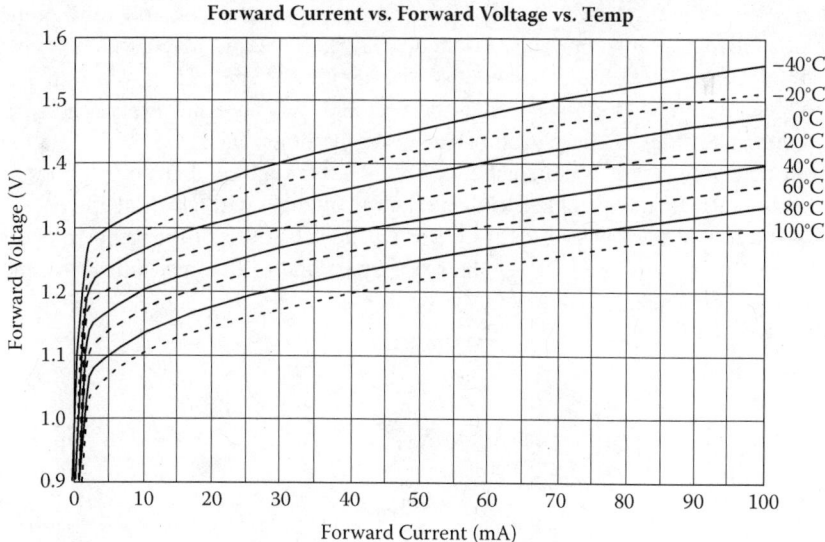

FIGURE 3.15 Forward voltage versus forward current for 0.011 × 0.011 in. GaAlAs IRLED. Changes in material or chip size will alter these characteristics. For example, a change to GaAs would lower the forward voltage. Temperature also alters these characteristics. Note that diode forward voltage shows a decrease as temperature increases.

These curves can be used for any IRLED by recording the V_F at 25°C ambient temperature at several different current levels, under pulse conditions that minimize junction heating (usually 100 µs at 0.1% duty cycle) and then sketching a new series of curves "with the same slope," as in Figure 3.15.

The contact to the N-type or top-side becomes increasingly critical if the device is to be used under pulse conditions at high currents. A GaAlAs IRLED with a 0.016 in. × 0.016 in. cross section is shown in Figure 3.16. The contact area is a compromise that allows V_F to be as low as practical and the output energy to be as high as practical. This is accomplished by making the variable resistance in the N-type region as small as possible (spreading the contact area out in lines) while blocking as little radiating area as possible.

The second portion of this section discusses the emission of the photons from the packaged IRLED and presents the advantages and disadvantages of plastic versus metal cans, energy output versus current, and energy output changes brought about by changes in temperature. Table 3.1 shows a comparison of IRLED characteristics of plastic and metal packages. Each of the points listed earlier will be considered in the following discussion.

A cross-sectional drawing of a package with the same dimensional outline in both plastic and metal is shown in Figure 3.17. This package will be referred to throughout this portion of the text to illustrate the points made in Table 3.1.

3.4.1 Package Side Emission

The side of the metal package is a tubular piece of thin nickel or Kovar™ (tradename for a nickel–iron alloy), which is opaque to the infrared energy or photons emitted from the IRLED. This material is as smooth as possible and may be plated with a good reflecting material such that energy striking it will be reflected. A large portion of this energy ends up going through the lens and thus becomes useful energy for detection. The sides of the plastic package, on the other hand, are made of a thermosetting epoxy. If this is not coated with a reflective surface, then any photons that strike this surface at any angle less than the critical angle will penetrate the package walls and go out the sides of the package. As much as 50% of the radiated energy may be emitted from the sides of the plastic part due to this mechanism. In most cases, this side emission would be useless.

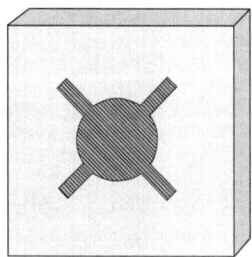

FIGURE 3.16 N-type region contact on a 0.016 × 0.016 in. GaAlAs IRLED chip. The variable resistance that is caused by the buildup in current density is minimal with this pattern as the forward voltage is increased, yet the emitting area is only slightly affected.

IRLED Packaging

TABLE 3.1
IRLED Package Comparison between Plastic and Metal Packages

Characteristic	Plastic	Metal
Side emission	Yes	No
Package height	Taller	Shorter
Lens quality	Usually superior	Usually inferior
Chip centering	Usually superior	Usually inferior
Heat dissipation	Usually superior	Usually inferior
Degradation	Usually superior	Usually inferior
Package cost	Significantly lower	Significantly higher
Storage temperature	Usually lower	Usually higher
Operating temperature	Usually equal	Usually equal
Temperature shock	Usually inferior	Usually superior
Solvent resistance	Usually inferior	Usually superior
Hermeticity	Usually inferior	Usually superior
Mechanical shock and vibration	Usually superior	Usually inferior

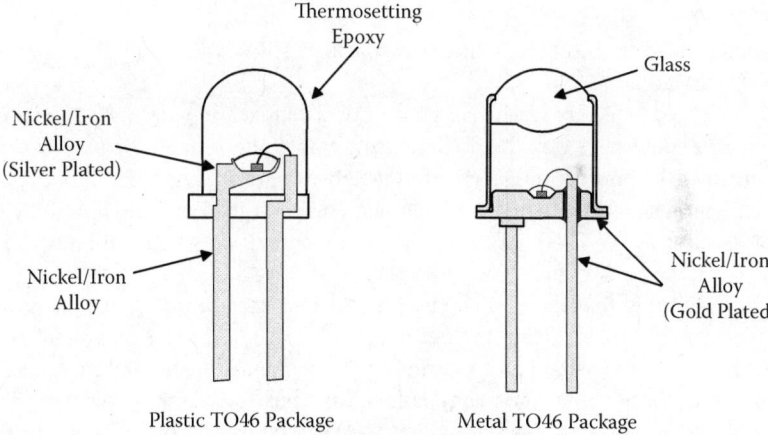

FIGURE 3.17 Cross-sectional drawing of plastic and metal packages. These two package outlines were chosen for the comparison because of the similarity in size, and they are available in both plastic and metal. As a result, the comparison of form, fit, and function can be more easily understood.

3.4.2 Package Height

The drawings shown for the two packages show identical heights. However, the lens magnification, or focusing ability, is higher on the metal package than on the plastic part. The plastic part requires a flag on the leads (below the IRLED) that prevents movement of the lead frame during insertion into the thermosetting epoxy and subsequent curing. As a result, the IRLED chip must be mounted further away from the base in the plastic package than in its metal counterpart. This simply means that if the exact mechanical outline is required in both packages, then the plastic part will have a wider beam or radiating pattern. If the same optical pattern is required, then the plastic package must be taller to accommodate the chip being higher in the plastic package. The plastic package will be taller by this difference in the distance from the chip position to the package bottom. The double-sided lens in the metal package also improves the focusing capability over the single-sided lens in the plastic package.

3.4.3 Lens Quality

The lens on the metal package is usually made from a hard glass whose thermal coefficient matches that of the nickel–iron alloy (Kovar) package material. In the fabrication of this lens, a graphite boat holding the molten glass while the other surface contour is controlled by the surface tension of the molten glass controls the shape or contour of the upper or lower surface. The lens is then flame-polished. The optical quality of these lenses is quite good, considering their relatively low cost, but leaves much to be desired when compared to precision ground lenses. The included beam angle between the 50% or half-power points is about 16° on the metal package and 26° on the plastic package. Because the outline dimensions are designed for an exact mechanical fit, the chip on the plastic package rides higher and as a result gives a wider beam angle.

The plastic package generally has a more consistent beam pattern when compared to its metal counterpart. Another difference not apparent in these beam plots is the total spread of the production distribution. Analysis of a large number of both plastic and metal parts shows that a much higher percentage of the distribution will be very similar to those shown for the plastic, whereas the metal parts will exhibit both more variability in the peak pattern and location.

It must be noted that there are a large number of variables that can cause these beam patterns to vary. These include chip placement, chip perpendicularity to mechanical axis, well shape, n_1–n_2 interfaces, lens quality, and distance from the lens tip. Figure 3.18 shows the beam pattern for a narrow radiating pattern in both the metal and plastic type package. These are typical radiation patterns for these parts. The slight dip in output at the center of the pattern is due primarily to the top metal contact in the center of the chip that blocks part of the light emission.

The TO46 plastic part is a popular package that first was widely used in visible LEDs and then later for IRLEDs. The length-to-diameter ratio of this package allows a relatively narrow beam angle to be obtained. As the chip location is moved from the top or lens end of the package toward the base, the emitting angle between half-power points decreases. Figure 3.19 shows the power coupling characteristics as the distance from the LED lens tip to the sensor is increased.

The wide-angle plastic and metal TO-46 parts are typically substituted for one another based on optical rather than mechanical interchangeability. The plastic package is taller, but the two have similar radiating patterns. Figure 3.20 shows a photograph of each package and the corresponding relative radiant intensity versus angular displacement.

IRLED Packaging

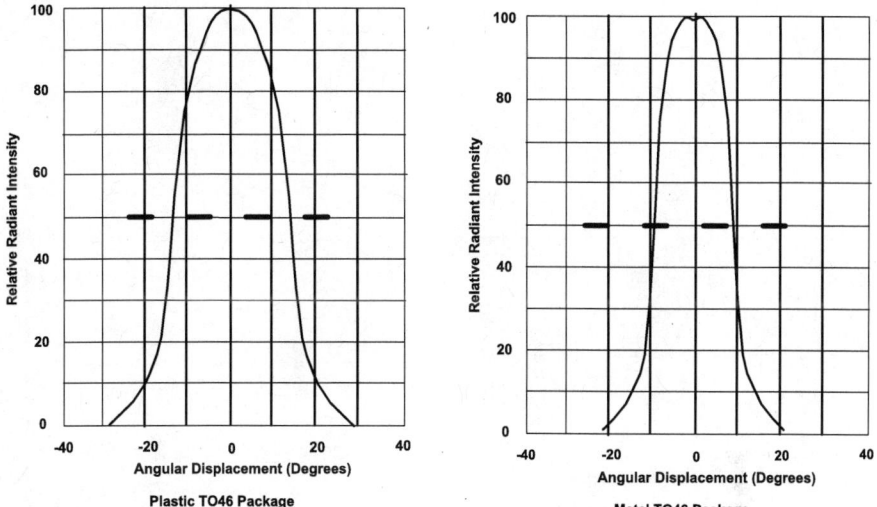

FIGURE 3.18 Beam pattern for narrow-beam plastic and metal TO-46.

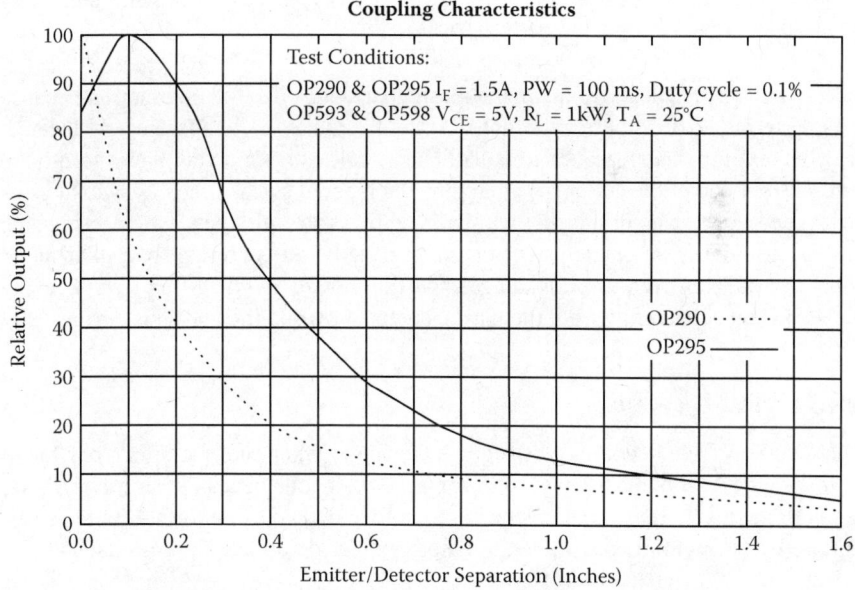

FIGURE 3.19 Relative radiant intensity versus angular displacement for plastic T-1 3/4 packages. The OP290 has the chip closer to the tip of the lens and has an emission angle of 50°. The chip in the OP295 is further back from the lens and has a narrower beam angle of 20°.

3.4.4 Chip Centering

IRLED chip centering is much more a function of manufacturing processing than package variation, because the tapered cup is usually identical in both the plastic and

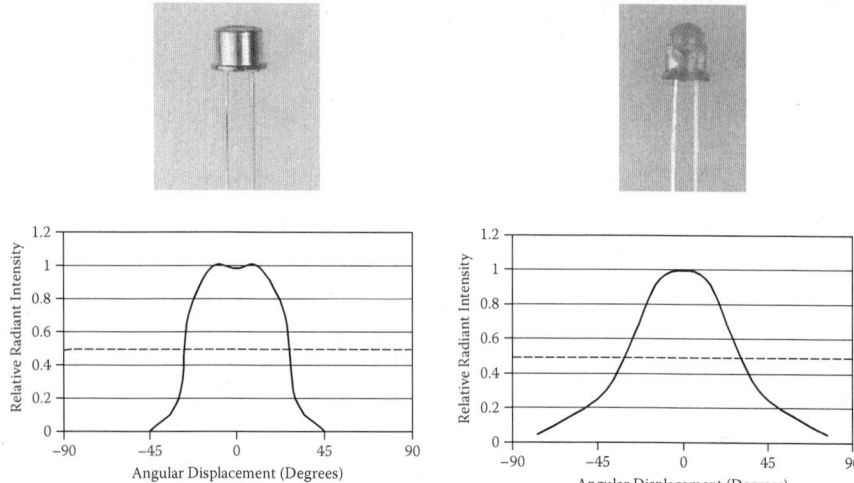

FIGURE 3.20 Radiation patterns on both the metal and plastic TO-46. Note the wider separation between half-power points for the plastic lens. Also note the higher uniformity of intensity for the plastic lens. Beam plots for metal devices do not show as much uniformity from one lot to another as do the plastic devices.

metal package. The lead frame on the plastic part lends itself to automation because the cup in the lead frame for the IRLED can mechanically be more easily located. A collet similar in appearance to a drill chuck holds the chip and places it into the cup. The outside of the collet is centered by the tapered sides of the cup, which gives very precise centering of the chip within the cup. As a result, the plastic parts, which are usually highly automated, will typically have better centering than their metal counterparts. However, this advantage is partially offset by the difficulty in keeping the lead frame centered within the plastic enclosure while the plastic is curing.

3.4.5 Heat Dissipation

The metal package becomes a better heat dissipater when the side or can portion of the package is mechanically placed in contact with a good heat sink. However, this is rarely done in actual applications owing to both added cost and the additional space required. Thermal ratings will be discussed in more detail later in this chapter.

3.4.6 Reliability

In optoelectronic technology, the two main reliability considerations are long-term LED degradation and catastrophic failure of LEDs or sensors due to thermal and mechanical stress. In the case of long-term LED degradation, the plastic device has an advantage because of its improved power dissipation characteristics and, as a result, lower junction temperatures. Figure 3.21 shows life test data for metal and plastic equivalent parts.

IRLED Packaging

Catastrophic failure due to thermal or mechanical stress, which usually occurs early in the operating life of a device, results from forces on the chip or bond wire that can dismount or delaminate the chip, disconnect the wire bond, or break the bond wire. The design of the metal part gives it the advantage, as there are no such forces on the chip or bond wire. However, the machine fabrication of the plastic part is very repeatable and mechanically accurate so that there are fewer failures due to assembly variables. In the end, neither part has a clear-cut advantage with respect to catastrophic failures.

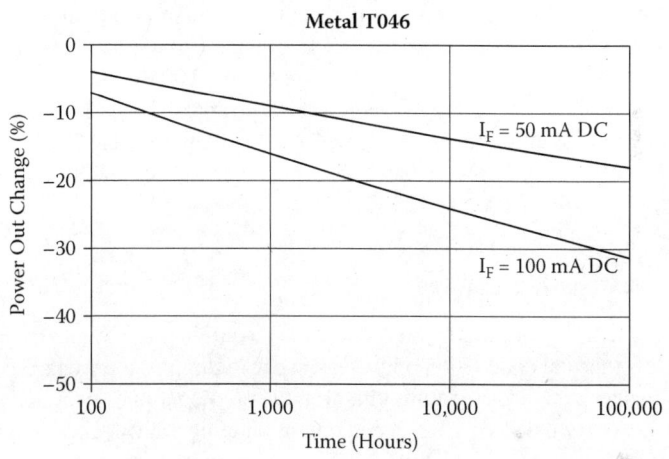

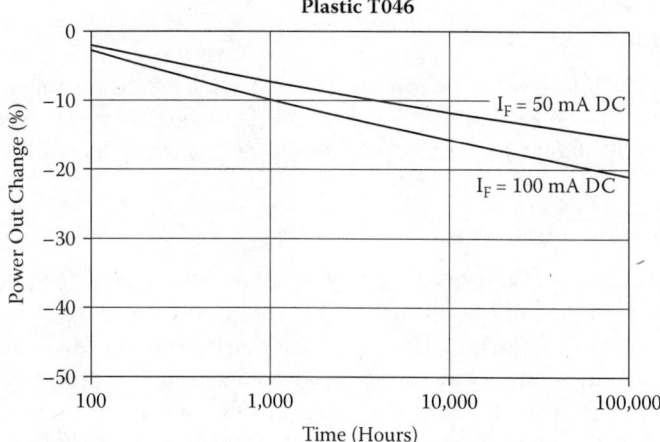

FIGURE 3.21 Operating life test data on metal and plastic TO-46 packages. The superior output degradation data for the plastic device results from the heat dissipation capability of its silver-plated copper leads.

3.4.7 STORAGE TEMPERATURE RANGE

The storage temperature range for metal IRLEDs is usually specified from −55°C to +150°C. These limits are somewhat arbitrary but primarily come from limitations specified for silicon transistors carried over to IRLEDs. In a silicon phototransistor, h_{FE} (current gain) decreases with decreasing temperature, and I_{CEO} (leakage current) increases with increasing temperature. These two factors lead to the arbitrary limits of −55°C and 150°C, with h_{FE} controlling the −55°C and leakage current controlling the 150°C.

The primary stress mechanism in plastic parts is the result of glass transition. This is the temperature at which plastic starts a recure cycle. The stresses that result are caused by thermal expansion mismatches, which can compound or stress the IRLED chip or shear the bond contacts or wire. In early plastics utilized in optoelectronic components, this glass transition occurred in the 100 to 110°C. The maximum storage temperature was specified at about 85°C. Improvements in plastic technology have allowed the glass transition rating to be raised to the 125°C to 130°C range. As a result, recent device ratings have been raised to 100°C for the package while allowing the chip to be raised to 125°C.

3.4.8 OPERATING TEMPERATURE

The operating temperature usually allows the chip to attain a 125°C maximum temperature. The low thermal conduction of the plastic package keeps the package well below the 125°C danger area. In the future, this trend should continue, eventually allowing plastic parts to carry the same storage and operating temperature range as metal parts. At the present time, however, the broader ranges remain with the metal package.

3.4.9 THERMAL SHOCK

Thermal shock follows the same pattern as the storage temperature range. The plastic packages must be kept below the glass transition point, or mechanical damage will occur as the plastic goes beyond the glass transition temperature.

3.4.10 SOLVENT RESISTANCE

Solvent resistance to chemical exposure can be graded into two basic categories. The thermosetting plastics or epoxies that are used in cast parts are less resistant than the temperature/pressure plastics used in transfer-molded parts or the metal can parts. Most acids, hydroxides, soaps, and detergents do not generally harm the thermosetting materials. Exposure to alcohol, gasoline, and most industrial solvents is also nondetrimental. However, acetone and xylene are two common solvents that should be avoided.

For purposes of cleaning or similar short-term exposures, the thermosetting plastic devices can be considered tolerant of any standard chemical that does not show obvious attack on a test sample. For long-term exposures, such as immersed applications, contact the manufacturer for more information.

The temperature–pressure characteristics of the metal parts make them even more resistant to chemical solvents. In general, the two weakest portions of the device are the lead egress points and the marking stamped on the device. If common sense and sample exposure to the particular solvent do not readily supply the correct results, then contact the manufacturer for more information.

3.4.11 Hermeticity

Hermeticity is a term that describes the ability of a package to resist the penetration of material from the outside to the inside. The metal packages have an inside cavity and can usually be leak-tested by helium or radioactive systems. The plastic packages have no inside cavity and thus must be leak-tested in a destructive mode by either a pressure cooker with a steam ambient or by being placed at an elevated temperature in a high-humidity environment such as 85/85 (85% relative humidity, 85°C). The seal or leak rate path on the plastic parts is primarily a function of the length of the leak path and the tightness of the bond between the plastic material and the regressing lead. The moisture or other harmful substance must traverse along the lead–plastic interface from the outside of the package to the junction of the IRLED. The "cast" plastic parts perform better on hermeticity tests than their molded counterparts. This is due to the reduced internal stress in the cast structure. The small leakage occurring in a nonhermetic IRLED is not a big problem, because these devices are operated in the forward mode and increased leakage due to contamination will appear as a very slight reduction in transmitted energy. If the penetrating contaminate is able to attack the IRLED chemically, then this argument is not valid. In general, however, the hermeticity advantage rests with the metal package, although the plastic package will usually continue to perform adequately.

3.4.12 Mechanical Shock and Vibration

This is usually considered to be a boundary condition for the IRLED in a hostile mechanical environment. Impact shock could occur in an application such as a fuse for an artillery shell. The quick acceleration in this application could create severe shock forces. A hostile vibration ambient could occur in an application involving the rotor blades on a helicopter. The plastic package will survive better in these environments because of the containment of the bond wire. Because bond wire movement is *ruggedized,* or constrained, throughout its entire length, the plastic part will have fewer failure modes and thus be more reliable.

3.4.13 Optical Consideration

The beam pattern for a wide-beam angle radiating plastic and metal package was shown in Figure 3.20. The corresponding beam patterns for the narrow-beam angle radiating plastic and metal package was shown in Figure 3.18. In either the plastic or metal package, if the assumption is made that the chip size and shape of the chip well are identical, then the radiated energy should be equal. The wide-beam package simply disperses this energy over a wider area. The wide-beam angle would be used where the application dictated a wide dispersion of energy. This would occur when the application required an accessory focusing lens such as surveillance (long-range detection)-type applications. Figure 3.22 shows the variation in apertured output radiant energy with respect to distance from the lens side of the mounting flange. The aperture used is 0.250 in. diameter and is normalized as shown.

The IRLED cannot be treated as a point source of energy when the distance between the IRLED package and receiver is short. On the broad-beam TO-46, this critical distance is approximately 2 in., whereas on the narrow-beam TO-46, the distance becomes approximately 6 in. Once these distances are exceeded, the inverse square law of a point source can be used. The inverse square law means that as the distance between the point source and the sensing area increases by a factor of two, then the energy per unit area decreases by a factor of four. This will be discussed in more detail later on in the chapter. Figure 3.23 shows this falloff graphically by illustrating the relative coupling on a production spread of photosensors and IRLEDs in the TO-18/TO-46 packages. Note the change in slope on the broad-angle parts at approximately 2 in. and on the narrow-angle part at approximately 6 in.

The radiant energy of an IRLED is not linear versus forward diode current. This was discussed in the first portion of Section 3.4. The output versus current is a function of the metallization geometry of the top or N-type side of the IRLED and the cross-sectional area of the chip. At high currents, the shape and location of this N-type metallization are particularly critical. The only major difference between GaAs and GaAlAs is the amount of radiated energy (GaAlAs has the larger radiated energy). The relative change is the same. Figure 3.24 shows this relative change

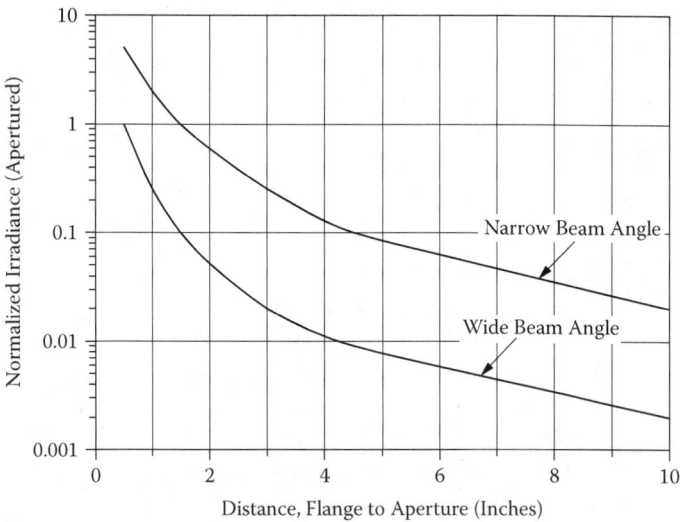

FIGURE 3.22 Change in apertured power output versus distance. Note that the inverse square law does not apply until the sensor is further than 2 in. from the lens on the wide-beam-angle unit and 6 in. on the narrow-beam-angle unit.

IRLED Packaging

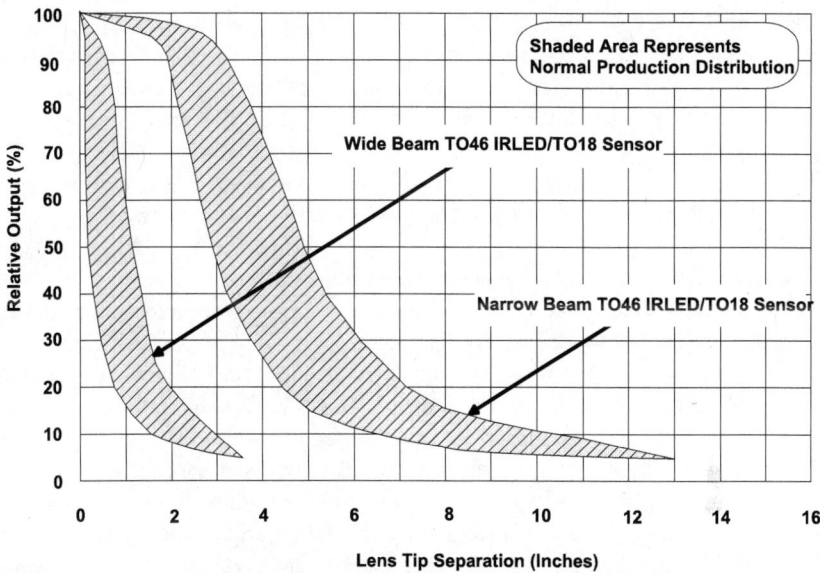

FIGURE 3.23 Coupling characteristics of broad- and narrow-beam TO-18 sensors and TO-46 IRLEDs. The slope of the coupling characteristic curve changes dramatically as the inverse square law takes over.

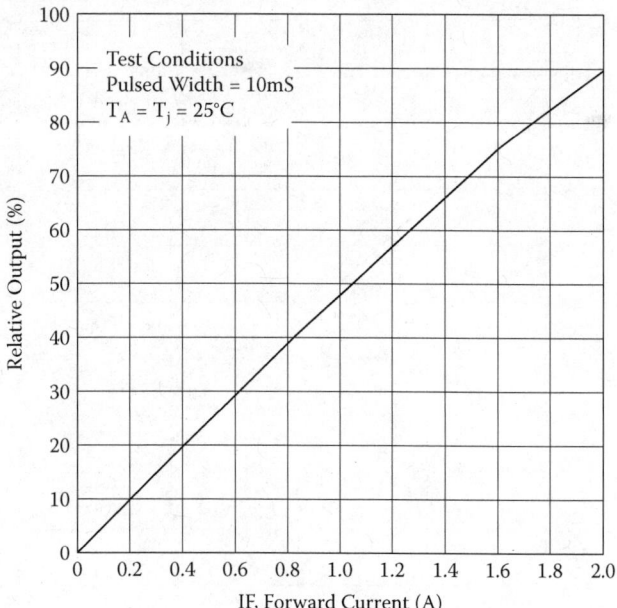

FIGURE 3.24 Relative change in apertured power versus forward current. Relative changes in emissions as a function of current are independent of package designs but dependent on chip size and metallization geometry.

versus current for a 0.016 in. × 0.016 in. IRLED. This change is also independent of the type of lens utilized in the package (the wider the dispersion angle, the faster the falloff per unit of distance). Note that there is less of a noticeable slope change at 2 and 6 in. on the broad and narrow radiating angle parts, respectively, in Figure 3.23 than in Figure 3.22.

The radiant energy emitted from an IRLED has a negative temperature coefficient. The curve shown in Figure 3.25 shows this relative change between −50°C and +125°C. The change is normalized to the reading at 25°C and is the same for GaAs and GaAlAs. Note that forward current is held constant.

The energy emitted from an IRLED is not all of a given wavelength. The IRLED emits a band of energy that is centered at a particular wavelength. This center point is near 875 nm for GaAlAs and 930 nm for GaAs at 25°C. Measurement of a number of different production lots shows the center emission at 25°C temperature will vary ±20 nm on the GaAlAs and ±15 nm on the GaAs. This follows logically from the earlier discussion in Chapter 1, which explained the effects of the additional variable of the aluminum in the GaAlAs process. This process requires both more sophisticated equipment and process control. Even with the controls, more variation in the IRLEDs also occurs. Figure 3.26 shows the spectral emission as a function of wavelength. Note that the GaAlAs emission band extends into the visible red spectrum between 700 and 800 nm.

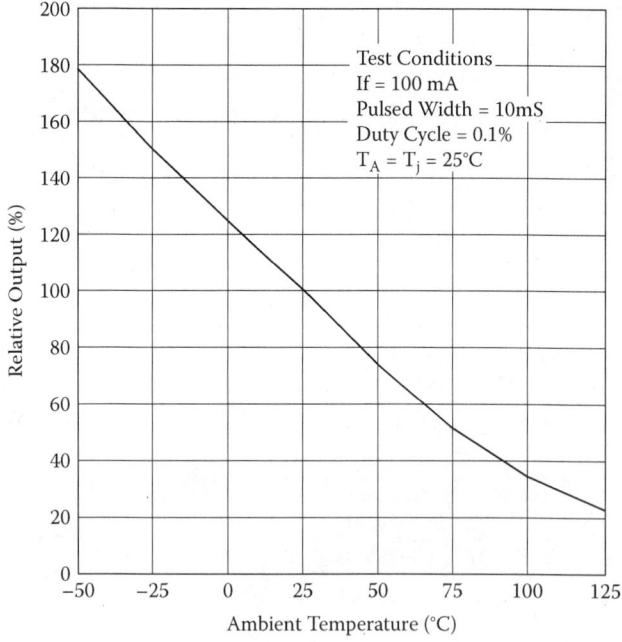

FIGURE 3.25 Percentage change in radiant energy versus ambient temperature. If forward current is held constant, radiant energy will decrease as temperature increases.

IRLED Packaging

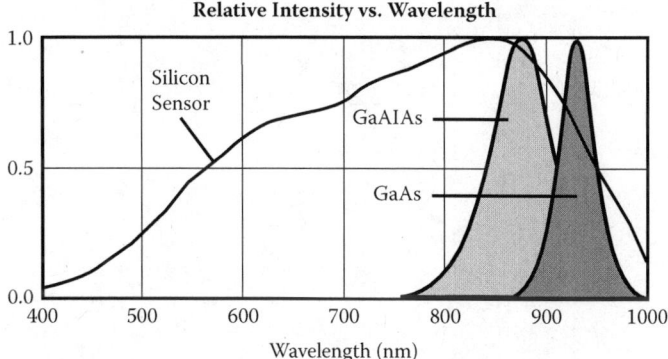

FIGURE 3.26 Spectral emission versus wavelength for GaAlAs and GaAs IRLEDs. As discussed in Chapter 1, the GaAlAs epitaxial growth process is more difficult to control, with the concentration of Al being related to the peak wavelength of the finished diode. This explains the wider variance in wavelength for GaAlAs devices.

This spectral emission also shifts with temperature. As the temperature increases, the wavelength of emission gets longer or shifts further into the infrared. Figure 3.27 shows this peak spectral emission shift versus temperature for both GaAlAs and GaAs. Note that the GaAs shifts more than the GaAlAs.

3.5 UNDERSTANDING THERMAL IMPEDANCE

The maximum power dissipation rating for a semiconductor device is usually defined as the largest amount of power that can be dissipated by it without exceeding the safe operating conditions. This quantity of power is a function of the following:

1. Ambient temperature
2. The maximum junction temperature considered safe for the particular device
3. The increase in junction temperature above ambient temperature per unit of power dissipation for the device package in a given mounting configuration

Item 1 results in lower power dissipation ratings at higher ambient temperatures as described by derating curves, described in the following paragraphs. Item 2 is determined from reliability experiments and is usually considered to be 150°C, although it may be lower owing to temperature limits imposed by the package material. Item 3 is called thermal impedance and is determined in the laboratory. The techniques used in this determination are also discussed in the following paragraphs.

3.5.1 Thermal Impedance Calculations

The formula for calculating thermal impedance is

$$R_{\text{THJA}} = \frac{T_J - T_A}{P_D} \tag{3.4}$$

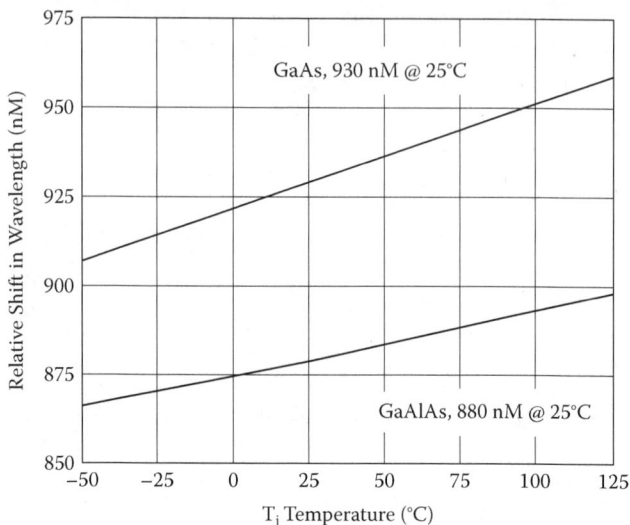

FIGURE 3.27 Shift in peak wavelength versus junction temperature for GaAlAs and GaAs. The GaAs device peak wavelength will shift more with temperature than the GaAlAs IRLEDs.

where

S_{THJA} = thermal impedance, junction to ambient (also called θ_{JA}); units are °C/W
T_J = junction temperature of the device under test
T_A = ambient air temperature
P_D = device power dissipation

R_{THJA} refers to the thermal impedance of a device with no heat sink, suspended in still air on thermally nonconductive leads. This is the worst case (highest value) for thermal impedance.

To calculate the maximum allowable power dissipation $(P_{D(MAX)})$, substitute numbers for R_{THJA} (measured in the laboratory) and T_J (using the maximum value determined from reliability experiments), then rearrange terms to get

$$P_{D(MAX)} = \frac{T_{J(MAX)} - T_A}{R_{THJA}} \tag{3.5}$$

This results in a linear power dissipation rating curve that intercepts zero power dissipation at $T_A = T_{J(MAX)}$, and with a slope which is $-1/R_{THJA}$, as shown in Figure 3.28.

The usual (and conservative) method of rating power dissipation is to limit the curve to the safe value for normal room temperature, which is 25°C. The result is a curve shaped as shown in Figure 3.29.

As there are voltage, current, and ambient temperature limitations not related to chip temperature, the final power dissipation rating curve (often called a *derating curve*) for a given device will be similar to the curve shown in Figure 3.30.

IRLED Packaging

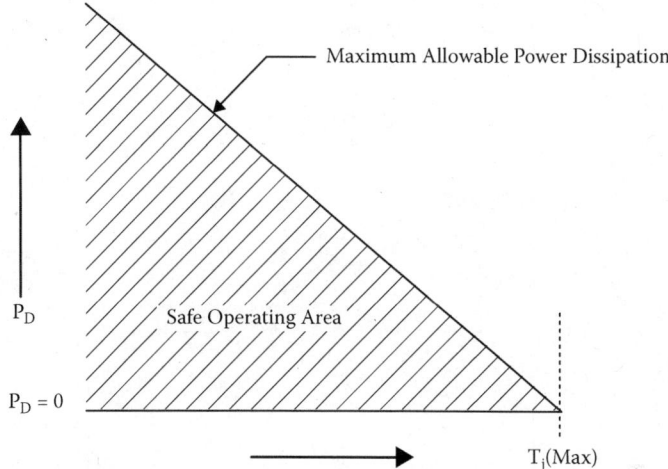

FIGURE 3.28 Initial thermal derating curve. The curve intercepts zero power dissipation at $T_A = T_{J(MAX)}$, and the slope equals $-1/R_{THJA}$.

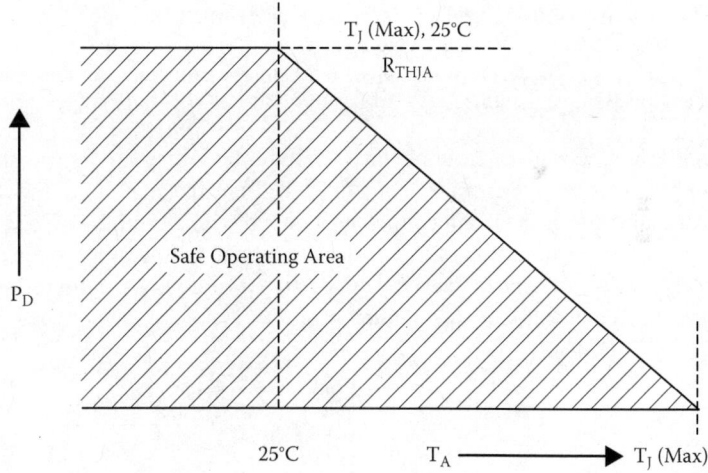

FIGURE 3.29 Thermal operating curve from 25°C. This curve limits power dissipation to safe values for generation at room temperature ($T_A = 25°C$).

Because thermal impedance is very nearly constant for different levels of power dissipation, the junction temperature is measured at a known quantity of power dissipation, and then substituted into the right-hand side of the formula:

$$R_{THJA} = \frac{T_J - T_A}{P_D} \tag{3.6}$$

to find the thermal impedance of the device.

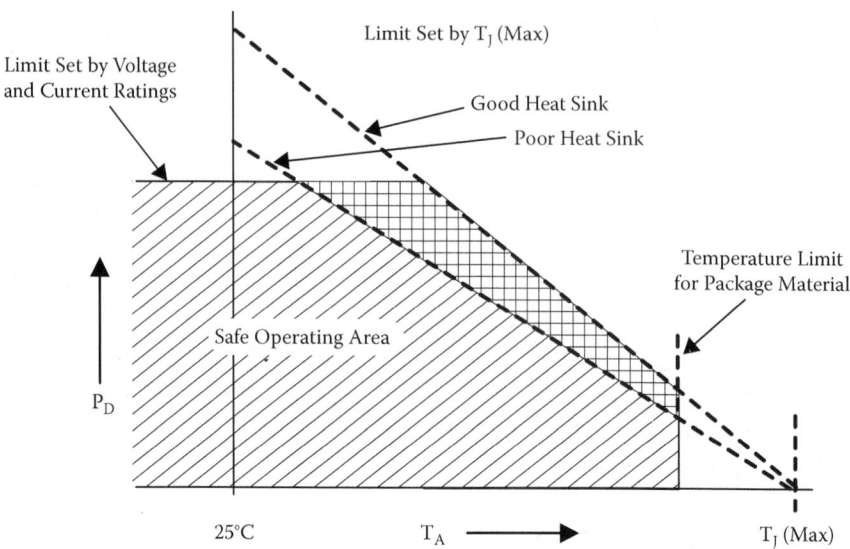

FIGURE 3.30 Final thermal derating curve. The final thermal derating curve or operating "envelope" is determined by several limits. Voltage and current limitations determine the maximum power dissipation when the ambient temperature is low enough so that no derating is required. The package material is normally limited to operating temperatures lower than the maximum junction temperature.

It is important to define the ambient conditions, because air movement, lead length, and contact with thermal conductors all affect the measured T_J. The best case (lowest value) of thermal impedance is obtained with an infinite heat sink, that is, by keeping the entire outside of the device at ambient temperature. Because case temperature equals ambient temperature under these conditions, infinite heat sink thermal impedance is called R_{THJC}, defined as

$$R_{THJA} = \frac{T_J - T_C}{P_D} \tag{3.7}$$

where T_C = case temperature.

To find R_{THJA} (thermal resistance junction to ambient), R_{THJX} (thermal resistance junction to a finite heat sink), and R_{THJC} (thermal resistance junction to an infinite heat sink), the device is placed in the desired mounting configuration and a specific amount of power is applied to the device to provide significant chip heating. The junction temperature is monitored by interrupting the power, and substituting the low forward-bias current (thermometer), 100 μA for the IRLED described in Figure 3.31.

The voltage drop must be measured before the junction has time to cool significantly. A 100 μs interruption is used, which is consistent with the thermal time constant of the devices being measured; a sample-and-hold circuit maintains the reading

IRLED Packaging

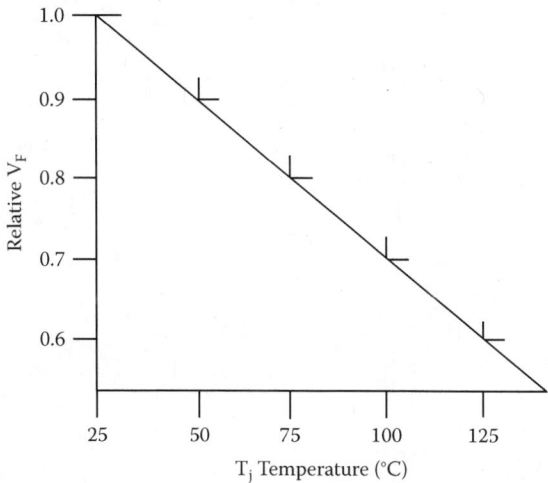

FIGURE 3.31 Voltage drop versus junction temperature for an IRLED. This curve offers the key to accurate measurement of junction temperature at various power levels, necessary for determining proper thermal deratings for IRLEDs.

so it can be recorded with a voltmeter. The applied waveform for the above IRLED would appear as shown in Figure 3.32.

Because of the sample-and-hold circuit, the voltmeter reading reflects the junction temperature of the chip. For a typical plastic package IRLED, the junction temperature rises after application of DC power for several minutes, as shown in Figure 3.33.

When the voltmeter reading has stopped changing, (1) substitute the reading back into the graph to get the actual T_J; (2) multiply the large forward current, in this case 100 mA, by the voltage drop on the diode with 100 mA applied to get the power dissipation; (3) measure the actual T_A; and (4) substitute into the R_{THJA} formula to get a value for thermal impedance. See Table 3.2 for an example of a typical T-1 3/4 packaged 0.016 × 0.016 IRLED.

The unit is then connected to a test circuit and immersed in agitated silicone dielectric fluid at a temperature of 25°C. This is a good approximation of an infinite heat sink for a low-power device. An I_F of 100 mA is applied. Every 100 ms, the I_F is reduced to 100 µA for a period of 100 µs, after which the I_F returns to 100 mA (see Figure 3.32).

Using a sample-and-hold circuit, it is observed that the device V_F during the low-current intervals starts out at 1.080 V but rapidly decreases, eventually stabilizing at 1.050 V. Interpolating between 1.080 V (25°C) and 1.030 V (50°C), it is found that the junction temperature is now 40°C.

The V_F is measured during the 100 mA I_F period and found to be 1.50 V. Thus, the power dissipation is 150 mW (99.9% of the time). Substituting into the formula:

$$R_{THJA}(\text{infinite heat sink}) = R_{THJC} = \frac{40 - 25}{0.150} = 100° \text{ C/W} \qquad (3.8)$$

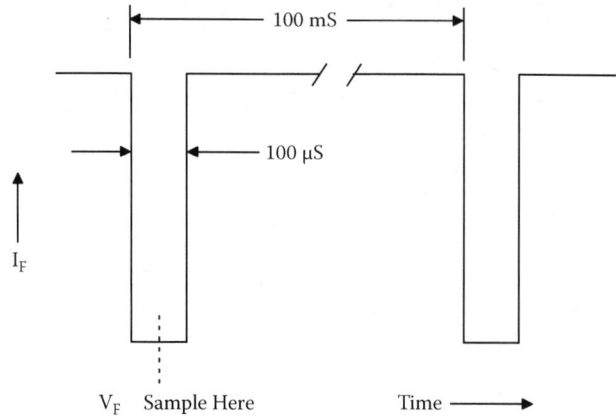

FIGURE 3.32 Timing cycle for device heating and monitoring of junction temperature. The objective is to measure junction temperature quickly by taking a voltage drop reading, with minimal interruption of the power level being tested.

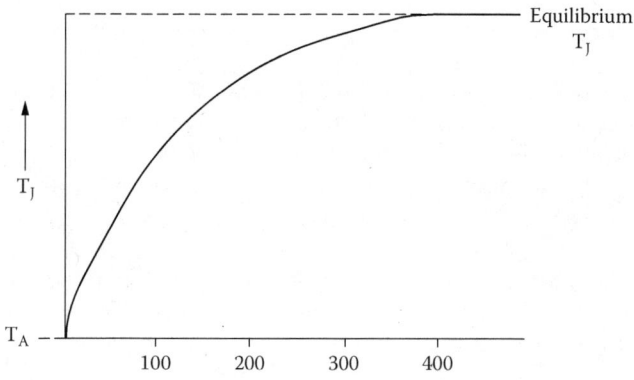

FIGURE 3.33 Equilibrium of junction temperature. The application of DC power causes a rise in junction temperature, which tends toward a new equilibrium, or steady-state, temperature.

TABLE 3.2
Voltage Drop versus Temperature

T_A	V_F (volts, at I_F = 100 µA)
25	1.080
50	1.030
75	0.980
100	0.930

IRLED Packaging

When the same test is conducted with the device in still air, mounted in a PC board socket, the final values of V_F are 1.024 at 100 µA and 1.40 at 100 mA. Thus, $T_J = 53°C$ and

$$R_{THJX} = \frac{53-25}{0.140} = 200° \text{C/W} \qquad (3.9)$$

The power derating curves are

$$P_D = \frac{T_{J(MAX)} - T_A}{R_{THJA}} = \frac{125 - T_A}{100} \qquad (3.10)$$

with infinite heat sink, and

$$P_D = \frac{125 - T_A}{200} \qquad (3.11)$$

with no heat sink.

Graphing the derating curve gives two lines, as shown in Figure 3.34.

Because the plastic package can withstand only 100°C (because of the glass transition temperature), the device is limited to 250 mW. Thus, the final power derating curve is shown in Figure 3.35.

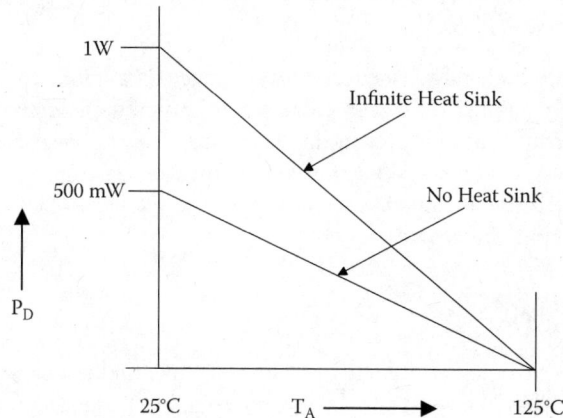

FIGURE 3.34 Thermal derating for "infinite" and "no" heat sink. With no heat sink, logic predicts that the P_O rating will be lower; hence, the bottom line represents operation with no heat sink. In practice, the optimal derating would lie somewhere between the two curves.

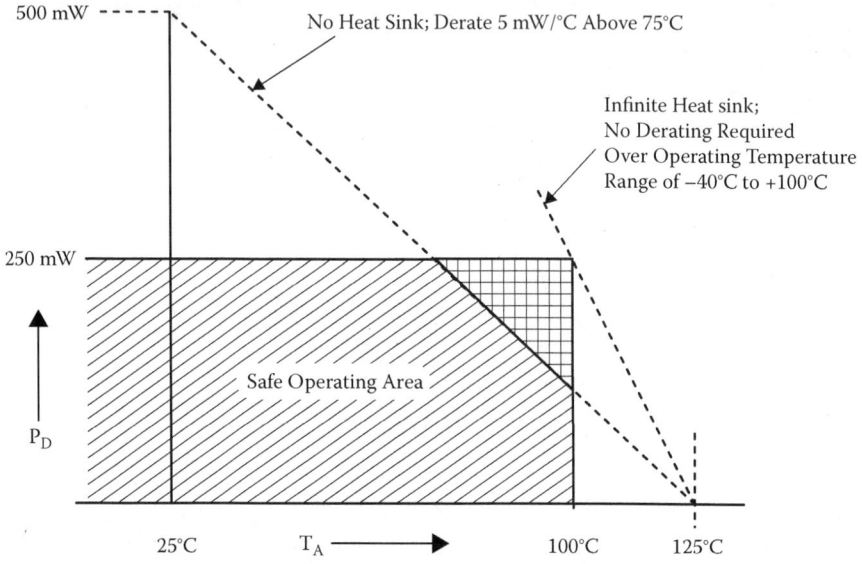

FIGURE 3.35 Final thermal derating. Only a small portion of the safe operating area extends into the region between zero and infinite heat sinking. The vast majority of applications will facilitate operation in the portion below the line representing no heat sink.

The entire shaded area can be used with an infinite heat sink; the cross-hatched area is forbidden for a device with no heat sink.

3.6 UNDERSTANDING THE MEASUREMENT OF RADIANT ENERGY

Infrared-emitting diode power measurement depends on a number of variables that must be precisely defined for design engineers to utilize manufacturers' data sheet information. Manufacturers differ in the techniques used in measuring power and also in their interpretations of the definitions of measured parameters. This section should help clarify the differences, especially those related to GaAs and GaAlAs solution-grown epitaxial devices.

3.6.1 General Discussion

Power is measured in units of energy per unit of time, and the conventional MKS unit is the watt. Some factors that must be controlled to make accurate power measurements are now discussed.

The energy an IRLED emits is in the form of photons, and the photon's energy is inversely proportional to its wavelength. To measure the power emitted, the technique must take into account both the rate of photon emission and the average

IRLED Packaging

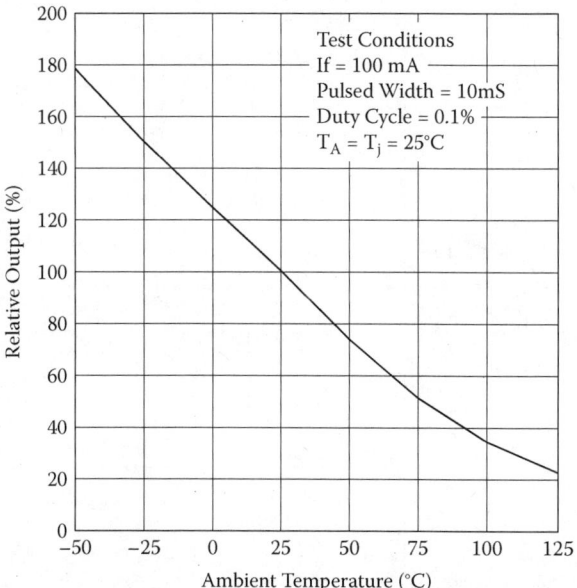

FIGURE 3.36 Output power versus ambient temperature for both GaAs and GaAlAs IRLEDs. Output decreases as temperature increases for these devices.

wavelength of the emitted photons. Both of these vary as functions of chip temperature. See Figures 3.36 and 3.37 for examples of these changes.

Stress on the chip will cause any defects in the chip to expand along the planes of the crystalline structure in a process called *dark line defect formation*. This degrades the chip, causing the power output to decrease. These dark line defects effectively destroy the crystalline formation and thus reduce the ability to create photons. Throughout the chip's operating life, these dark line defects continue to propagate, resulting in time-dependent output degradation. Measurements made after the chip has been stressed mechanically, thermally, or electrically will be lower than initial readings. If the operating and environmental elements are accounted for, then IRLED degradation becomes a predictable function of time. Many manufacturers provide information on output reduction versus time for fixed current, temperature, and heat sink conditions. Figures 3.38, 3.39, and 3.40 illustrate the magnitude of these output changes due to applied DC current for variations of ambient temperature, current level, and different materials used as emitters.

The response of most detectors is also wavelength and temperature dependent. One reason for these variations is that the surface of the detector can reflect photons depending on the wavelength, the angle of incidence, and the type of protective coating on the surface. Some emitting devices can also exceed the range of linearity in power detection. Finally, there are other minor characteristics of detectors to be considered. Obviously, the accuracy of the detection system is critically important to measuring the output of an IRLED accurately.

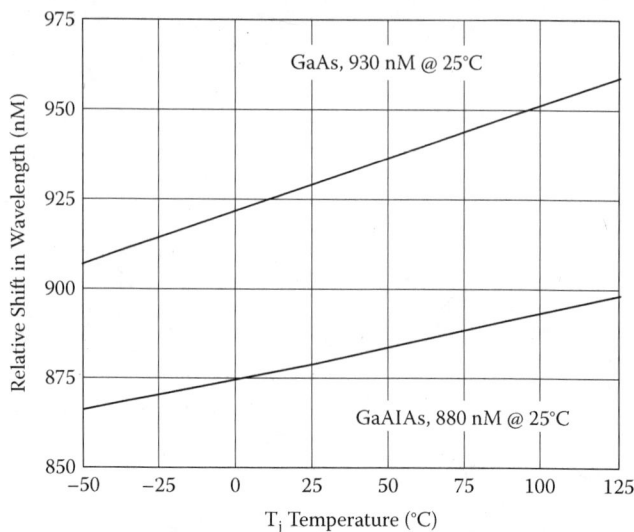

FIGURE 3.37 Peak wavelength versus ambient temperature for both GaAs and GaAlAs IRLEDs. Peak wavelength increases as temperature increases for these devices.

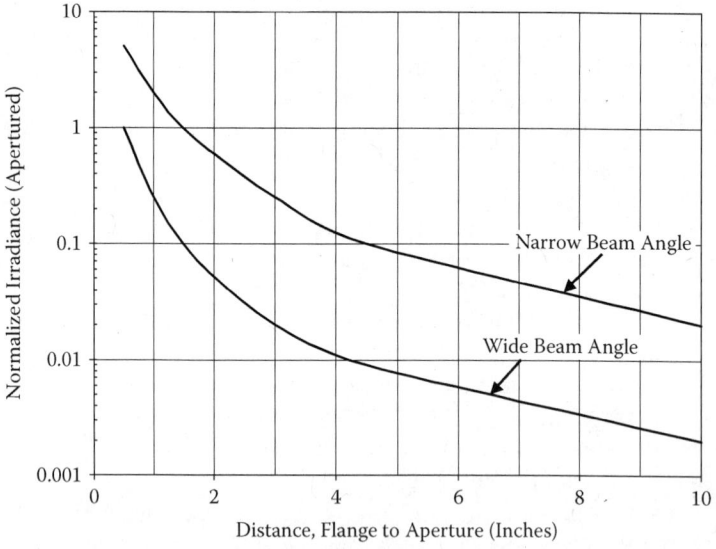

FIGURE 3.38 Percentage change in GaAs IRLED mounted in metal TO-46 package versus time at 25°C and 55°C. Note the increase in degradation caused by the higher ambient temperature.

IRLED Packaging

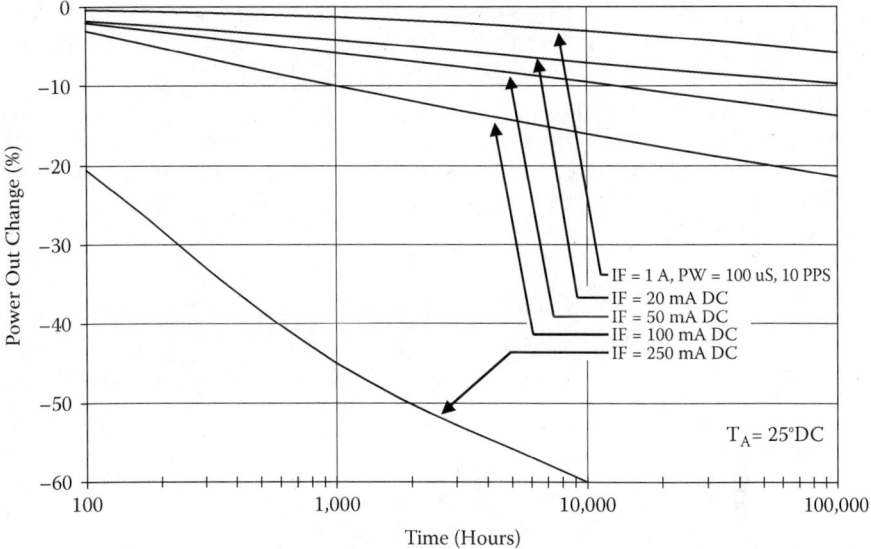

FIGURE 3.39 Percentage change in GaAlAs IRLED mounted in plastic TO-46 package versus time at various current levels. Note the increased rate of degradation caused by higher current.

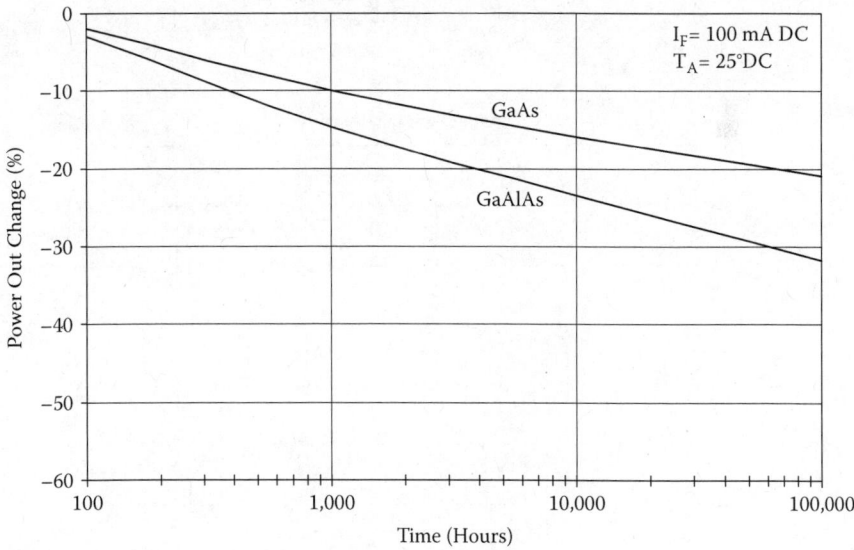

FIGURE 3.40 Percentage change in GaAs and GaAlAs IRLED mounted in metal TO-46 package versus time under same conditions. Note the increased susceptibility to degradation effects of GaAlAs under the same forward current and temperature conditions. Remember that the GaAlAs is emitting more usable energy than the GaAs for the same current. The peak wavelength of the GaAlAs IRLED is also more closely matched to photosensor peak spectral response.

Any measurement of directed output is dependent on complex optics, which include package design, chip position in the package, lens design, energy-blocking components within the packaging, and the fact that only a small percentage of the emitted photons exit the chip.

Many devices have radiation patterns that change as the distance from the device to the detector is varied, so this distance can be important in directed output measurement. Logically, this distance also becomes critical in slotted switch design, as discussed in Chapter 8 (See Figure 3.41).

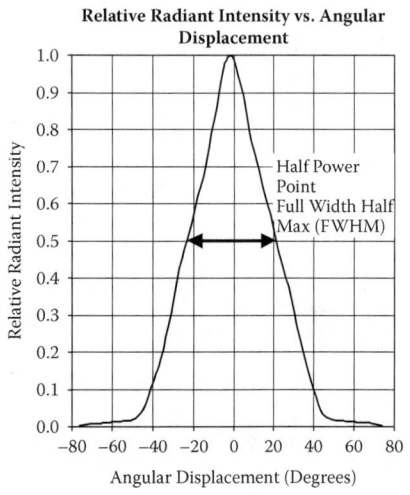

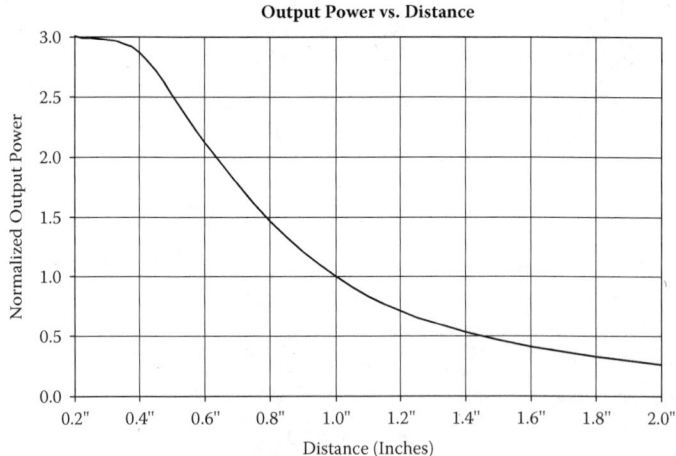

FIGURE 3.41 Relative radiant intensity versus angular displacement and distance for T-1 3/4 package. The *beam angle* is generally measured between the half-power points and is sometimes referred to as the *full width half max* (FWHM).

IRLED Packaging

3.6.2 Parameter Definitions and Measurement Techniques

There have traditionally been two methods of defining power measurement, but they have been interpreted differently.

The first method is radiant power output (P_O or E_e), sometimes called *total power*. A strict interpretation of P_O is that the total amount of radiation exiting the package, regardless of direction, should be measured. Some manufacturers have interpreted radiant power output to be only that radiation which exits the package in a direction useful to most designers.

The measurement may include only that radiation collected by a flat surface detector near the lens tip and orthogonal to the lens axis. Radiation emitted from the sides or back of the package and surface reflections from the detector are not collected. Therefore, the devices from these manufacturers are conservatively rated (sometimes by as much as a factor of two, depending on the device type). Their output may actually be quite high when compared to devices measured differently by other manufacturers. For instance, P_O readings for the narrow (15° between half-power points) radiation pattern plastic packages utilizing the 0.016 in. × 0.016 in. GaAlAs IRLED are typically 60% higher when using a parabolic reflector than when using a flat P_O test fixture. This is due primarily to the collection of radiation from the sides of the package. P_O measurements are normally useful only for devices with wide radiation patterns, because the primary application is in providing a relatively even intensity over a large area. Radiation that exits the side or back of the package is not useful without external reflectors; and if external reflectors are added, there are intensity peaks in the radiation pattern that are detrimental in most applications.

The second way to measure power is on-axis intensity. Measuring the power incident on a specified area does this. The most common method is to utilize a fixture that controls the distance from the device to a measured aperture on the detector. This measured power can then be specified as average power per unit area (both $E_{e(APT)}$ and P_A are equivalent, and the measurement is usually expressed as mW/cm^2) or as I_e average power per unit of solid angle (i.e., where the measure is expressed as mW/sr or milliwatts per steradian). Note that measurements expressed in milliwatts per square centimeter should also include information on the distance from the device to the aperture and the size of the aperture, for them to be useful to design engineers.

The calculated value of I_e is also dependent on distance for most applications, and a design engineer can be misled by the mathematical model into assuming that I_e is a constant regardless of distance. Most IRLEDs cannot be modeled as a point or discrete source except at distances that are very large compared to the package dimensions and/or optical dimensions. Thus, the foundation assumption in spherical calculations (using mW/sr) is invalid, and attempts to use this model may lead to errors. Note in Figure 3.42 how mW/sr becomes consistent after approximately 6 in. separation.

Some manufacturers have chosen to use $E_{e(APT)}$ or P_A rather than I_e for devices that do not have a virtual source that is distance independent. This is the preferred

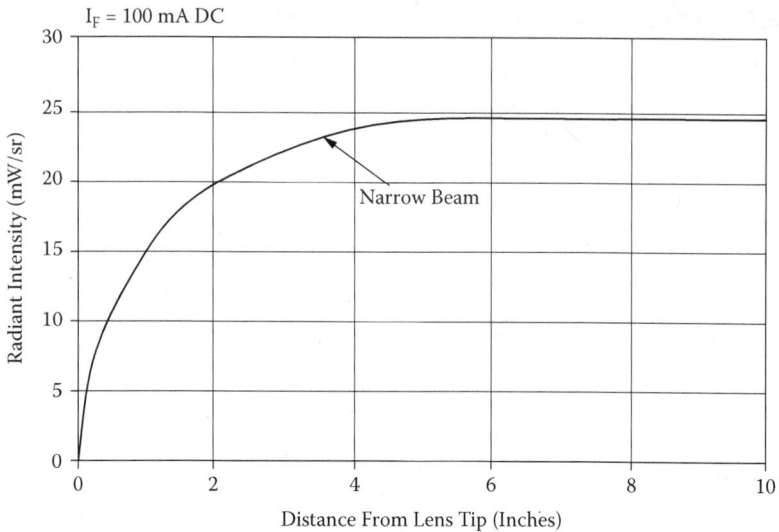

FIGURE 3.42 Output intensity in mW/sr versus distance from lens tip on T-1 3/4 package. A consistent function of separation distance only occurs after approximately 6 in.; therefore, use of geometrical models to predict irradiance can lead to design errors at close proximity.

parameter because a simple performance graph can then show how $E_{e(APT)}$ varies with distance, as shown in Figure 3.43.

$E_{e(APT)}$ measurements have historically been made only for narrow-radiation-pattern devices because their major application is to have a high on-axis intensity for good coupling efficiency with a small-sensing-area photodetector (see Figure 3.44).

However, some manufacturers are beginning to use the measurement parameter with wide-radiation-pattern devices also. $E_{e(APT)}$ is a key design parameter when the distance and aperture are chosen to give maximum useful information. The distance is normally chosen so that two criteria are met: (1) all intensity peaks should fall within the aperture opening for devices with normal optics and (2) the distance should be at a maximum, with the constraint that the intensity does not vary more than 10% from point to point within the aperture opening for normal devices. This gives the maximum useful information to the systems designer. Aperture size is typically chosen so that it is slightly larger than the lens diameter of a detector, which is mechanically matched to the dimensions of the IRLED. This provides the user with a mechanical alignment tolerance as well as the average power intensity within the aperture.

3.7 RELIABILITY

In optoelectronic technology, the two main reliability considerations are long-term IRLED degradation and catastrophic failure due to thermal and/or mechanical

IRLED Packaging

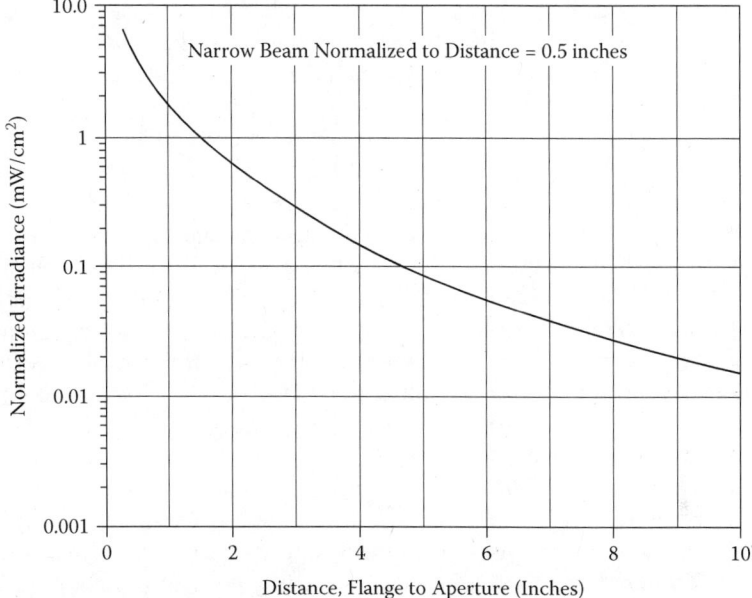

FIGURE 3.43 Output intensity in mW/cm² versus distance from lens side of mount area. This method of specifying device output is convenient for design engineers, because most photosensors are tested and specified in terms of radiant intensity expressed in mW/cm², making sensor output current prediction an easy calculation.

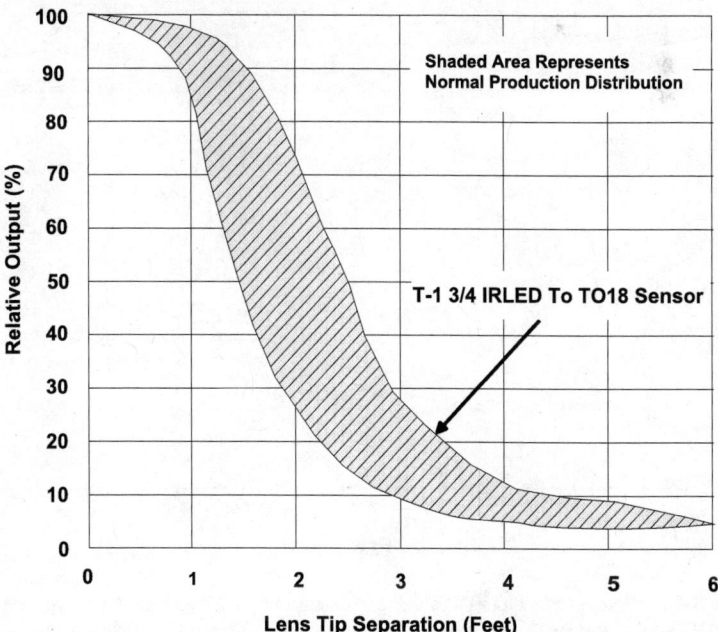

FIGURE 3.44 Coupling characteristics of plastic TO-18 phototransistor and GaAlAs IRLED versus separation between lens tips. Virtually all of the complex optical calculations can be avoided by the use of simple coupling characteristic curves provided by many manufacturers.

stress. Stress on the chip will cause any crystalline defects in the chip to expand along the planes of the crystalline structure in a process called *dark line defect formation*. This causes a degradation of power output as these crystalline defects result in a reduction in photon formation. As these defects expand, the total output for the chip is measurably decreased. Measurements made after the chip has been stressed will be lower. Once the chip has been mounted and encased by the manufacturer, the primary degradation is controlled by those stresses created by current density, temperature, and time. Most manufacturers will accumulate data on different package styles and different chips under varying stress levels. This data is then plotted, showing a linear decrease in relative output when time is plotted logarithmically. Some manufacturers feel that there is an annealing effect on the defect formation and that this linear line will become asymptotic to a fixed degradation percentage at some time. As this time period is too long to be tested empirically, the information is presented as a linear decrease versus logarithmic time in order to offer a conservative approach for the systems designer.

Initially (the first 120 hr), degradation data is very sporadic. Some units improve, some units remain flat, but most units will decrease at varying rates. Once these initial variations are complete, the devices will assume a linear decrease. The observed data is usually plotted as an average of the distribution. Sigma, or the standard deviation, is then calculated for each measurement point. The one sigma, two sigma, or three sigma plots can be added, and the system designer can then calculate the degradation rate for the system stress conditions.

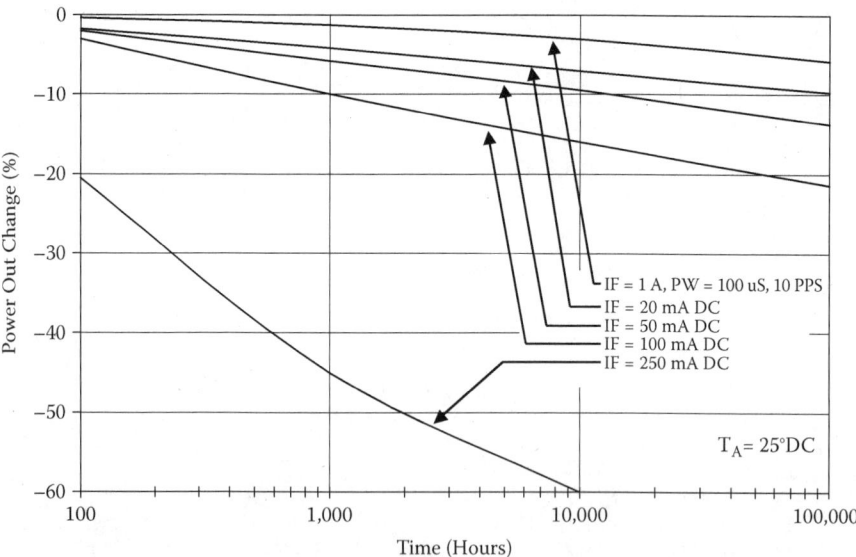

FIGURE 3.45 Percentage change in power output versus time for 0.016 in. × 0.016 in. GaAlAs IRLED in plastic T-1 3/4 package. Note that operation at higher forward current will increase the rate of degradation.

IRLED Packaging

One of the methods of minimizing this early variation is to pre-age or burn in the IRLEDs. There are two important points to be considered. If the system design is so critical that a pre-age or burn-in is required, then the length of time needs to be established. Twenty-four hours will remove most of the variation; 96 hr will essentially remove all of it. If the system is operated for testing or stabilization, then the burn in may be accomplished during this time. It is almost always bad to pre-stress, pre-age, or burn-in the device under accelerated conditions. Because of the normal way the current distribution in the IRLED behaves, an over stress at system operating levels occurs when accelerated aging is employed. This simply means that if the device is to be operated at some lower forward current than the level used for burn-in, then the emitted energy will be abnormally degraded at the lower forward current. For example, if the unit is to

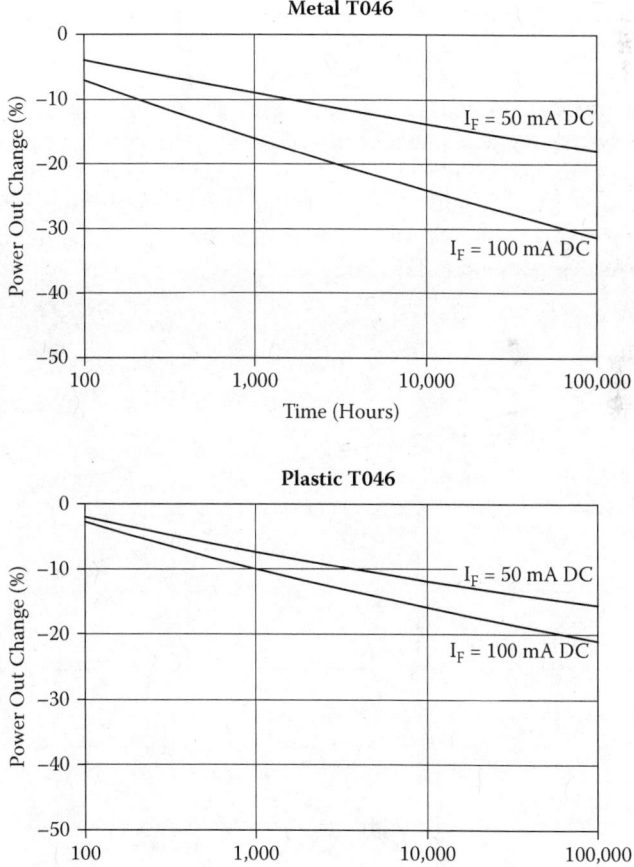

FIGURE 3.46 Percentage change in GaAs IRLED mounted in metal TO-46 package and plastic TO-46 package versus time.

be operated at 10 mA and aged at 50 mA, the percentage drop in energy output measured at 50 mA test condition will be lower than that measured using a 10 mA test condition. The accelerated burn-in destroys a useful portion of the 10 mA performance of the device.

Use of thermal derating curves requires an understanding of thermal properties as well as a starting point. The thermal derating implies that the stress level is a linear derating function from 25°C (as temperature is increased) to the temperature at which the stress level is zero. For ease of calculation, we will assume that V_F is constant. If a device is rated for 150 mW at 25°C and V_F at 100 mA is 1.5 V, then I_F maximum is 100 mA. If the device derates to zero at 125°C, then the device will be rated for 100 mA at 25°C, 75 mA at 55°C, 50 mA at 75°C, etc. The 100 mA degradation curve at 25°C would be identical to the 75 mA degradation curve at 50°C, which would be identical to the 50 mA at 75°C, etc. This also is a conservative approach because V_F is decreasing with both decreasing current and increasing temperature. In fact, a unit operated in the previous example will exhibit improved degradation rates as the temperature is increased.

The net result is that a family of curves for various forward currents at a 25°C ambient with associated sigma is all that is necessary for the designer to be able to calculate the system degradation rate. Such a family of curves is shown for a 0.016 in. × 0.016 in. GaAlAs IRLED in a plastic T-1 3/4 package in Figure 3.45.

The curves shown in Figures 3.45, 3.46, and 3.47 show that increasing current causes a faster degradation rate, increasing temperature causes a faster degradation rate, and GaAs degrades more slowly than GaAlAs.

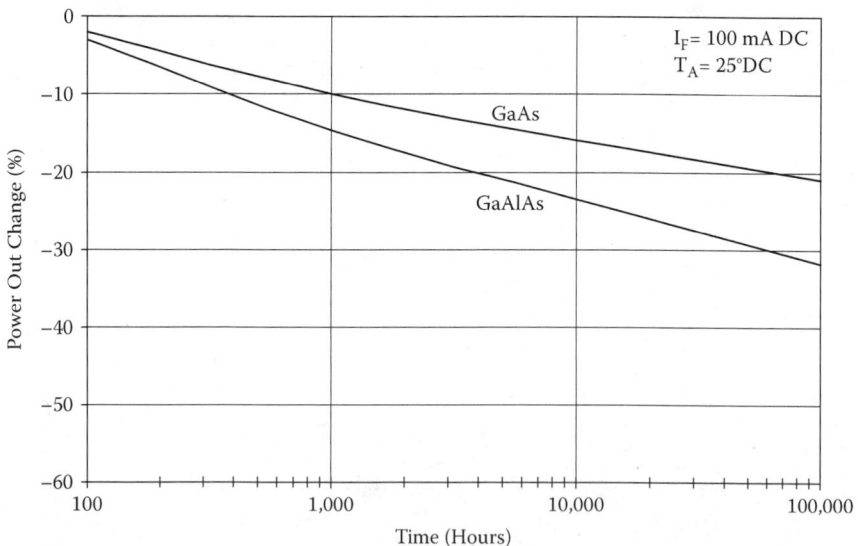

FIGURE 3.47 Percentage change in GaAs and GaAlAs IRLED mounted in metal TO-46 package versus time under same stress conditions. Note that GaAs shows less degradation than GaAlAs at identical forward currents.

Catastrophic failure due to thermal or mechanical stress results from forces on the IRLED chip or wire bond that in turn shear or break the bond wire. If such failures occur, they usually occur early in the operating life of a device. Because the machine fabrication of the plastic part is very repeatable and mechanically accurate, there is a lower probability of obtaining a weak bond. Machine fabrication of plastic devices tends to offset the low-shear-stress environment of the air chamber inside a metal package. The net effect is equal reliability. Catastrophic failures are very infrequent in plastic and metal package styles as long as the manufacturer properly controls production. As a result, neither plastic nor metal can be said to be less reliable than the other. The greater variation is found among the different manufacturers.

3.8 CONCLUSION

Power measurement of IRLEDs varies more than any other parameter among different manufacturers. Part of the difference is in interpretation of the definitions of the parameters measured, and part is in the technique used.

Part 2

The Receiver (Silicon Photosensor)

4 The Photodiode

4.1 BASIC THEORY

The formation of a P-N junction in silicon and the corresponding built-in potential barrier is identical to the process described in Chapter 1, Section 1.2, for IRLEDs. However, in the receiver or silicon P-N photodiode, the main objective is the generation of photocurrent. Figure 4.1 shows a typical SMD-packaged photodiode. A typical photodiode chip is shown in Figure 4.2.

Semiconductor material will generate a photocurrent on absorption of incident radiation. This is known as the *photoelectric effect*. Once the photoelectric effect takes place within the semiconductor material, photons excite electrons and move them from the valence band into the conduction band. This process is explained in more detail in the following text.

4.1.2 Photoelectric Effect

Semiconductor material contains energy states that can be occupied by electrons that are limited to bands of energy, as shown in Figure 4.3. When all electrons are in their lowest energy states, they are occupying states in the valence band. The gap between the valence and conduction band is known as the band gap of the semiconductor material. Typically, $E_6 = 1.12$ eV. This is the required energy the material must absorb to release one electron from the valence band to the conduction band. In the conduction band, electrons are free to move about the crystal. A hole in the valence band is left for each electron that moves to the conduction band. Photoexcitation in P-type material results in the creation of additional minority carriers (electrons), which generate a photocurrent on crossing the P-N junction.

With zero-bias voltage applied to the semiconductor material, the minority carriers must reach the P-N junction by diffusion. When the electron–hole pair is formed more than a diffusion length from the junction, simple recombination is likely to occur with no contribution to photocurrent. Diffusion current flows without any electric field, which is generated once an external bias is applied across the P-N junction.

The diffusion current is given by:

$$J_n = qD_n \frac{dn}{dx} \qquad (4.1)$$

where

J_n = Current density, amperes/cm^3
D_n = Diffusion constant
n = Electron density, electrons/cm^3

FIGURE 4.1 Example of a PLCC (SMD) package. Photograph showing the package, chip, and the bonding wire.

FIGURE 4.2 Example of a typical photodiode chip. The large area is the active P-N junction. The two areas on the left and right are the anode contacts of the photodiode. The backside of the chip is the cathode.

Electrons and holes in the semiconductor are in constant thermal motion. In the diffusion current equation in the preceding text:

$$D_n = \mu_n V_T \text{ and } D_p = \mu_p V_T \tag{4.2}$$

$Vt = KT/q$ is the thermal voltage and is equal to approximately 26 mV at room temperature

K is Boltzmann's constant

q is electron charge 1.6×10^{-19} C

T is the absolute temperature, and μ_n and μ_p are the electron and hole mobilities, respectively

The Photodiode

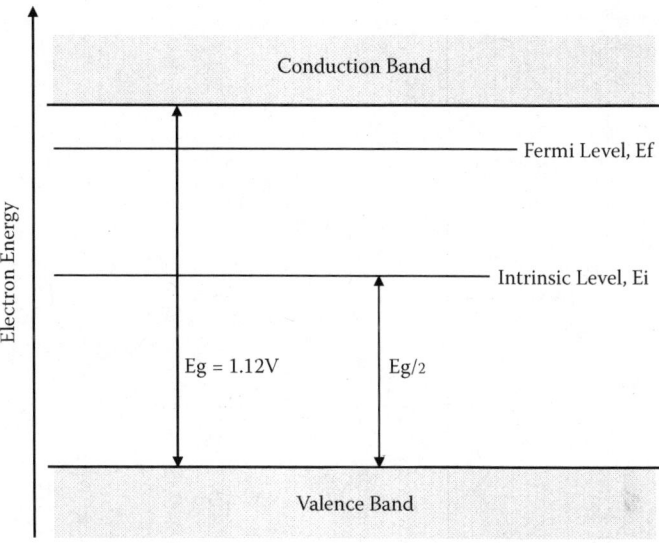

FIGURE 4.3 Electron energies in silicon. Electrons free to move about the crystal occupy states in the conduction band. Valence electrons attached to silicon atoms occupy the valence band. The intrinsic level is approximately halfway between the conduction and valence bands. The Fermi level shown corresponds to N-type silicon.

Response time for a device is relatively slow when there is no electrical field present to accelerate newly formed minority carriers toward the P-N junction. This mode of operation is known as *photovoltaic operation*. When an external bias is applied to the semiconductor, the response time speeds up dramatically with the addition of a depletion region near the junction. This region has a field gradient sufficient to accelerate the newly formed electrons across the junction. This type of operation is referred to as the *photoconductive mode*. Figure 4.4 shows the electric field with the recombination of electron–hole pairs and the separation leading to photocurrent. Analysis of Figure 4.4 shows that the electric field is not uniform. The field is much stronger in the depletion region than it is in the adjacent N-type or P-type regions. The field shown applies to the diffused junction shown in Figure 4.5.

Figure 4.5 shows the depletion region in a diffused structure with respect to penetration into the N-type starting material and the P-type diffused region. It should be remembered that in the P-type region, the majority carriers are holes and the minority carriers are electrons. In the N-type region, on the other hand, the majority carriers are electrons and the minority carriers are holes.

As the reverse voltage across the P-N junction is increased, the depletion region will widen. For a given voltage, the minority carriers within that region govern the diffusion into the P- or N-type regions. For charge neutrality, the depletion region has to deplete the same number of minority carriers from the N-type region as are depleted from the P-type region for a given voltage. The number of minority carriers within the P-type region is significantly higher than the number within the N-type region. Therefore, the depletion region is wider in the N-type region than in the P-type region. The depletion region exists even with zero applied bias. The "built-in" potential

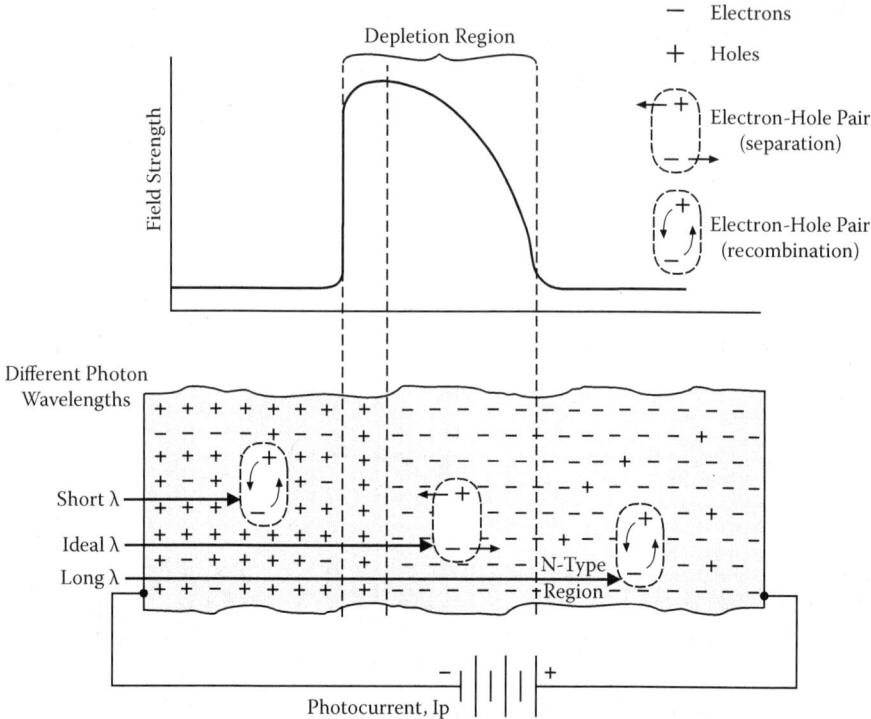

FIGURE 4.4 P-N photodiode junctions. Photocurrent created within internal field. Photons traverse the crystal a distance that is dependent on their wavelength. Electron–hole pairs that are far away from the junction or not generated in the presence of a strong field usually recombine before a photocurrent can be generated.

discussed in Chapter 1, Section 1.2, is the same for the P-N junction of a silicon photodiode. Reverse bias aids this built-in field and thus expands the depletion region.

A photodiode has the best performance when the largest number of photons is absorbed in the depletion region. An ideal photodiode would have a depletion region wide enough to allow absorption of the desired photon wavelengths within this depletion region. The depth which a photon travels before it is absorbed is a function of its wavelength. Short photon wavelengths (such as ultraviolet or visible energy) are absorbed near the surface, whereas those having longer wavelength (such as infrared) may penetrate the entire thickness of the semiconductor crystal.

4.2 OPTIMIZATION

The optimization of response time and bandwidth between a P-N junction photodiode and a PIN photodiode is somewhat analogous. Both devices require minimizing capacitance and controlling the width of the depletion layers.

The response time for all practical purposes of the junction photodiode is largely dependent on the transit time of the carriers across the depletion layer. The wider

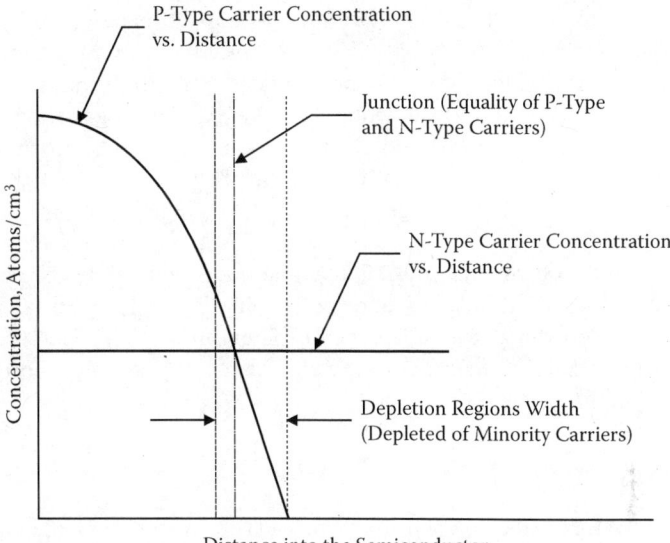

FIGURE 4.5 Depletion regions in a diffused P-N junction. The wider depletion region in the N-type material results from the fact that N-type material has a lower concentration of minority carriers than P-type material. Equal numbers of minority carriers must be depleted on either side of the junction in order to build an electrical potential equal to the reverse-bias voltage.

the depletion layer, the longer it takes the carriers to cross. Applying a reverse bias across the P-N junction to create an electric field, which accelerates the carriers across the junction, and/or decreasing the depletion width can decrease this time.

In the PIN photodiode, the response time is proportional to the i-layer, which for this report is analogous to the depletion layer. By decreasing this i-layer, the transit time can be reduced proportionally. Further reduction in transit time can be accomplished by applying moderate reverse bias to allow the carriers to drift across the i-layer.

For optimal design, it is normal to have

$$t_{tr} = \frac{1}{2} \text{ (modulation period) and } t_{tr} = \tau_{RC} \text{ the RC constant} \quad (4.3)$$

To further improve the speed of the photodiode, processes are modified to lower the internal resistance of the bulk material. An accepted practice is to thin the photodiode to reduce the impedance from the backside contact to the junction. However, in processing, this will cause a yield reduction due to breakage. Another method is to grow a lightly doped epitaxial layer on a highly doped substrate. This lowers the internal resistance, improving the speed of the photodiode.

Often times in photodiodes, there is a compromise between responsivity of the device and the bandwidth. Responsivity is a measure of the photocurrent generated in response to an input light source. Bandwidth is a measure of the time it takes the output to respond to the input light source (transit time).

Wider depletion or i-layers allow for more photons to be absorbed by the photodiodes, thus increasing responsivity. However, this slows down the transit time, reducing bandwidth. Decreasing the depletion or i-layer speeds up the transit time and allows for an increase in bandwidth. This approach does not allow all of the photons to be absorbed by the photodiode and lowers the device's responsivity.

Additional improvement to the response of the photodiode to a particular wavelength is achieved by adding an antireflective (AR) coating. This coating adjusts the refractive index from air to the crystal. The layer is a multiple of a quarter wavelength of the light source, which yields maximum absorption into the photodiode, whereas the half-wave will maximize the reflection.

Using the same theory, several layers of different films with different refractive indexes can be added to the surface of the photodiode to pass only a portion of the light's wavelength.

4.3 CHARACTERIZATION

Operation of the P-N or the PIN photodiode may either be in the forward or the reverse mode. Operation in the forward mode with zero bias is called the *photovoltaic mode*. The photons impinging on the photodiode generate a voltage potential. Common applications using this mode employ a P-N photodiode operated as a voltage source, also commonly called a *solar cell*. Operation with reverse bias or in the reverse mode is called the *photoconductive* or *photocurrent mode*. When comparing the photovoltaic mode to the photoconductive mode, the photoconductive mode offers the following advantages:

1. Higher speed.
2. Improved stability.
3. Larger dynamic range of operation.
4. Lower temperature coefficient.
5. Improved long-wavelength response within the planar diffused region.
6. Response to the short wavelengths in the edge region adjacent to the planar diffusion. (The formation of the depletion region at the surface allows hole–electron pairs to be created from the short wavelengths in the depletion region rather than in the N-type or P-type regions.)

Figure 4.6 shows the relative response against wavelength for a PIN photodiode made with starting material of 1000 Ω/cm N-type material or a P-N photodiode made with 10 Ω/cm N-type material. The geometry of the devices is identical.

Note that the P-N photodiode peak relative response is at a shorter wavelength. This is due to the narrower depletion region located closer to the surface, and hence, the loss of useful hole electron pairs created by the longer-wavelength photons. Overlaying this with the GaAs and GaAlAs spectral emission versus wavelength plots in Chapter 3, Figure 3.29 clearly shows the better compatibility of the PIN photodiode to GaAs and GaAlAs IRLED emission.

The total capacitance versus reverse-bias voltages on the same two photodiodes is shown in Figure 4.7. Note the improvement in capacitance inherent in the PIN photodiode. This sharply illustrates the wider depletion region for a given voltage

The Photodiode

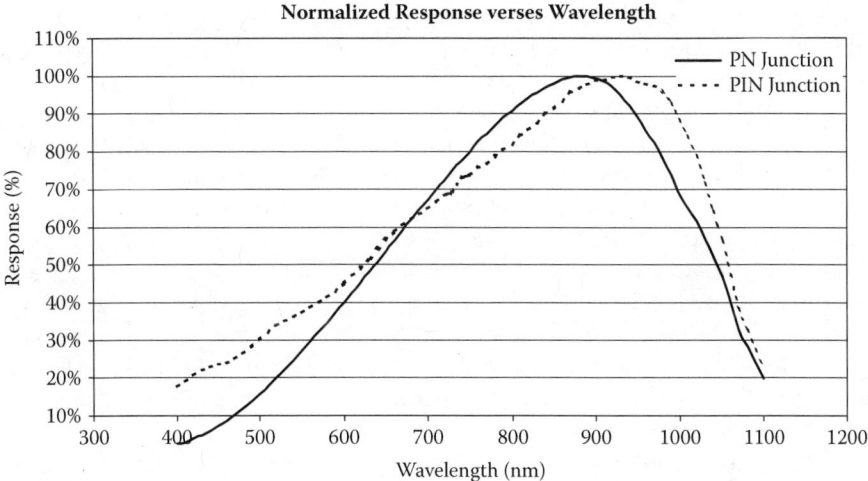

FIGURE 4.6 Relative response versus wavelength. The wider depletion region of the PIN photodiode results in longer wavelengths being absorbed within the region and forming useful electron–hole pairs. The net result is a relative response peak at a longer wavelength than that of a typical P-N photodiode.

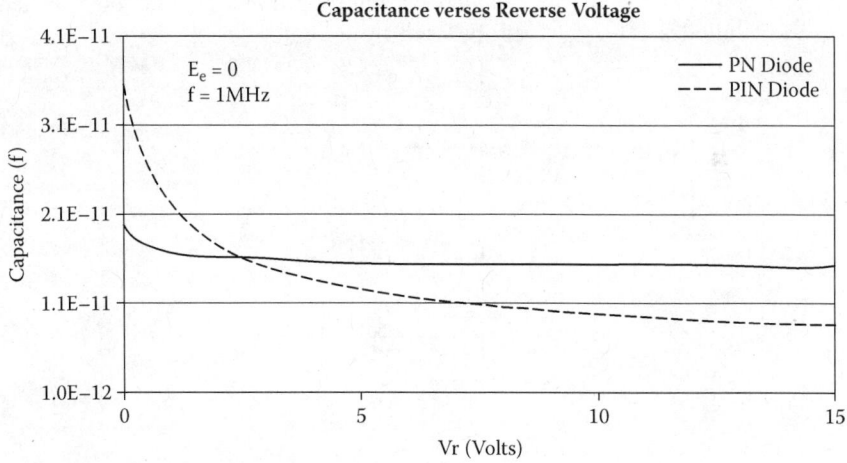

FIGURE 4.7 Capacitance versus reverse-bias voltage. The physical separation of charge layers at the edge of the depletion region is significantly greater for the P-N photodiode. This results in a lower capacitance, making a higher operating speed possible.

across the PIN structure and, hence, the lower capacitance. The plates of the capacitors are simply the cross-sectional area of the P-type and N-type regions adjacent to the depletion region with the width of the depletion region being the separation or distance between the plates. The low capacitance of the photodiode when operated in the depleted mode allows the unit to operate at extremely high speeds. The response time for a given PIN photodiode depends on the areas of the photodiode that are irradiated, the amount of reverse bias applied, the capacitance, and the value of the effective load resistance.

Photocurrent begins to flow within a few picoseconds after the photons impinge on the photodiode, but there is junction, package, and stray wiring capacitance to be charged. This then makes the rise or fall time largely dependent on the load resistance unless the load resistance is so small that the internal resistance of the photodiode limits the speed. When the effective load resistance approaches 100 Ω, then this effect becomes pronounced. A PIN photodiode with effective load resistance in the 100 to 500 Ω range should switch in the 0.5 to 5 ns range for both rise and fall time. Because the normally available IRLEDs discussed in Part 1 will operate only up to a 1 MHz rate, it is readily seen that almost all P-N and PIN photodiodes are more than adequate to use as receivers. Figure 4.8 shows a recommended amplifier circuit for a photodiode operated in a linear mode.

The negative-going input is very close to ground potential. The dynamic resistance seen at this negative-going input by the photodiode is R_1 divided by the loop gain. If the operational amplifier has extremely high input resistance, then the loop gain closely approximates the forward gain of the operational amplifier. This type of application under open-air communications will be discussed in more detail in Chapter 5.

Another major advantage of the P-N or PIN photodiode is the linearity of response with respect to the quantity of photons impinging on the surface. These photodiodes can effectively be used from photon levels that will produce

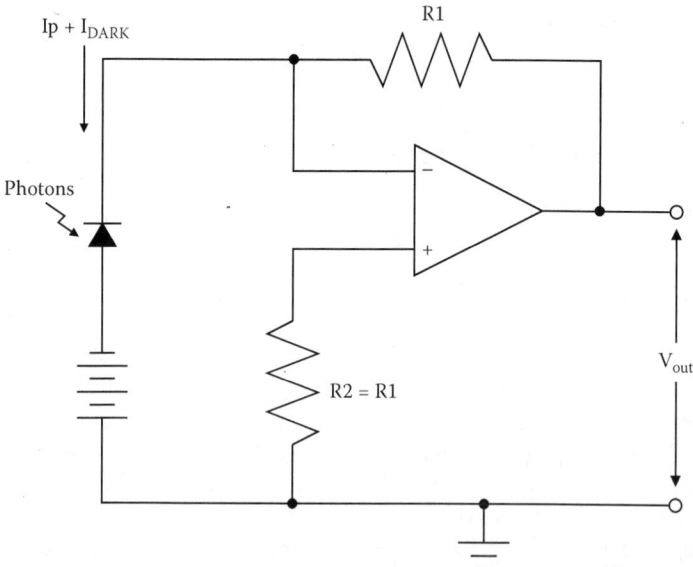

FIGURE 4.8 Linear mode circuit for photodiode and amplifier. Although offering the advantage of fast rise and fall times, photodiodes almost always require the addition of an amplifier to be useful as infrared detectors.

The Photodiode

photocurrent slightly above the dark current level 50 to 100 pA (e.g., 50×10^{-12} A) to direct sunlight in the 80 mW/cm² level. This corresponds to approximately nine orders of magnitude. Figure 4.9 shows this linearity graphically for both a P-N or PIN photodiode.

The photocurrent of a P-N or PIN photodiode depends on a number of variables. Ideally, each photon (quantum of energy) should cause one electron to be added to the photocurrent. "Quantum efficiency" is therefore dimensioned as "electron per photon." Most manufacturers express this in terms of the *flux responsivity* (R_ϕ). This takes into account the photon energy, and represents the ratio of photocurrent to the amount of spot flux:

$$R_\phi = n_q \left(\frac{\lambda}{1240} \right) = \frac{I_p}{\phi_e} \tag{4.4}$$

where
R_ϕ = flux responsivity in amps per watt
I_p = photocurrent in amperes
ϕ_e = radiant flux in watts
n_q = quantum efficiency in electrons per photon
λ = photon wavelength in nanometers

R_ϕ will approximately follow the curve shown in Figure 4.6, where the relative response normalized to 1.0 is changed to flux responsivity with a maximum of approximately 0.60 A/W. This is shown in Figure 4.10.

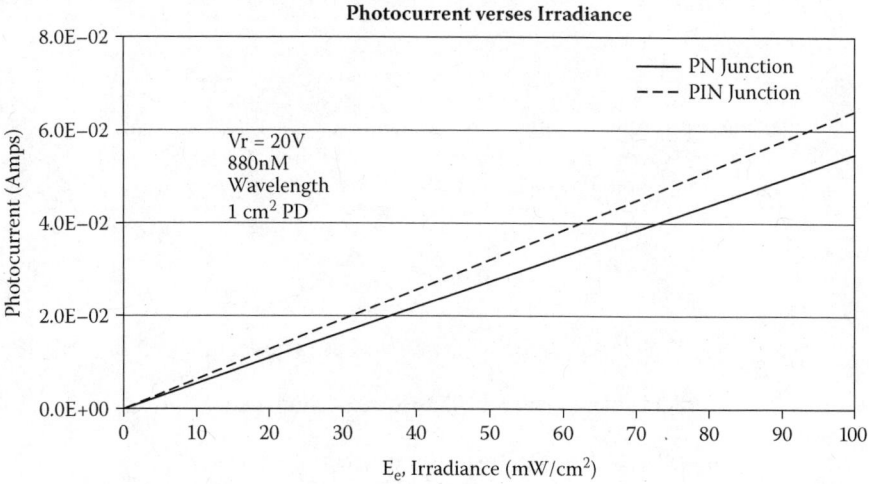

FIGURE 4.9 Photocurrent versus irradiance. The graph is for a P-N and a PIN photodiode. Linearity over nine decades is achievable for photodiodes. Irradiance levels range from virtually dark to sunlight (10^2) and higher.

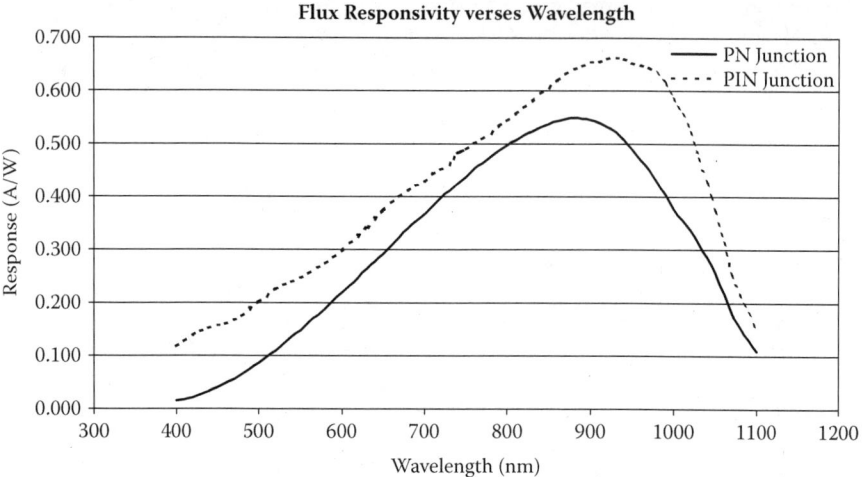

FIGURE 4.10 Flux responsivity versus wavelength. Flux responsivity is analogous to the relative spectral response. Controlling variables include the width of the depletion layer itself and the distance from the surface of the crystal to the depletion layer.

The quantum efficiency on the PIN photodiode at 900 nm with R_ϕ equal to 0.6 A/W is

$$0.6 = n_q \left(\frac{900}{1240} \right)$$

$$n_q = 0.83 \qquad (4.5)$$

$$n_q = 83\%$$

Another measure of performance is the incidence response, R_E, which adds the variable of the photosensitive area (or the magnified area if a lens is used). This is equal to the ratio of photocurrent to incidence:

$$R_E = \frac{I_P}{E_e} = \int \left[R_\phi (A_D) \right] d A_D \sim R_\phi A_D$$

R_E = incidence response in amps/watts/square centimeters

E_e = radiant incidence in watts/square centimeters   (4.6)

A_D = effective photosensitive area in square centimeters

I_P, R_ϕ (from previous equation)

Assume a radiant incidence of 2 mW/cm² at 900 nm (peak response) and a diode effective area of 0.065 cm² (~0.100 in. × 0.100 in.):

The Photodiode

$$I_p \sim E_e R_\phi A_D$$
$$\sim 2\,\text{mW/cm}^2 (0.6\,\text{A/W}) 0.065\,\text{cm}^2 \quad (4.7)$$
$$\sim 0.13 \times 10^{-3} (0.6\,\text{A/W})$$
$$\sim 78\,\mu\text{A}$$

(assumes radiant incidence at 900 nm or peak response),

There is noise associated with operation of the photodiode. The two types of noise to be considered are *flicker* noise and *shot* noise. The two components of current that contribute to this noise are junction current and leakage current. The junction current causes *shot* noise, whereas the leakage or dark current causes thermal noise from the leakage resistance and *flicker* noise. A worst-case value of the leakage current noise is obtained by applying the *shot* noise formula to the entire dark current. This thermal noise is usually dominant below 20 Hz.

When the frequency is increased to greater than 20 Hz, the shot noise becomes dominant. The signal-to-noise ratio is defined as the ratio of the photocurrent, when signal is applied, to the noise current, when there is no signal. This signal-to-noise ratio increases as the square root of the photodiode area because the signal rises linearly with the area, whereas noise current varies as the square root of the area. The noise equivalent power (NEP) is defined as the signal flux level when the signal-to-noise ratio is one and the bandwidth is narrow (1 to 10 Hz). The NEP varies inversely with the responsivity (see Figure 4.6). It will thus be minimum at the peak wavelength and increase slightly as the wavelength of the photons increases or decreases.

The formula for calculating the thermal noise is

$$\frac{I_{N(\text{Thermal})}}{\sqrt{B}} = 25.3\sqrt{I_s(\text{nA})} \quad (\text{fA}/\sqrt{\text{Hz}}) \quad (4.8)$$

where

$I_N/\sqrt{B}$ = bandwidth-normalized noise current in femtoamps per root hertz
I_s = reverse saturation current in nanoamps

A sample calculation of thermal noise from the stated formula will be useful in understanding the magnitude of the noise. A PIN diode with an effective area of 0.065 cm² (0.100 in. × 0.100 in.) will be used. Leakage current is assumed to be 25×10^{-9} A.

$$\frac{I_N}{\sqrt{B}} = 25.3 \sqrt{25 \times 10^{-9}}$$
$$= 25.3 \times 1.58 \times 10^{-4}$$
$$= 400 \times 10^{-5}\,\text{fA/root hertz} \quad (4.9)$$
$$= 4.0 \times 10^{-18}\,\text{A/root hertz}$$

The formula for calculating the shot noise is

$$\frac{I_N(\text{shot})}{\sqrt{B}} = 17.9\sqrt{I_{dc}(\text{nA})} \quad (\text{fA}/\sqrt{\text{Hz}}) \tag{4.10}$$

I_{dc} = total dark current in nanoamps
$I_N(\text{shot})/\sqrt{B}$ = Bandwidth-normalized noise current in femtoamps per root hertz

A sample calculation of shot noise from the stated formula will be useful in understanding the magnitude of the noise. A PIN diode with an affective area of 0.065 cm² (0.100 in. × 0.100 in.) will be used. Leakage current is assumed to be 25 × 10⁻⁹ A.

$$\begin{aligned}
\frac{I_N}{\sqrt{B}} &= 17.9 \ \sqrt{25 \times 10^{-9}} \\
&= 17.9 \times 1.58 \times 10^{-4} \\
&= 28 \times 10^{-4} \ \text{fA/root hertz} \\
&= 28 \times 10^{-19} \ \text{A/root hertz}
\end{aligned} \tag{4.11}$$

The formula for calculation of NEP is

$$\text{NEP} = \frac{I_N/\sqrt{B}}{R_\phi} \quad \left(\text{fW}/\sqrt{\text{Hz}}\right) \tag{4.12}$$

where
NEP = the radiant signal flux at a specified wavelength required for unity signal-to-noise ratio normalized for bandwidth
$I_N/\sqrt{B}$ = bandwidth-normalized noise
R_ϕ = flux responsivity in amps/watt at a given wavelength

A sample calculation for NEP from the stated formula will be useful in understanding the magnitude of NEP. A PIN diode with an effective area of 0.065 cm² (0.100 in. × 0.100 in.) will be used. A flux responsivity of 0.6 A/W will be used.

$$\begin{aligned}
\text{NEP} &= \frac{28 \times 10^{-19}}{0.6} \\
&= 47 \times 10^{-19}
\end{aligned} \tag{4.13}$$

The leakage current or dark current (I_{dc}) is the major drawback to operation in the photoconductive mode. This current refers to the current that flows when no radiant flux or photons are applied to the photodiode. Dark current consists of two components:

1. Surface leakage
2. Bulk leakage

The surface leakage of a diode approximates a fixed number, whereas the bulk leakage approximately doubles for each 10°C increase in temperature. When this is plotted on log-linear paper, it results in a curve similar to Figure 4.11. At 25°C and below,

The Photodiode

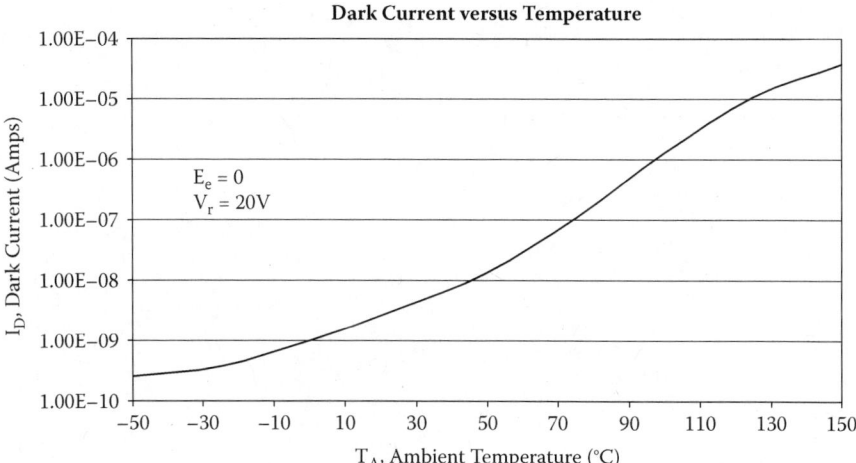

FIGURE 4.11 Dark current versus ambient temperature. The shape of this curve is similar to a plot of bulk leakage versus temperature. The dark current is composed of a fixed amount of surface leakage plus bulk leakage, which approximately doubles for each 10°C increase in temperature.

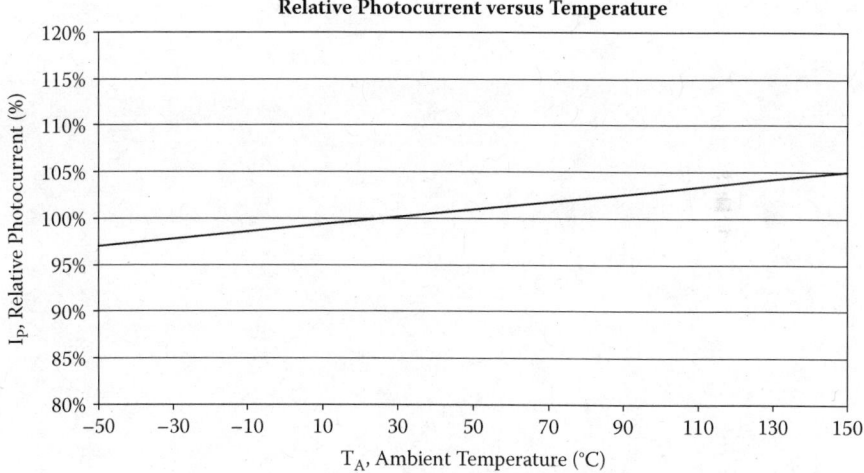

FIGURE 4.12 Relative photocurrent versus ambient temperature. As with dark current, only less dramatically so, relative photocurrent increases with increasing temperature. In the case of phototransistors, on-state collector current varies with temperature to a much greater extent, and in direct proportion to, the h_{FE} (current gain) of the device.

the surface leakage will usually dominate. At high temperature, the bulk leakage will dominate. The photocurrent (I_p) of the photodiode will vary with temperature. This positive temperature coefficient is shown in Figure 4.12. This change is a significantly lower percentage than is found in multiple junction photosensors units such as a phototransistor. The leakage in a phototransistor is the diode leakage multiplied by the h_{FE} (current gain) corresponding to that collector current and bias voltage level.

5 The Phototransistor and Photodarlington

5.1 BASIC THEORY

A phototransistor operates in a manner similar to a conventional small signal transistor except that the base or control current can come from both impinging photons on the depletion region or traditional base current supplied to the base–emitter junction. The impinging photons utilize the depletion region formed by the reverse-biased collector–base junction to create photocurrent, which acts as base current or control current. Most phototransistors are NPN types because this material lends itself to both ease of manufacture and useful electrical parameters. The hole–electron pairs created in the collector–base depletion region cause photocurrent. The electrons move toward the collector or N-type region, whereas the holes move toward the base or P-type region. As described in the basic theory of the P-N junction covered in Chapter 4, Section 4.1, Figure 5.1 shows the distribution of impurity concentrations in a diffused NPN transistor.

The series of statements made in the following text summarizes the basic dimensions and concentration levels within a typical phototransistor. This information corresponds to the data presented.

Single-crystal silicon has a density of 5×10^{22} atoms/cm^3.
N-type emitter impurity concentration at the surface is 5×10^{20} atoms/cm^3.
The emitter–base junction is approximately 2 µm deep (dependent on beta).
The base width is approximately 2 µm (dependent on beta).
The collector–base junction is approximately 4 µm deep.
P-type base impurity concentration at the surface is 1×10^{19} atoms/cm^3.
N-type starting material impurity concentration is 1×10^{15} atoms/cm^3.
The finished NPN transistor is usually 250 µm thick.
The collector region is approximately 240 µm thick.

Figure 5.2 shows the collector current versus collector emitter voltages for varying levels of irradiance (or photons) impinging on the base of the NPN transistor.

With zero photons impinging on the base region, the collector-to-emitter voltage is gradually increased, and only leakage current (dark current) will flow until breakdown is reached. If the quantity of photons is increased to a certain level and stabilized, and the photon level is then increased again and stabilized with a corresponding increase of collector-to-emitter voltage, a family of characteristic curves will be formed. The current gain of the transistor is the collector current divided by the base current at a particular voltage. The base current corresponds

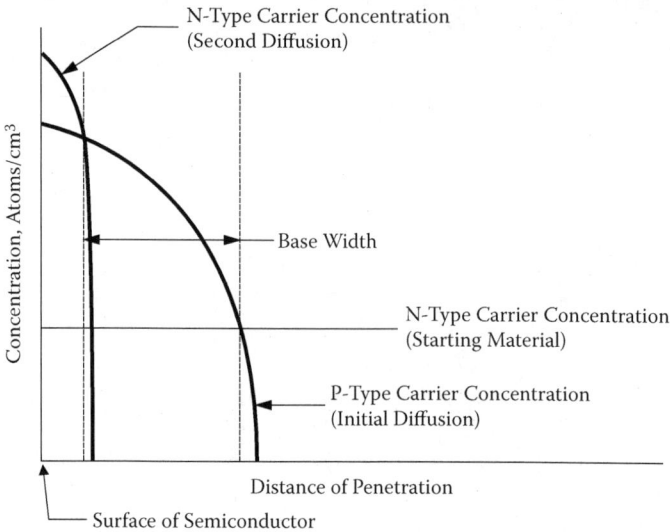

FIGURE 5.1 Impurity concentrations for NPN transistor. To form a transistor, P-type impurities are diffused into N-type starting material. N-type material is then diffused into the newly formed P-type material, leaving two junctions for the NPN transistor.

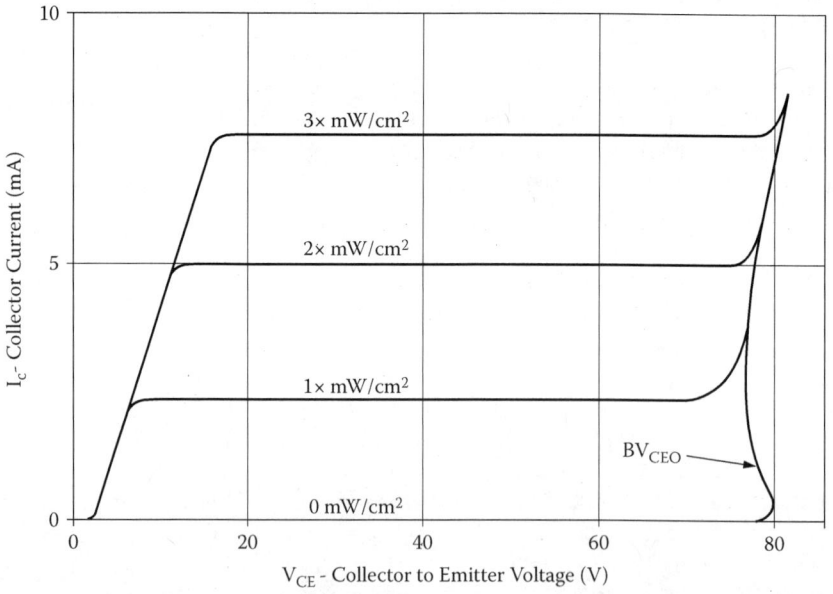

FIGURE 5.2 Collector-to-emitter voltage versus collector current. This family of curves is analogous to traditional transistor I_C versus V_{CE} curves except that incident radiation on the base substitutes for base current. Each step in the light level is the analog to an incremental increase in base current, resulting in a family of curves.

to photocurrent ($I_b = I_p$), as discussed in Chapter 4, Section 4.2, on photocurrent for P-N diodes. These curves will be discussed in more detail in the characterization section.

5.1.1 Phototransistor

Figure 5.3 shows the top of a phototransistor chip. The emitter diffusion is the chevron shape in the upper-left corner of the top view. The phototransistor is optimized for a large exposed base area to increase the quantity of photons that can strike it. This creates a higher-than-normal capacitance that compromises the frequency performance of the unit and will be discussed later in Section 5.2. The current gain, or h_{FE}, of a phototransistor typically can be varied from 50 to 2000 by controlling the base width and doping concentrations in the base and emitter. Figure 5.4 shows the side view of the same phototransistor. The view is along the dashed line in Figure 5.3. This view shows the diffusions and contact metallization.

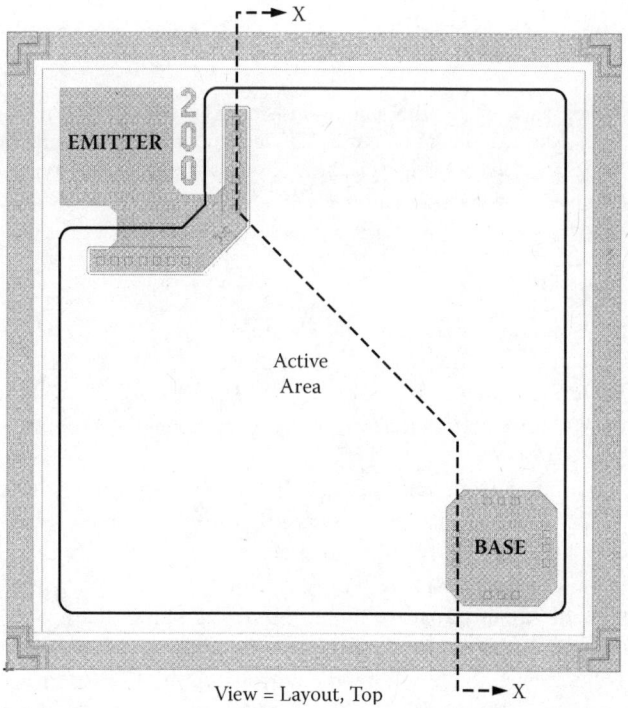

FIGURE 5.3 Top view of an NPN phototransistor. In an NPN phototransistor, the maximum top area possible is allotted to base diffusion to provide as much light-sensitive area as possible. The larger base area increases proportionally the base–collector capacitance, decreasing the speed of the transistor.

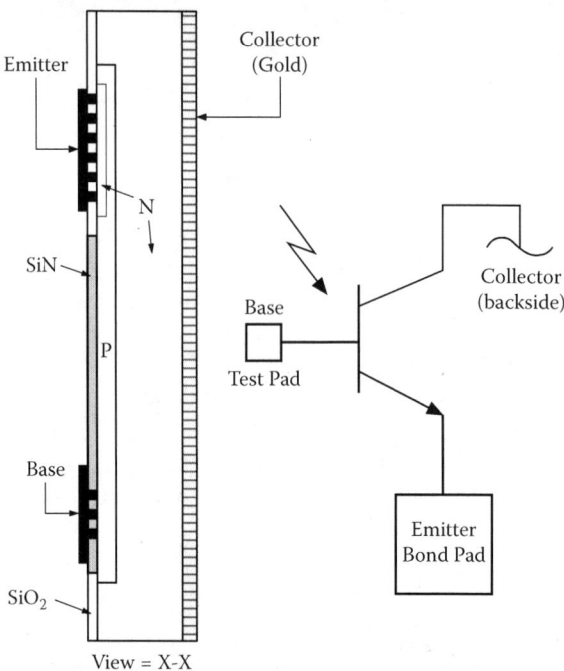

FIGURE 5.4 Side view of an NPN phototransistor. The cross-sectional view shows the diffusions along the dashed line from Figure 5.3. The emitter and base contacts are typically aluminum, whereas the collector contact is the backside of the chip and is typically a sintered gold contact. The electrical symbol is for the NPN phototransistor.

5.1.2 Photodarlington

Figure 5.6 shows the schematic of a photodarlington transistor. The unit has a common collector with the emitter of the input transistor driving the base of the output transistor. This effectively squares the current gain or sensitivity, but significantly decreases the switching performance. This type of device is sometimes referred to as *cascaded transistors*.

The photodarlington transistor gain usually varies from 1000 to 100,000. When gains are needed below 1500, the phototransistor is used, whereas gains above 25,000 usually become unstable (particularly at high temperatures). The photodarlington transistor can be constructed to be highly sensitive to low light levels by making the base area of the input transistor large. Figure 5.5 shows the top-view layout of a photodarlington transistor. The output transistor occupies the upper-left corner of the chip. This photodarlington will handle currents in the 10 to 20 mA range before having a high current h_{FE} roll-off. By enlarging the emitter (Q_2) of output transistors, improvements can be obtained in the Vce/Ic characteristics. Some photodarlingtons can handle up to 200 mA of IC_{ON}.

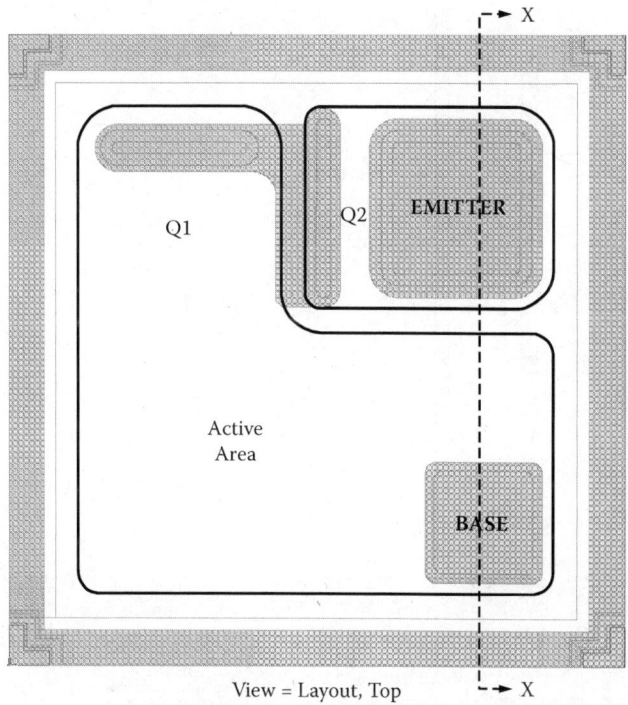

FIGURE 5.5 Top view of a darlington or cascaded transistor. The emitter of the photosensitive transistor drives the base of the second transistor, approximately squaring the h_{FE}. Current gains from 1000 to 100,000 are possible. Operating speed is traded to achieve this very high gain. VCE_{sat} are much higher ($T_2\ VCE_{sat} + T_1\ V_{be}$).

5.1.3 R_{BE} Phototransistor

Many applications of the phototransistor are used in a high ambient light condition. The transistor cannot differentiate between the ambient light and the light from the signal source, so it will amplify both equally. The R_{BE} phototransistor is the answer to this problem. On the chip with a phototransistor is a high-value resistor connected from the base to the emitter. A typical R_{BE} is shown in Figures 5.7 and 5.8. The resistor that is in parallel with the base–emitter junction will allow photocurrent to be generated (ambient light) without the transistor's gain. This lower photocurrent is leakage current. Once the photocurrent is great enough to generate a voltage across the base–emitter junction equivalent to a V_{BE}, the transistor becomes forward-biased and will amplify the rest of the photocurrent (the signal).

The basic construction method for these types of transistors employs standard bipolar technology techniques and will be discussed later in Chapter 6.

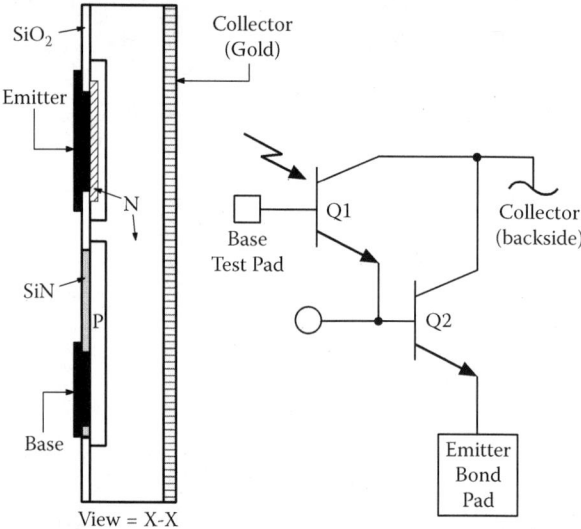

FIGURE 5.6 Side view of the darlington transistor. The cross-sectional view shows the diffusions along the dashed line from Figure 5.5. The emitter and base contacts are typically aluminum, whereas the collector contact is the backside of the chip and is typically a sintered gold contact. The electrical symbol is for the NPN darlington phototransistor.

5.2 CHARACTERIZATION

The NPN transistor has two junctions, the collector–base and the base–emitter. The reverse breakdown of the collector–base junction is shown in Figure 5.9. As the voltage is increased above zero, a small leakage current identical to the dark current in the P-N diode is observed. This leakage current will increase slightly with increasing voltage. At some point determined by the impurity concentration and the gradient of this impurity concentration versus distance, the depletion region becomes wide enough that a carrier (hole or electron) traveling through it will strike and dislodge a like carrier from the lattice structure of the atom. Increasing the voltage further will cause high currents to flow at the *breakdown* or *avalanche voltage*. Under this condition, the heat generated can cause damage to the device.

If a voltage is applied between the collector and emitter, a similar phenomenon happens but at a lower voltage. This is shown in Figure 5.10. The observed leakage current is higher due to h_{FE} caused by the forward bias on the base–emitter junction, but because h_{FE} is low at low currents, that is only slightly higher than that observed on the collector–base diode.

As the voltage approaches the avalanche voltage (the sum of the voltage drop across the forward-biased emitter–base diode, and the voltage drop across the reverse-biased collector–base diode), a similar phenomenon occurs. Due to the narrow base width created by the wide depletion region at high voltages, when avalanche

The Phototransistor and Photodarlington

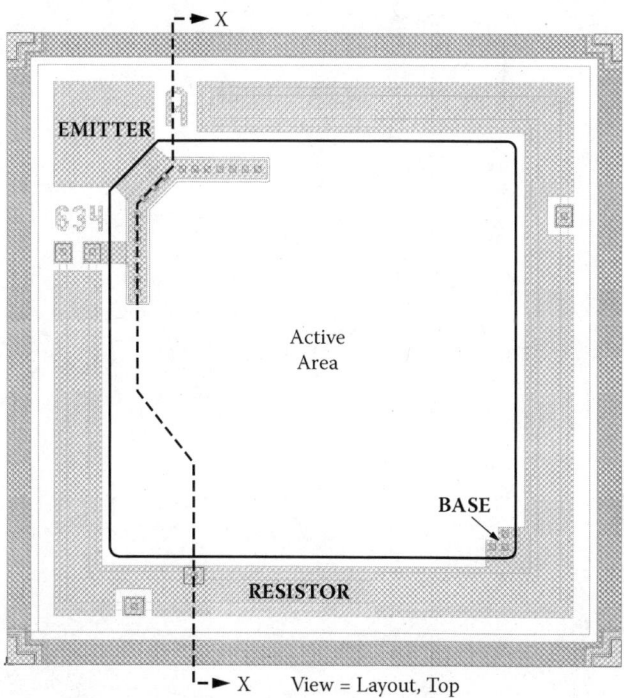

FIGURE 5.7 R_{BE} transistor top view. The large value of the resistor across the base–emitter junction will allow high ambient light operation and let the transistor differentiate between ambient light and the signal.

occurs, alpha goes to unity and beta approaches infinity (beta or h_{FE} which is equal to $\alpha/1-\alpha$). Carriers are actually being created. Because this is an unstable condition, the depletion region narrows (voltage drop decreases) until alpha drops back to below unity. The collector–base junction leakage current is normally specified with a low leakage (~100 nA) at the operating voltage and a higher value of leakage (~100 µA) to show breakdown (avalanche). The collector–emitter junctions are also normally specified with a similarly low leakage (~100 nA at operating voltage) but a significantly higher current (~1 to 10 mA) to ensure the voltage that is read is in the stabilized region after alpha has dropped back to below unity.

If the emitter–base junction is reverse-biased, a normal avalanche will occur at a low voltage (~5 to 10 V). If the reverse bias is applied emitter to collector, the snap-back phenomenon will not be seen. The inverse alpha is significantly lower and the depletion region is much more narrow (owing to the lower applied voltages).

The collector–emitter leakage current constitutes one boundary of the unit when it is operated as a switch. When the transistor is off, the value of the leakage or dark current flowing through the load resistor prevents the unit from being totally "off." This leakage current is usually specified at room temperature. As the temperature is increased, the leakage current will increase. The leakage current of the

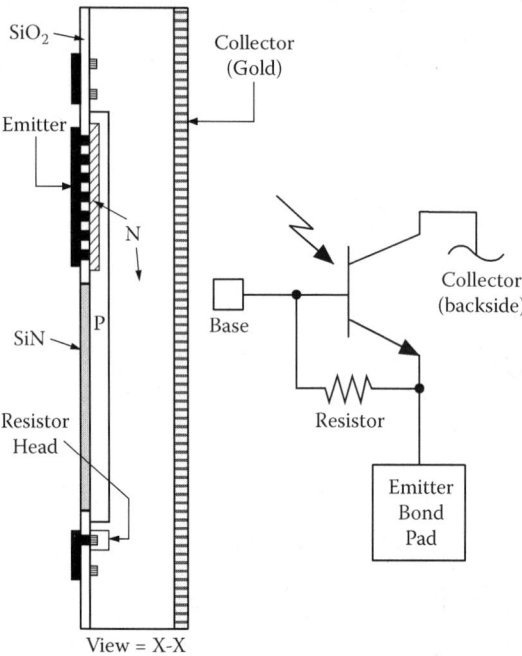

FIGURE 5.8 Side view of a R_{BE} transistor. The cross-sectional view shows the diffusions along the dashed line from Figure 5.7. The emitter and base contacts are typically aluminum, whereas the collector contact is the backside of the chip and is typically a sintered gold contact. The electrical symbol is for the NPN R_{BE} phototransistor.

collector–base junction is multiplied by beta (h_{FE}). Because beta is increasing with both increasing collector current and increasing temperature, the observed leakages are significantly higher than those observed for a P-N or PIN diode. Figure 5.11 shows this increase on a log-linear graph. Increasing the voltage will cause a corresponding increase in leakage current (refer to Figure 5.10).

5.2.1 Switching Characteristics

The phototransistor operated as a switch has two boundary conditions: the "off" position (controlled by leakage current) and the "on" position (controlled by saturation voltage). Figure 5.12 shows these two boundary conditions when the load is 1000 Ω and the supply voltage is 5 V.

In a typical example, if the leakage current is 1 µA and the saturation voltage is 0.2 V, then the "off" condition would be 4.999 V and the "on" condition would be 0.2 V. The voltage drop across the load would be 0.001 V in the "off" condition and 4.8 V in the "on" condition. The unit would not be a perfect switch (5 to 0 V) by 1 mV in the "off" condition and 200 mV in the "on" condition. Further analysis of the saturation characteristics is required to understand this phenomenon. Figure 5.13 shows

The Phototransistor and Photodarlington

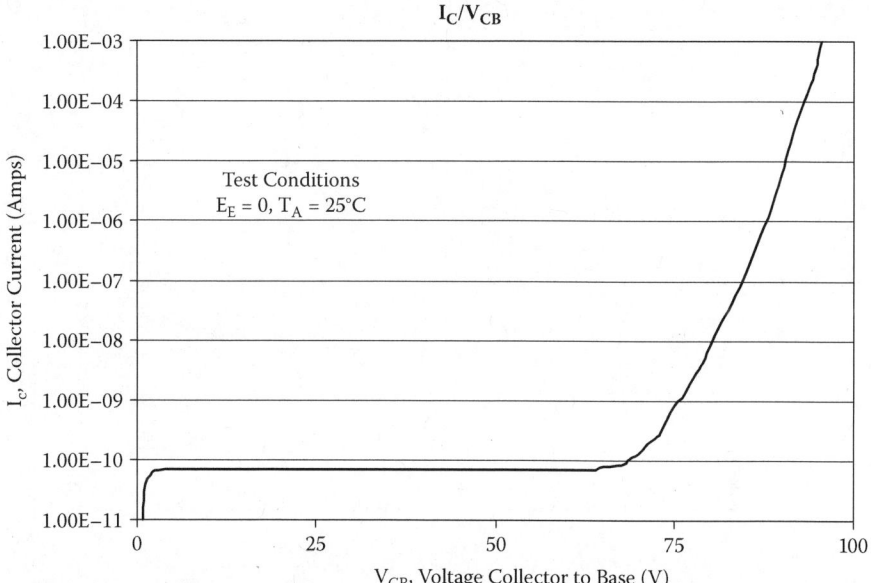

FIGURE 5.9 Reverse breakdown of collector–base junction. Exactly analogous to diode behavior, reverse breakdown of the collector–base junction is characterized first by the occurrence of a small leakage current as the collector–base voltage rises above zero. The depletion region becomes wider as more voltage is applied, until at breakdown, carriers passing through this region will dislodge additional like carriers, causing high currents to flow. Any additional increase in voltage at this point would heat and damage the device.

an exaggerated view of the saturation voltage versus collector current for two different levels of base current. If we trace the $I_b = 1$ line from 0 to 0.75 V, we observe three different regions of saturation voltage. The initial offset voltage is approximately 30 mV and is due to the inverse h_{FE} being lower than the forward h_{FE}. (If inverse and forward h_{FE} were identical, then there would be no offset.) The offset voltage is fairly consistent between units made by the same techniques. Because most commercial devices are planar diffused (causing the emitter to be more heavily doped and having a graded junction), a value of 30 to 50 mV is usually considered typical. The second region has a resistive slope controlled by the device geometry and the bulk resistance in the collector region. A common technique for lowering this resistance is to use epitaxial material. Figures 5.14 and 5.15 show how this significantly lowers the collector bulk resistance.

The third region of saturation voltage is the normal linear operating region, and it will be discussed in more detail in the h_{FE} section. The second saturation voltage line of $I_B = 1.5x$ is similar to the $I_B = 1x$ line. Its normal linear operating region is at a higher collector current, which is reasonable if one assumes h_{FE} is relatively constant. The changes in h_{FE} will be discussed later in this chapter. The important point is that the higher the base current for a given collector current, the lower the

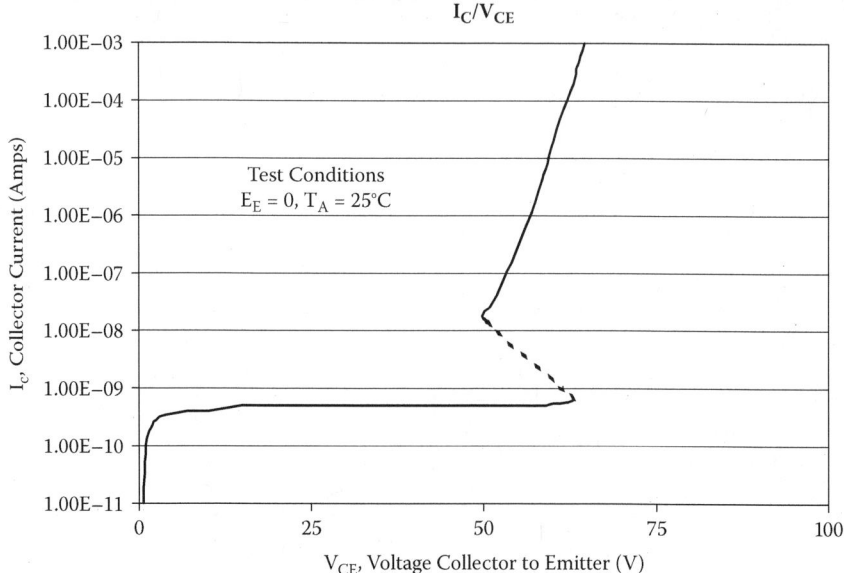

FIGURE 5.10 Reverse breakdown of collector-to-emitter junctions. The collector–base junction is reverse-biased; however, the emitter–base junction is simultaneously forward-biased. At high voltages, the wide collector–base depletion region causes the base to become very narrow, and current gain (h_{FE}) then approaches infinity. New carriers are actually created until the depletion region narrows (the base widens) sufficiently to cause h_{FE} to drop back. V_{CE} also drops back when this stabilization process occurs. Continued high voltages will heat and damage the device.

saturation voltage. This is an important factor because a production spread of h_{FE} is normal. Figure 5.16 shows these three regions in a planar epitaxial phototransistor.

The photodarlington will behave in a similar fashion. The leakage (dark) current is multiplied by the cascaded transistors' h_{FE}, so the value will be proportionally larger and will increase more rapidly with temperature. Figure 5.17 shows this increase graphically.

The saturation voltage will also increase because a forward-biased diode is added to the saturated circuitry. The offset voltage will increase by approximately 0.6 V, and the slope of the saturation voltage will also increase. The net result will be an increase of the saturation voltage to approximately 0.63 V at zero collector current and to 0.9 V at 5 mA. These values will vary with the geometry of the devices. The saturation resistance has more impact on the total voltage drop at high currents than the forward voltage drop of the base–emitter diode.

The current, gain or β, has a number of variables that cause it to change. In the section on quantum efficiency of the P-N diode (Chapter 4, Section 4.2), it was shown that for a PIN diode with a peak response at 900 nm, the maximum quantum efficiency (electrons per photon) was 83%. As the wavelength increased or decreased, the quantum efficiency decreased. Because the phototransistor has a

The Phototransistor and Photodarlington

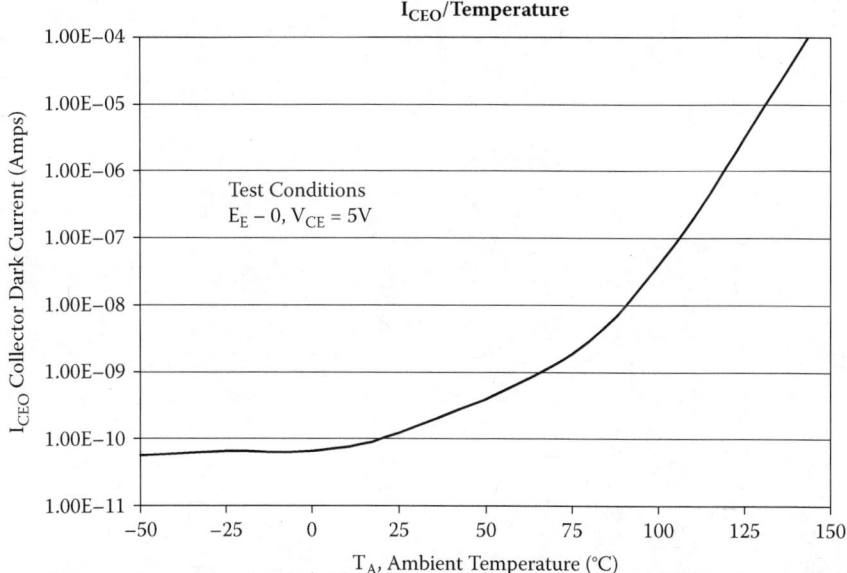

FIGURE 5.11 Collector–emitter leakage (dark) current versus ambient temperature. When the phototransistor is used as a switch, collector dark current becomes the "off" state or lower limit. Dark current is typically higher than for photodiodes because leakage current is multiplied by the current gain (h_{FE}) of the device.

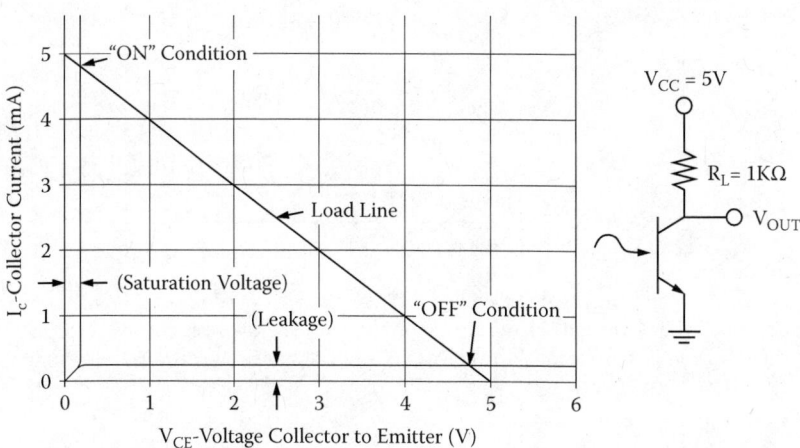

FIGURE 5.12 Boundary conditions for an optical switch. A "perfect" switch would be "off" at V_{CE} equal to 5.0 V and "on" at V_{CE} equal to 0 V. Leakage current and saturation voltage limit the device to less than 5.0 V off and more than 0 V on for the circuit shown.

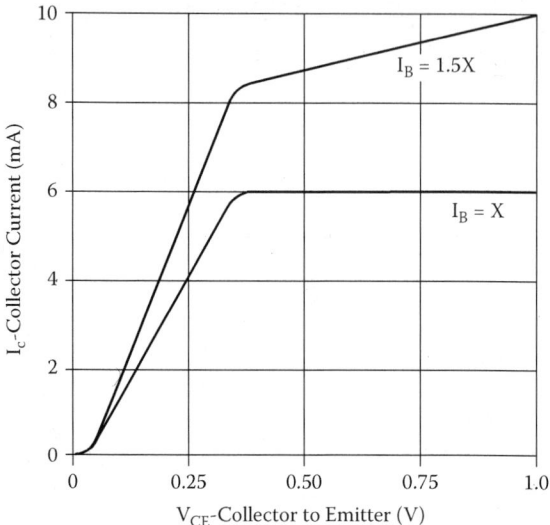

FIGURE 5.13 Saturation voltage versus collector current. For the $I_B = X$ curve, the initial offset is due to the difference between forward and inverse current gain. Device geometry and bulk resistance in the collector create the characteristic of the second region. The third portion is the normal linear operating region.

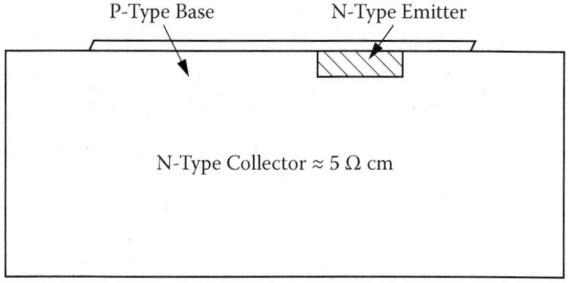

FIGURE 5.14 The comparison of collector bulk resistance—homogeneous material. The thickness of the homogeneous material is approximately the thickness of the chip: 250 μm. This thickness adds significantly to the overall collector resistance of the phototransistor chip.

narrower depletion region located closer to the surface, the peak response will be both lower and less efficient. This response plotted against wavelength is shown in Figure 5.18.

The response is modified by the addition of an antireflective coating deposited on top of the photosensitive area that is designed to enhance the IR response. This

The Phototransistor and Photodarlington

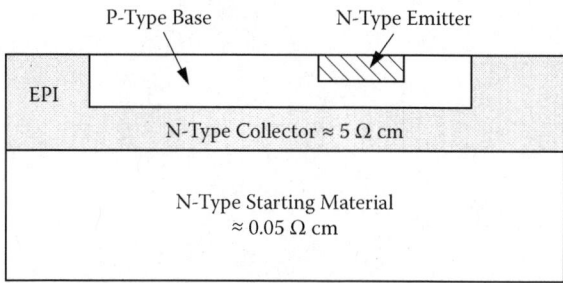

FIGURE 5.15 The comparison of collector bulk resistance—epitaxial material. In actual practice, the epitaxial deposition is approximately 20–30 μm thick and the total transistor chip is 250 μm thick. The resistance of the epitaxial layer improves the breakdown characteristics of the phototransistor. The lower resistance of the substrate allows a much lower collector resistance.

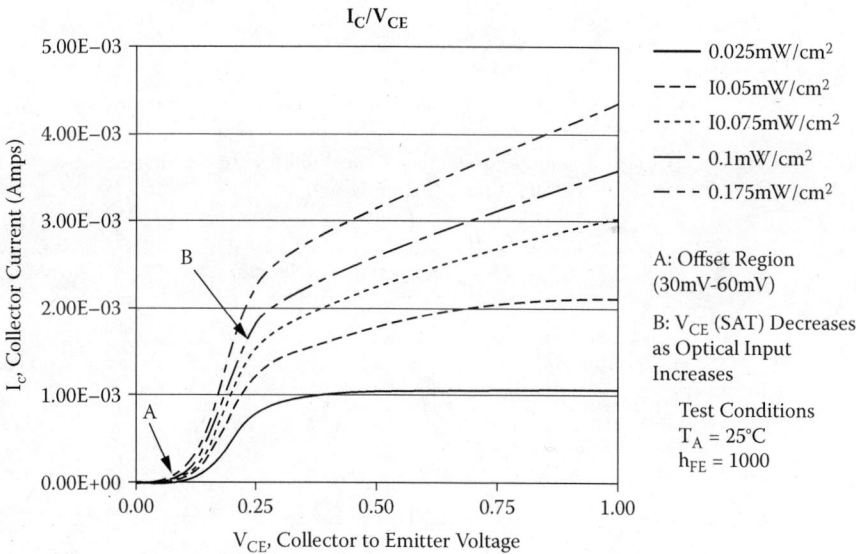

FIGURE 5.16 Collector current versus collector-to-emitter voltage. The offset region (A) is largely independent of the irradiance and is primarily a function of the difference between h_{FE} and forward h_{FE}. The irradiance level, as the source of base current, dramatically affects device performance in the operating range (B).

quarter-wave overcoat is designed to maximize device sensitivity for wavelengths of 880 nm. The two local maximums to the left of the 880 nm global maximum are also the result of the optical overcoating and are analogous to the secondary resonance peaks found in acoustics. Although theory predicts that these secondary

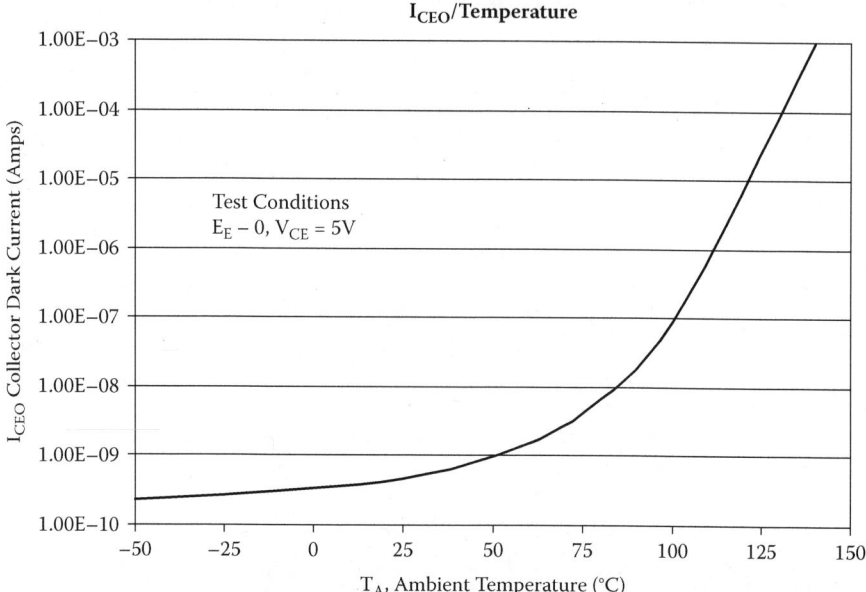

FIGURE 5.17 Collector dark current (photodarlington) versus ambient temperature. Just as in Figure 5.11, dark current increases with temperature and is then multiplied by the gain of the device. This effect is even more dramatic for the photodarlington (depicted in figure) owing to the additional current gain of the output transistor.

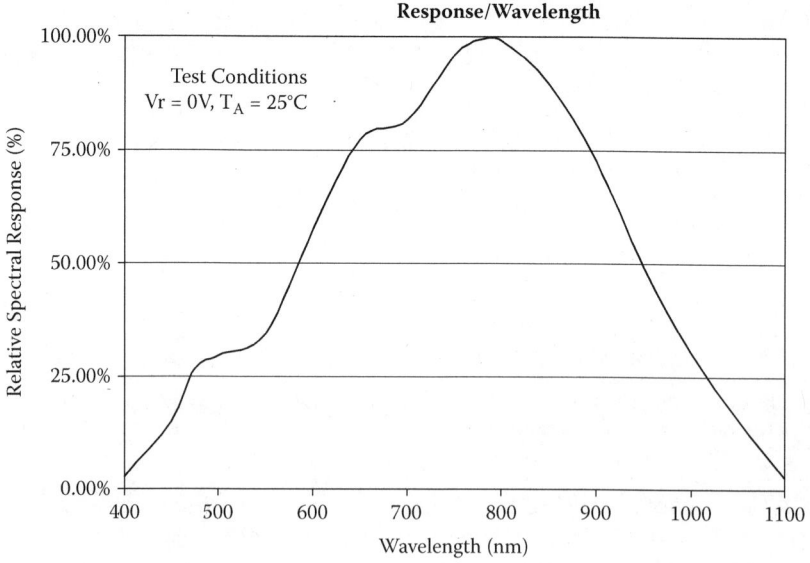

FIGURE 5.18 The spectral response for an NPN phototransistor. An optical overcoat is often applied to help center the peaks' sensitivity at 800 nm. The two secondary peaks are also created as a result of this overcoat.

The Phototransistor and Photodarlington

peaks should appear at 440 nm (quarter-wave) and 660 nm wavelengths, in reality, they are confirmed at 475 and 660 nm. Typically, the coating is deposited at a multiple of the quarter-wave to improve process control of the thickness of the film.

The peak spectral response is effectively lowered as the diffused junctions move closer to the surface. This is done in order to lower capacitance and improve the frequency response of the phototransistor. This further lowers the quantum efficiency. The resulting phototransistor has a reasonably linear response in the 100 µA to 15 mA collector current range. This is normally adequate because smaller geometries would not improve the response at the lower end and the photon–h_{FE} mix would not make the maximum end any more usable.

The lower light levels are usually detected by a photodiode–amplifier combination or, in some cases, by a photodarlington. The upper end of output current requirements can be extended by a photodarlington with a high-power output transistor or by feeding the output of the phototransistor into another silicon transistor. Figure 5.19 shows relative IRLED output versus collector current for a typical phototransistor.

Different light energy sources with their different spectral content cause a phototransistor to have different outputs. Figure 5.20 shows the $I_{C(ON)}$ of a narrow included acceptance angle phototransistor (25° between half-power points) encapsulated in a plastic TO-18 outline package. This graph will shift with different sensitivity levels of photosensors and variations in energy sources among different manufacturers, but the shape remains relatively consistent.

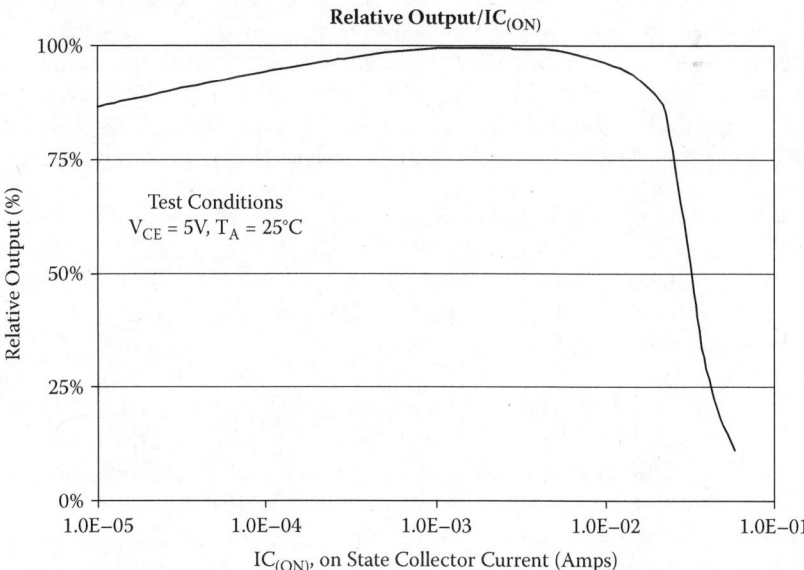

FIGURE 5.19 The relative output versus collector current. The key to successful use of phototransistors lies in the relationship between changing incident radiation and the on-state collector current. This curve assumes a given collector—emitter voltage.

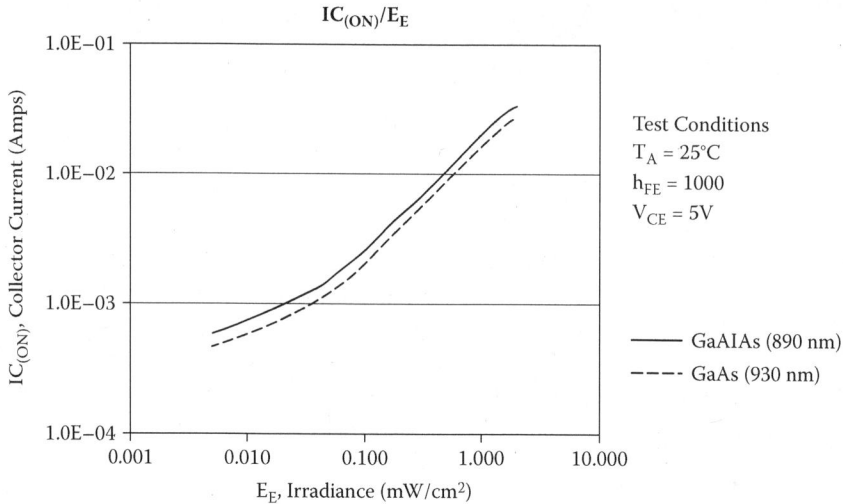

FIGURE 5.20 On-state collector current versus irradiance. Phototransistor collector current is typically a linear function of irradiance in the range shown. Design differences (e.g., lens changes) will shift the location of these curves but not their linear characteristics.

The phototransistor collector current has a positive temperature coefficient. As h_{FE} decreases, this curve will tend to flatten, and as it increases, it will become steeper. When h_{FE} is in the 400 to 600 range, the curve will nearly cancel the IRLED output–temperature curve, so the resultant or coupled output is nearly flat. As the summation of these curves approximates a multiplication rather than an addition, the curve will tail down at low temperature (transistor dominated) and tail down again at high temperature (IRLED dominated). This is the curve most often shown on manufacturers' data sheets. Figure 5.21 shows this change for a unit with h_{FE} in the 400 to 600 range.

The switching time of a phototransistor is significantly longer than its small signal transistor equivalent with equal current-carrying capability. The major reason is the significant increase in the capacitance of the collector–base diode. In a phototransistor, the area of the base exposed to impinging photons is greatly increased in order to enlarge the energy-gathering area and improve device sensitivity.

Figures 5.22 and 5.23 show the comparable sizes for a phototransistor and its equivalent current-carrying small signal transistor. The base area on the phototransistor is 300K sq. μm, whereas the base area on the equivalent transistor is only 71.5 sq. μm, a factor of 4200 larger. With equivalent diffusions, this ratio would be an accurate predictor of the difference in capacitance. In addition, the diffusion profile on the phototransistor is deeper in order to improve the photon wavelength response in the infrared region. This further aggravates the problem of bandwidth response for the phototransistor. Figures 5.24 through 5.27 show the results of this higher capacitance and resulting lower bandwidth response and the test circuit.

The Phototransistor and Photodarlington

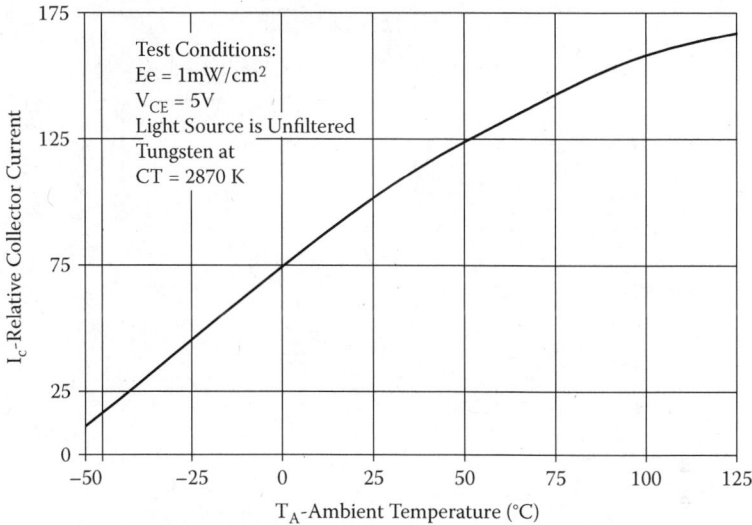

FIGURE 5.21 Normalized collector current versus ambient temperature. With a constant source of infrared energy, higher operating temperature results in higher relative output current. The opposite behavior occurs for relative output versus temperature for IRLEDs, making possible coupled configurations with lower overall temperature drifting.

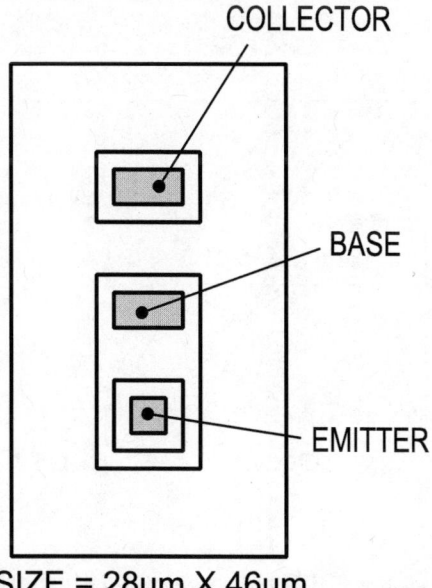

FIGURE 5.22 An NPN small signal transistor. Small base area allows for a lower base-to-collector capacitance and a higher-speed device.

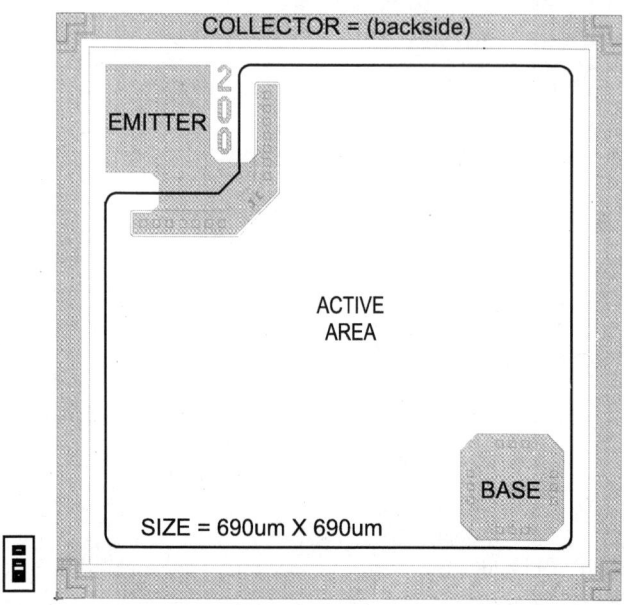

FIGURE 5.23 The active area of a NPN phototransistor is 4200 times larger than that of a standard NPN transistors. Although this allows for a much greater light-gathering area, base-to-collector capacitance is increased, resulting in slower speed than a conventional transistors.

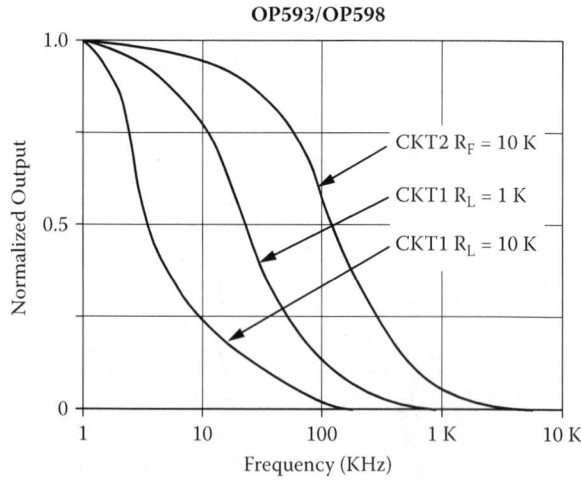

FIGURE 5.24 The normalized output versus frequency for an NPN phototransistor. Starting from the left, the lower load resistance (by an order of magnitude) results in a tenfold frequency response increase. The use of a linear amplifier in test circuit two affords the fastest frequency response.

The Phototransistor and Photodarlington

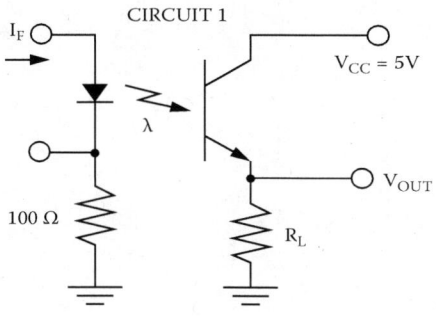

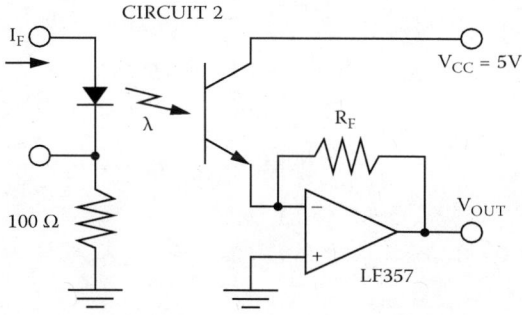

FIGURE 5.25 The test schematics for the graph in Figure 5.24.

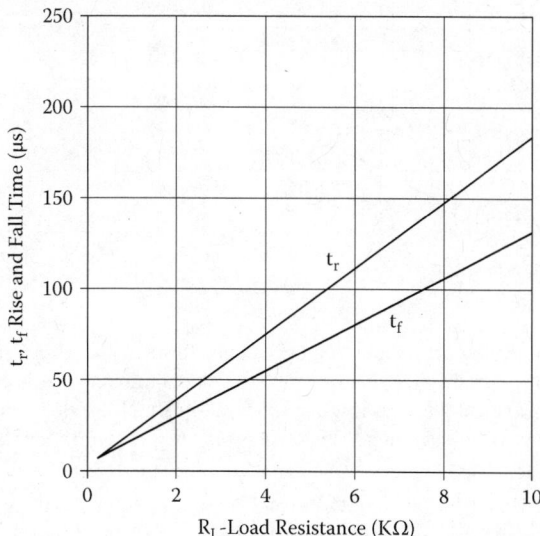

FIGURE 5.26 Rise and fall time versus load resistance for an NPN phototransistor. A model depicting charging or discharging a capacitor through a resistor (RC network) would predict faster speed at lower resistance, just as shown in the relationship between load resistance and rise/fall times or frequency response.

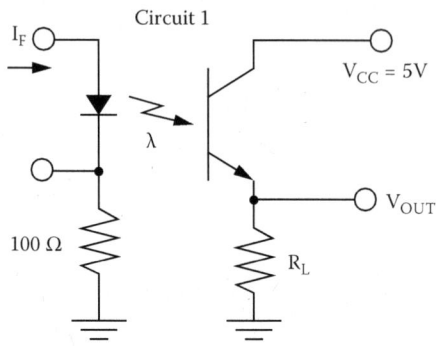

FIGURE 5.27 The test schematics for the graph in Figure 5.26.

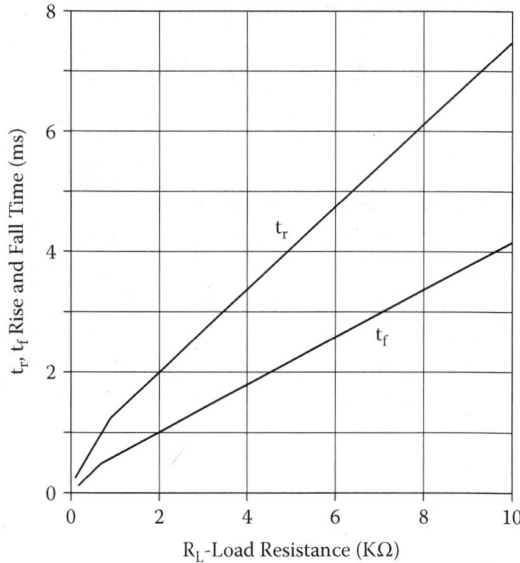

FIGURE 5.28 Switching time versus load resistance for a photodarlington transistor. More than an order of magnitude in switching speed over phototransistors is traded for higher output currents in a typical photodarlington; however, in simple motion-sensing applications where speed is not critical, the component savings may be well worth the trade-off.

The switching time of a photodarlington transistor is significantly slower than the transistor counterpart. Figure 5.28 (see also Figure 5.29) shows the switching time versus load resistance and the test circuit. Note that the times are given in milliseconds rather than microseconds.

The Phototransistor and Photodarlington

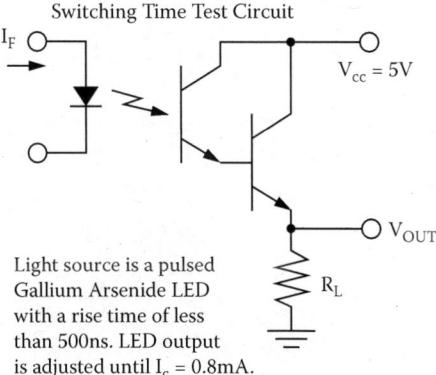

FIGURE 5.29 The test schematics for the graph in Figure 5.28.

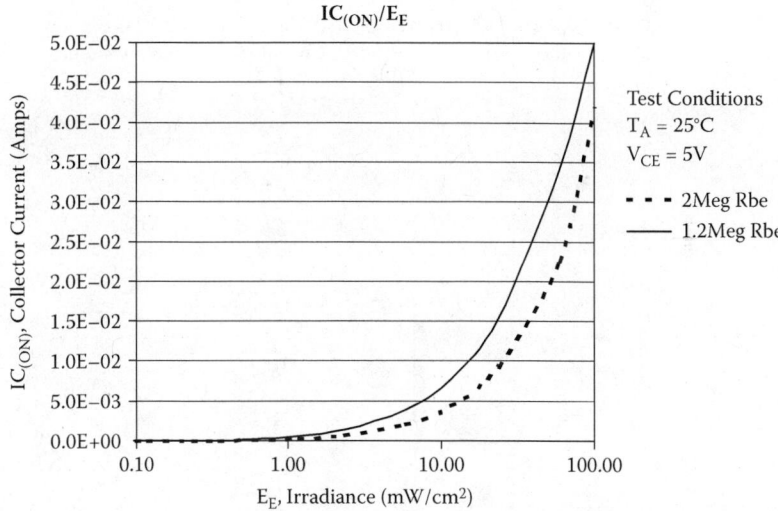

FIGURE 5.30 IC plots of a typical R_{BE} phototransistor. The R_{BE} transistor is switched off at low light levels because of the resistor in the base–emitter junction. This allows some ambient light level to be present before the phototransistor gains the light signal.

Figure 5.30 shows the I_C current of the R_{BE} phototransistor versus a standard phototransistor. The R_{BE} phototransistor has similar characteristics as the standard phototransistor except at low light levels. The $I_{C(ON)}$ at low light levels is in the

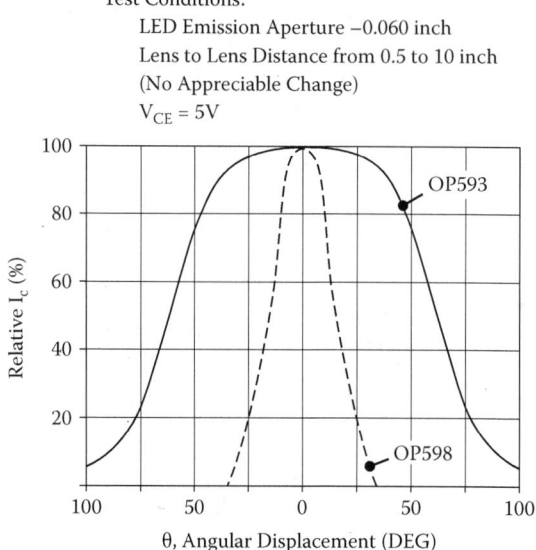

FIGURE 5.31 Receiving pattern plots of wide and narrow beam outline TO-18 plastic packages. These plots represent the inverse of the IRLED beam pattern plots shown earlier. To create these patterns, a collimated source is rotated through an arc in front of the device while $I_{C(ON)}$ is measured.

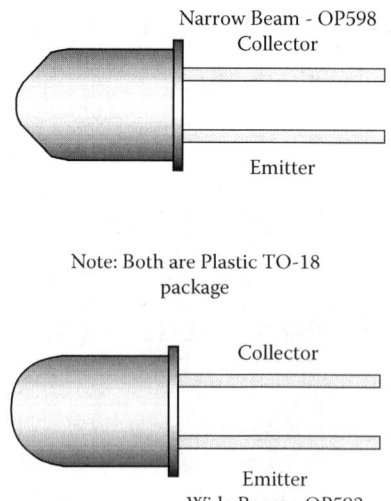

FIGURE 5.32 Drawings of the two packages from Figure 5.31.

The Phototransistor and Photodarlington

microampere (μA) range on the R_{BE}, whereas the $I_{C(ON)}$ for the standard transistor is in the milliampere (mA) range. The effective gain of the R_{BE} is defined as:

$$G_M = I_C \frac{I_C}{I_P - \left(\dfrac{V_{BE}}{R}\right)} \tag{5.1}$$

where
  G_M = Effective gain
  I_C = Collector current
  I_P = Photocurrent
  V_{BE} = Base–emitter voltage of the transistor
  R = Value of the base–emitter resistor

5.2.2 Package Lens Effects

The lens characteristics of the phototransistors are similar to their LED counterparts. The photosensitive area of the chip is generally larger than the emitting size

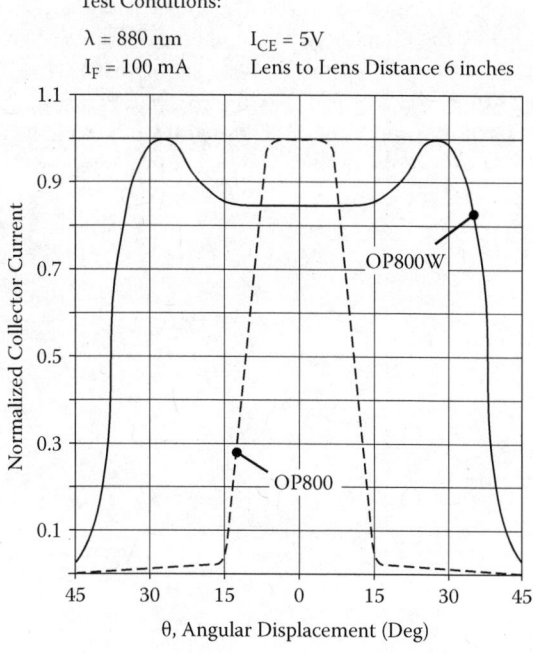

FIGURE 5.33 Beam pattern plots and side-view drawings of metal TO-18 outline package. The wide-angle "doughnut" effect results from reflections off the side of the metal can.

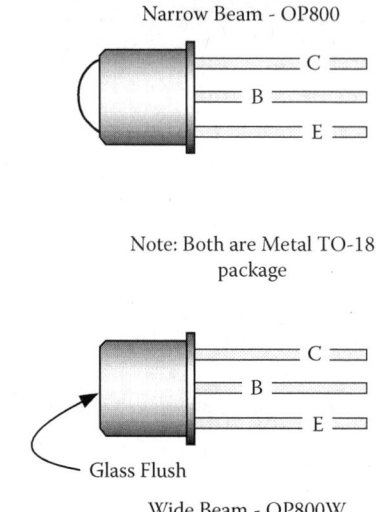

FIGURE 5.34 Drawings of the two packages from Figure 5.33.

of the IRLED, but the phototransistor normally does not set in a well. A typical well diameter is 0.030 in., whereas a typical photosensitive area diameter is 0.022 in. The net result is that the effective size of the emitting area of an IRLED is close to the receiving area of the phototransistor. The subtle differences, however, lead to physical differences in the outline of the two packages. Figures 5.31 and 5.32 show receiving angle pattern plots and side-view drawings of a wide beam angle plastic. Figures 5.33 and 5.34 show comparable drawings for the metal package.

6 The Photointegrated Circuit

6.1 BASIC THEORY

Chapter 3 showed that the P-N and PIN diodes were significantly faster in speed than the basic phototransistor. A review of Chapter 2 leads one to the conclusion that the phototransistor adds gain to the photodiode but compromises the speed performance, and the photodarlington advances the trade-off even further. In many applications, the speed–gain relationship of the phototransistor is the most cost-effective method of solving the system gain design problems in which speed is not an issue.

The photointegrated circuit (photo IC) offers other characteristics that can be very useful in certain applications. In many designs, the photoswitch must be in a remote location from the processing electronics. In order to achieve precise switching control (resolution), the photosensitive area may also be reduced. This type of application will be discussed in more detail in Chapters 8 and 9. The light source may not be supplying adequate signal to the phototransistor. This results in a small electrical signal that could cause reliability problems if the signal path back to the processing electronics is long and there is a high probability of erroneous signals caused by coupled noise from motors, relays, or other forms of electrical interference. If this occurs, the design will require special shielding or a higher signal level than is possible from an apertured phototransistor, and the photo IC becomes the most cost-effective solution to this type of problem. The high-resolution rotary or linear encoders use photo ICs to take advantage of the inherent speed, the reduced system space, reliability, and cost reduction attained over systems fabricated from discrete devices. At the present time, bipolar technology is widely used to fabricate simple photo ICs, whereas metal-oxide-silicon (MOS) technology can be used to include complex digital circuit designs. The more advanced photo IC designs will be discussed in Chapter 7.

The simple photo IC using bipolar technology is shown in Figure 6.1. The electrical schematic is shown in Figure 6.2. This circuit will be used in a discussion defining the basic construction used in bipolar technology. The photodiode is made as a P-N junction (similar to the collector–base diode) with the P$^+$-type material exposed for photon coupling. The N-type region is electrically connected to the collector of the first amplifying transistor (Q1). The output transistor (Q2) adds gain to the circuit and allows for an open collector output. The resistor (R1) must be electrically isolated from the other elements for the circuit to function properly. Figures 6.1 and 6.2 show the top-view layout of the photo IC and its electrical schematic.

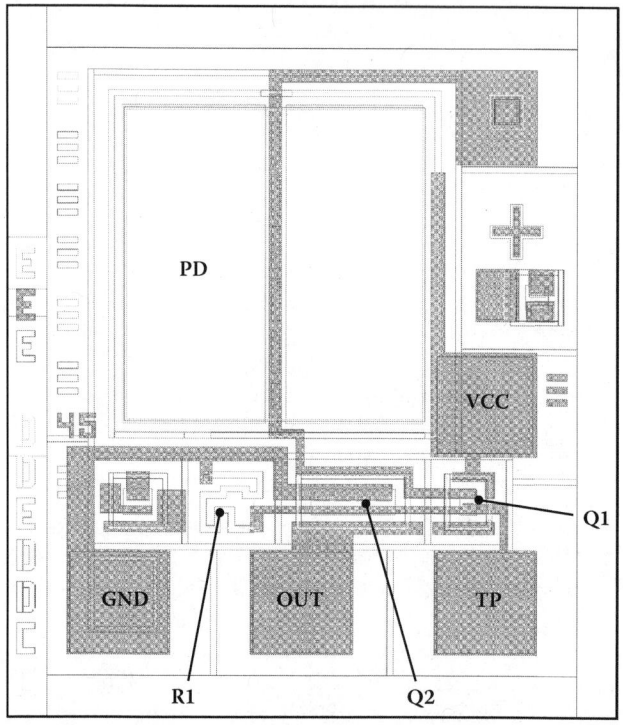

FIGURE 6.1 Layout of photo IC with transistors, resistor, and photodiode.

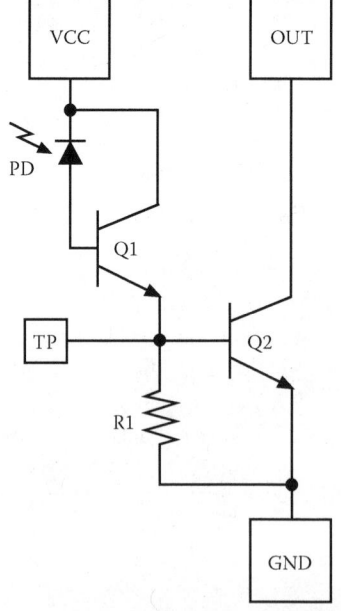

FIGURE 6.2 Schematic of photo IC with transistors, resistor, and photodiode.

The Photointegrated Circuit

6.2 IC PROCESSING STEPS

Understanding Figure 6.1 is difficult unless the reader is familiar with bipolar processing technology. In order to bridge the gap, it is necessary to review the construction of the IC itself. The approach used will be to follow a simplified overview of the various process steps utilized in the fabrication of the finished chip. The construction of the output transistor (NPN transistor Q2) will be used as a focus for the various steps.

6.2.1 STARTING WAFER

First, the starting silicon wafer of P-type silicon is oxidized. This is done by placing it in a furnace with a surplus of oxygen molecules. As the oxygen combines with the silicon, a layer of silicon oxide (SiO_2 or oxide) is grown on the wafer. A thin layer of photographic emulsion (photoresist) is then placed over the entire wafer surface. Selected areas of the photoresist are exposed to ultraviolet light, which causes the photoresist to harden. The unexposed photoresist is then removed, and the SiO_2 is chemically removed in the areas where no photoresist is located. The top illustration in Figure 6.3 shows a cross section of the silicon wafer depicting this oxide removal process step.

6.2.2 BURIED LAYER

After the exposed photoresist has been removed, the wafer is then placed in another furnace and coated with a layer of N-type impurities. At high temperature, these N-type impurities diffuse into the surface. This process is similar in principle to a drop of ink dropped into a glass of still water. The ink drop (high concentration) will gradually diffuse into the surrounding water. The diffusion rate will approximately double for each 10°C rise in temperature. Of key importance is

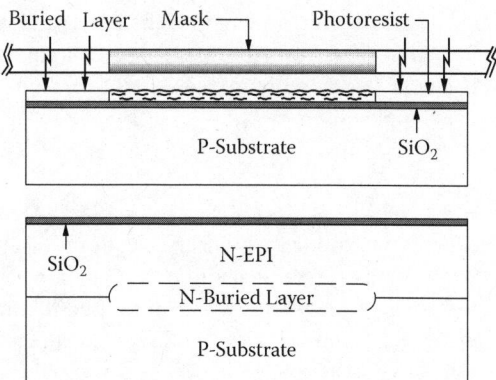

FIGURE 6.3 Photomask operation. Top: Illustrates the location of a photomask above a wafer for the purpose of exposing the photoresist prior to the oxide removal procedure. This allows for precise placement of geometries. Bottom: Cross-sectional view of an IC NPN transistor after completion of the buried layer diffusion and epitaxial growth.

the fact that the diffusion rate into the S_iO_2 is much lower or slower than the diffusion rate into the exposed silicon (Si). When the SiO_2 has been grown thick enough, it will act as a diffusion barrier, offering selective penetration of the N-type impurity atoms. The localized N-type areas (buried layer) will become the areas located beneath the active components. These buried layer areas will aid in reducing the collector's series resistance for the NPN device. Figure 6.3 shows the cross-sectional view of these areas. The oxide area over the N-buried layer will be removed before the deposition. The N-type nomenclature is used to show a heavier-than-normal concentration of N-type impurities and indicates lower resistance per unit area.

6.2.3 Epi

The wafer is then placed in an epitaxial (Epi) reactor, and an N-type region approximately 10 to 15 µm thick is grown on the wafer. The process is similar in function to the epitaxial layer grown on the GaAs substrate for IRLED fabrication, except that the method uses gas-phase rather than solution-phase material. The bottom illustration of Figure 6.3 shows a cross section after the Epi has been grown on top of the wafer and the buried layer regions. The wafer thickness is approximately 250 µm, the N-type material is approximately 3 µm, and the epitaxial layer is approximately 12 µm.

6.2.4 Isolation

The next oxide removal is performed selectively removing the oxide in a narrow region around the edges of the buried layer. A thin layer of photoresist is then placed over the entire wafer surface. Selected areas of the photoresist are exposed to ultraviolet light. The unexposed photoresist is then removed, and the SiO_2 is chemically removed in the areas where no photoresist is located. The wafer is then placed in another furnace and coated with a layer of P-type impurities. The P material is then diffused into the wafer. This will create a frame of the P^+-type material after the diffusion is performed (isolation). The P^+ diffusion creates separate N-Epi areas that are electrically isolated from one another. Figure 6.4 shows the results after the isolation diffusion step has been completed.

6.2.5 Deep N^+

The next oxide removal step is followed by an N^+ diffusion (deep N^+) that creates the collector of the NPN. A thin layer of photoresist is then placed over the entire wafer surface. Selected areas of the photoresist are exposed to ultraviolet light. The unexposed photoresist is then removed, and the SiO_2 is chemically removed in the areas where no photoresist is located. The wafer is then placed in another furnace and coated with a layer of N-type impurities. The N material is then diffused into the wafer. The net effect is to create a low-resistance path for the carriers traveling from the buried layer back to the top surface collector contact. Figure 6.5 shows the cross-sectional view of the added deep N^+ diffusion.

The Photointegrated Circuit

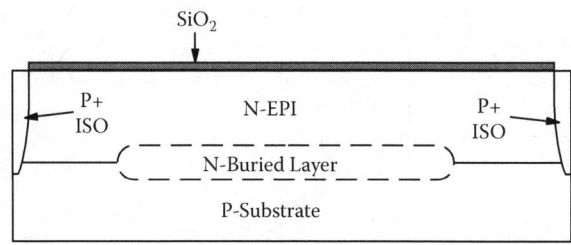

FIGURE 6.4 Cross-sectional view of an IC NPN transistor after P+ isolation step.

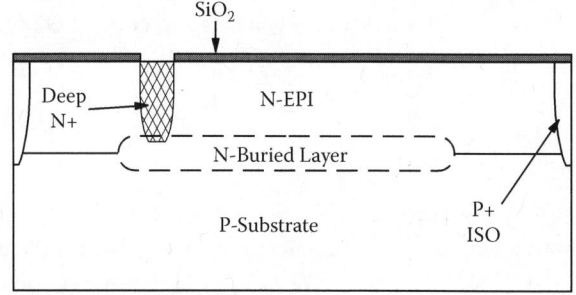

FIGURE 6.5 Cross-sectional view of an IC NPN transistor after deep N+ collector step.

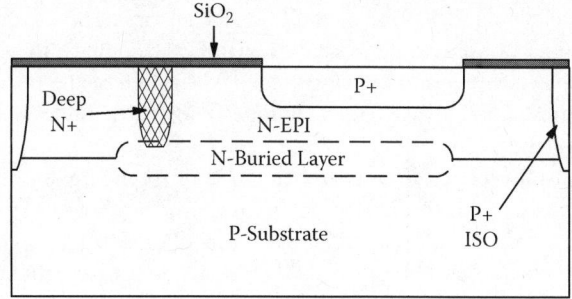

FIGURE 6.6 Cross-sectional view of an IC NPN transistor after P+ base step.

6.2.6 P+

The next oxide removal is performed to create the base region of the NPN. This area is slightly smaller than the isolated Epi areas. A thin layer of photoresist is then placed over the entire wafer surface. Selected areas of the photoresist are exposed to ultraviolet light. The unexposed photoresist is then removed, and the SiO_2 is chemically removed in the areas where no photoresist is located. The wafer is then placed in another furnace and coated with a layer of P-type impurities. The P material is then diffused into the wafer that will become the base region of the NPN transistors and the exposed photosensitive region of the photodiode. Figure 6.6 shows a cross-sectional view of the P+ step.

6.2.7 N⁺

The next oxide removal step will create N⁺ regions that will become the emitters of the NPN transistors. A thin layer of photoresist is then placed over the entire wafer surface. Selected areas of the photoresist are exposed to ultraviolet light. The unexposed photoresist is then removed, and the SiO_2 is chemically removed in the areas where no photoresist is located. The wafer is then placed in another furnace and coated with a layer of N-type impurities. The N material is then diffused into the wafer. At the same time, N⁺ will be diffused into the collectors to improve the contact to the metal. Sometimes, an output transistor (Q2) requires both low saturation voltage and good current-carrying capability in the "on" condition. To meet this requirement, multiple emitters are used, which in effect parallels four transistors. Figure 6.7 shows the side view of the transistor after the emitter diffusion. The deep N⁺ diffusion is cross-hatched for clarity.

6.2.8 RESISTORS

The resistor depicted in Figure 6.11 may be formed by one or more techniques. A thin layer of photoresist is placed over the entire wafer surface. Selected areas of the photoresist are exposed to ultraviolet light. The unexposed photoresist is then removed, and the SiO_2 is chemically removed in the areas where no photoresist is located. Diffusing P⁺ through a long thin opening in the S_iO_2 forms the P⁺-type of resistor. The P⁺-diffusion sheet resistance is normally 100 to 500 Ω/square (a square is defined as length/width = 1). A 1000 Ω resistor at 200 Ω/square would be 5 squares long (an example would be that a resistor 50 µm long × 10 µm wide or 100 µm long × 20 µm wide are equal in resistance). The width of the resistor is defined by the design rules of a process. This technique is normally used for wide-tolerance (±10 to 20%), relatively low-value (<10 kΩ) resistors.

A second resistor type utilizes ion implantation of P-type impurities deposited in a thin layer on the surface of the silicon. A thin layer of photoresist is then placed over the entire wafer surface. Selected areas of the photoresist are exposed to ultraviolet light. The unexposed photoresist is then removed, and the SiO_2 is chemically removed in the areas where no photoresist is located. This technique is similar to the

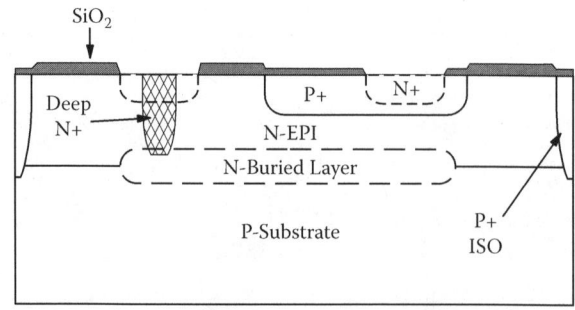

FIGURE 6.7 Cross-sectional view of an IC NPN transistor after N⁺ emitter step.

The Photointegrated Circuit

diffused resistors technique except that the deposited layers are higher in resistivity (5000 Ω/square) and shallower in depth, and better control of resistor value is usually possible. The process is a separate step just before the emitter step. In the circuit shown in Figure 6.1, the resistor is designed for 50 kΩ. At 2000 Ω/square, the resistor is 25 squares.

A third resistor type utilizes thin film deposited resistors. This technique is also similar to the ion implantation method except that the film that is deposited remains on top of the oxide, and the excess material is removed by selective etching. Manufacturers use this technique to lessen the effect of temperature on the resistance value, improve repeatability, and increase the ohms per square values. This resistor can also be adjusted by using a laser beam. Instead of opening a metal link, the laser trims the deposited resistor to a desired value.

6.2.9 Contact

Another oxide removal step is performed to define where electrical contact is made on the devices. A thin layer of photoresist is then placed over the entire wafer surface. Selected areas of the photoresist are exposed to ultraviolet light. The unexposed photoresist is then removed, and the SiO_2 is chemically removed in the areas where no photoresist is located. Openings are etched through the S_iO_2 to the N^+ and P^+ diffusions to allow for the metal to connect to the transistor. The NPN transistor is now complete, and Figure 6.8 shows a cross section after all diffusion steps and contact openings. Figure 6.9 has been drawn to depict the transistor in a three-dimensional view, and to illustrate the different oxide steps created during the process.

6.2.10 Metal

The metallization is then deposited using the evaporation, sputtering, or electron beam methods. An oxide removal step is performed, leaving the metal contacting and interconnecting the transistors, resistors, and other devices. A thin layer of photoresist is then placed over the entire wafer surface. Selected areas of the photoresist are exposed to ultraviolet light, which photochemically causes the

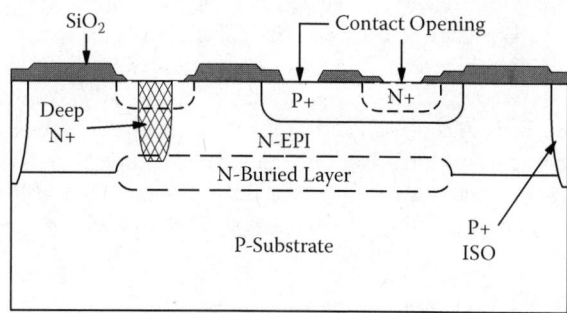

FIGURE 6.8 Cross-sectional view of an IC NPN transistor. All diffusions are completed and show electrical contact openings prior to metal deposition.

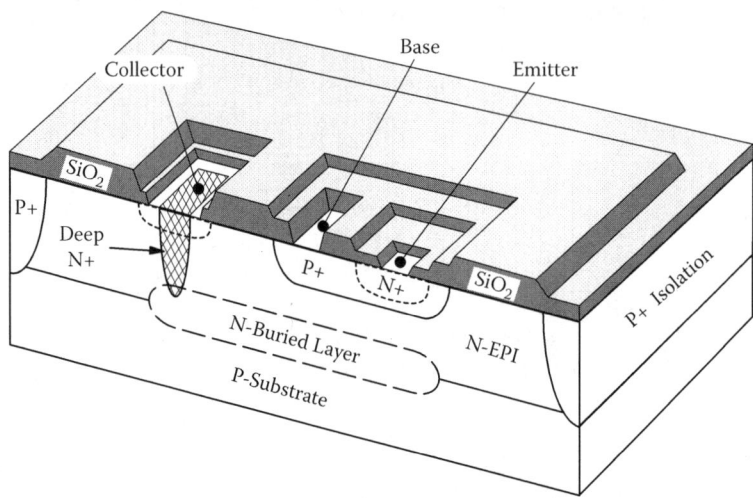

FIGURE 6.9 3-D Cross section of the IC NPN transistor.

photoresist to harden. The unexposed photoresist is then removed. The metal is chemically removed in the areas where no photoresist is located. The metallization step utilized may actually consist of more than one layer of metal, but is usually a single layer of aluminum with or without a small percentage of impurities to improve the metal current-carrying and processing characteristics. The deposition systems and the type of conductors can vary widely from one manufacturer to another. Figure 6.9 shows a three-dimensional view of the NPN transistor after contacts have been opened. The collector contact is on the left, the emitter contact is to the right, and the base contact is in the center.

6.2.11 Passivation

The last deposition puts a protective coating (passivation) on top of the metallization. This coating is typically a low-temperature oxide, nitride, or a combination of both. The final masking step opens the passivation over the metal contact pads and photodiode. A thin layer of photoresist is then placed over the entire wafer surface. Selected areas of the photoresist are exposed to ultraviolet light. The unexposed photoresist is then removed, and the passivation is chemically removed in the areas where no photoresist is located. The contact pads allow for wire bonding of the die to the package. The passivation lessens the mechanical damage to the substrate from subsequent operations and seals the surface from foreign material that could cause corrosion to the metal traces. Figure 6.10 shows the completed diffusion processes.

6.2.12 Backside Processing

The final steps use a mechanical lap or grind to thin the wafer. The processing of the wafer is much easier using a 500-μm-thick wafer, with less breakage in the

The Photointegrated Circuit

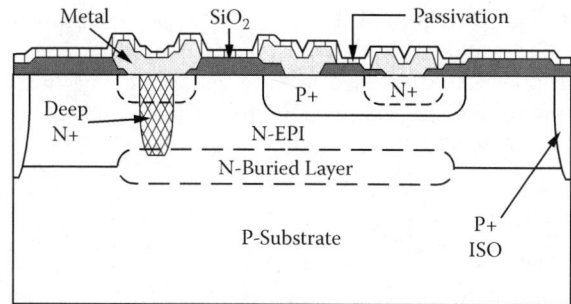

FIGURE 6.10 Cross-sectional view of the completed IC NPN transistor structure.

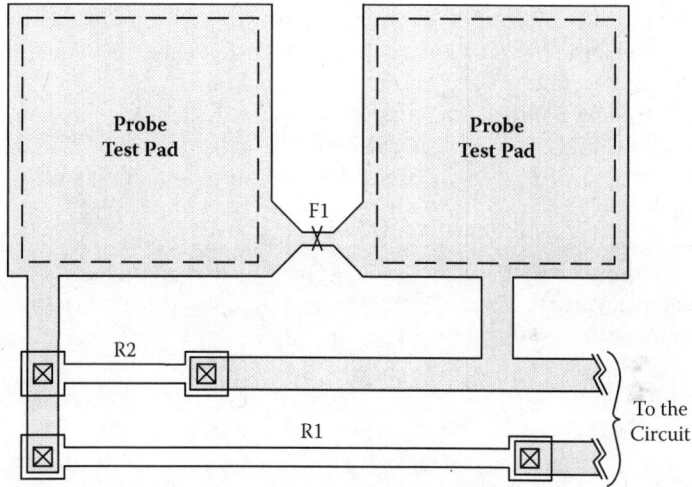

FIGURE 6.11 Layout of the diffused resistor with an adjustable fuse link.

process. However, most packages need a thinner wafer to optimize the optics in the package. Thus, the wafer is thinned to the appropriate thickness, with the typical thickness ranging between 200 and 400 μm. If the backside of the wafer needs to have electrical contact, a metallization is deposited to improve the contact and reduce the possibility that an oxide could grow on the back of the wafer, causing an open backside contact. Gold and silver are common metals used for the backside metallization.

6.3 OTHER IC DEVICES

The tolerance of resistors may be improved by fusible links, or shorting bars across the ends of the resistor. A fuse link acts, as its name implies, as a thin metal connection between probe pads that can be opened by an electrical pulse. Figure 6.11

shows a resistor (R2) with a fuse link shorted between two test pads and in series with resistor R1. This will give a 30% adjustment to the value of resistor R1. Several other methods can be employed such as laser energy, a metal mask change, or a zener diode. In the zener diode option, a high electrical pulse causes the P-N junction to go from an open to a short. The other options are shorted and open after the adjustment.

6.3.1 Photodiodes

The P-N photodiode is similar in structure to the P-N transistor, with two major exceptions. First, it is significantly larger to absorb more photons and, second, the N^+-type diffusion used for the emitter is eliminated. The diode or P-N junction utilizes the P^+-type base diffusion as the cathode and the N-type epitaxial layer as the anode. Peak wavelength for this type of photodiode is 650 to 750 nm. The wavelength is determined by several factors. The deeper a junction is from the surface, the longer the peak wavelength. The junction on this type of photodiode is within 2 to 3 μm from the surface. Figure 6.12 shows this type of photodiode.

Figure 6.13 shows an epitaxial-to-substrate photodiode. This junction, because it is deeper, has a higher wavelength response of 800 to 900 nm as compared to the base–epitaxial diode. However, it has much more capacitance for the same area and therefore is inherently slower.

The oxide over the P-N photodiode is a deposited oxide or nitride, the thickness of which is closely controlled. On the P-N photodiode, the film acts as a filter for GaAlAs energy, which peaks at 875 to 900 nm. The peaks were discussed in Chapter 5 and visually portrayed in Figure 5.13.

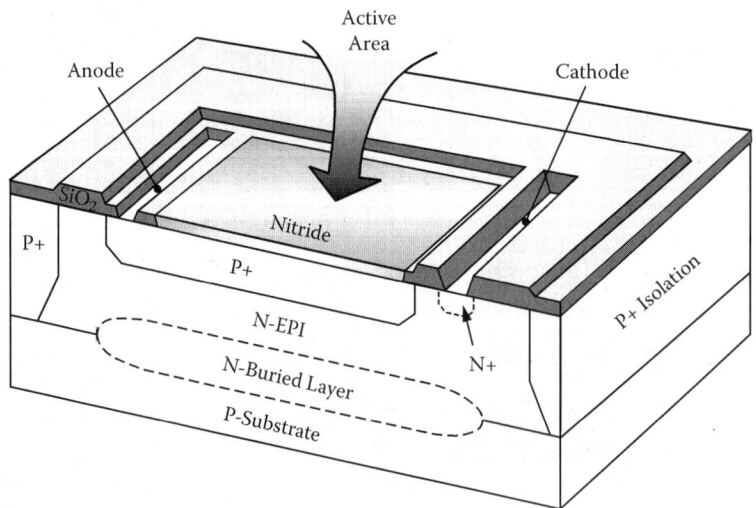

FIGURE 6.12 3-D Cross section of the IC P^+ photodiode structure. Peak wavelength response (λ) is 650 to 750 nm.

The Photointegrated Circuit

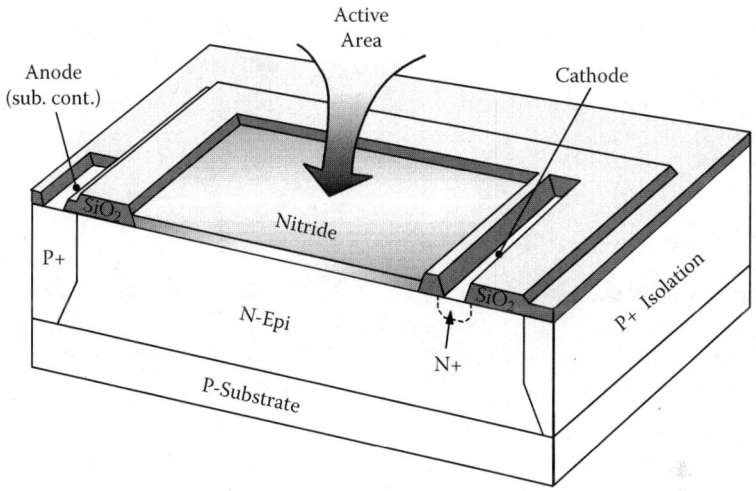

FIGURE 6.13 3-D Cross section of the IC Epi photodiode structure. Peak wavelength response (λ) is 800 to 900 nm.

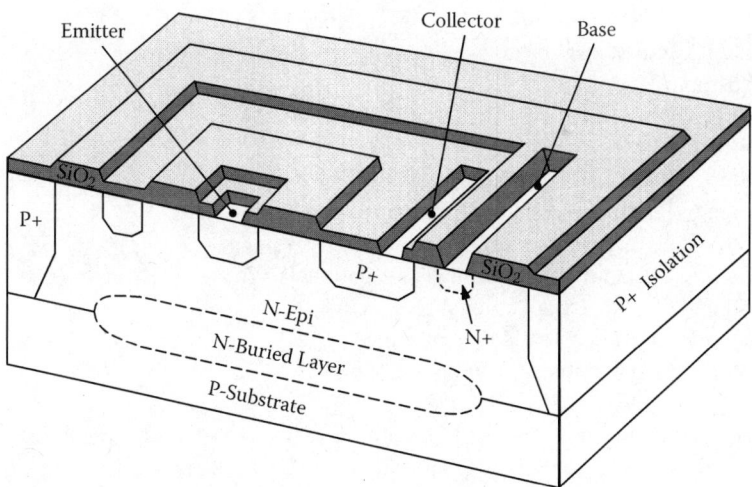

FIGURE 6.14 3-D Cross section of the IC lateral PNP structure.

6.3.2 Lateral PNP

Utilizing the N-type Epi region as the base and diffusing two P$^+$-type regions (emitter and collector) into the Epi region will create a lateral PNP transistor. The Epi region (base) is between these two diffused P-type diffusions, and h_{FE} for this structure depends mainly on the spacing between the emitter and collector P$^+$diffusions. A lateral PNP structure is shown in Figure 6.14.

6.3.3 Vertical PNP

Another type of common transistor is the vertical PNP. The transistor behaves the same as the lateral with the exception that the collector is the substrate, not another P⁺diffusion. Typically, the transistor has lower frequency response and current-carrying capabilities.

6.3.4 Capacitor

If necessary, a capacitor may be formed by utilizing the N^+-type emitter diffusion as one plate and a thin deposited oxide or nitride as the dielectric with a deposited metal overcoat acting as the other plate. This is called an *MOS* (metal-oxide-silicon) or *MNS* (metal-nitride-silicon) *capacitor*. The thinner the dielectric, the higher the value of the capacitor. However, thinner dielectrics have a lower breakdown voltage.

6.3.5 Summary

The completed IC has been processed through the following steps:

Starting wafer
    Step 1: N diffusion (buried layer Figure 6.3)
    Step 2: N-epitaxial layer
Deposition (Epi)
    Step 3: P⁺diffusion (isolation Figure 6.4)
    Step 4: N⁺diffusion (deep N⁺Figure 6.5)
    Step 5: P⁺diffusion (base Figure 6.6)
    Step 6: P implant (resistor—optional)
    Step 7: N⁺diffusion (emitter Figure 6.7)
    Step 8: Capacitor (oxide or nitride—optional)
    Step 9: Contact oxide removal (Figure 6.8)
    Step 10: Metal deposition and removal
    Step 11: Passivation deposition and removal

Figure 6.9 shows how the oxide thickness varies over the entire surface of the photo IC due to additional oxide being removed and regrown during the process. Because the subsequent diffusions are done at a lower temperature to minimize additional drive of the previous diffusions, the grown oxide becomes progressively thinner. When viewing a photo IC under magnification, the diffused areas are seen as different colors, and the color of the oxide is a function of its thickness.

6.4 CHARACTERIZATION

A number of different simple photo IC circuits are discussed later that are similar in function to the phototransistor and photodarlington discussed in Chapter 5. A comparison of gain versus speed shows the relative advantage of the photo IC compared

The Photointegrated Circuit

to phototransistors. Figures 6.15 and 6.16 show the schematic for a phototransistor and its photo IC equivalent, where the gain of each device is considered equal.

When the load resistance is 1000 Ω, the phototransistor will reach its half-power point at 25 kHz, whereas the photo IC could attain speeds of 500 kHz. The addition of a second gain stage to the photo IC would lower the frequency response of the photo IC to 25 kHz or equal that of the phototransistor. The gain, however, would be increased by two orders of magnitude. These changes are summarized in Table 6.1.

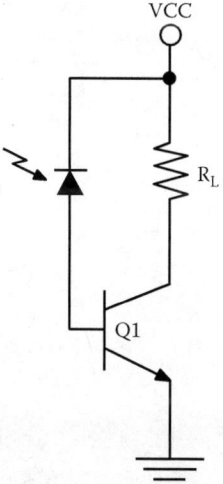

FIGURE 6.15 Schematic of a simple photointegrated circuit. This is a simple photo IC with a photodiode, transistor, and pull-up resistor.

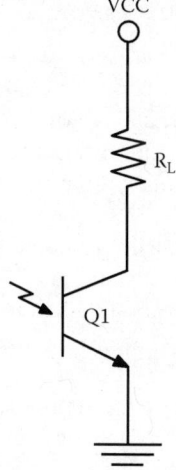

FIGURE 6.16 Schematic of the equivalent phototransistor circuit. A phototransistor chip with external load resistor.

TABLE 6.1
Relative Gain and Frequency Response of Photo IC versus Phototransistor (1 kΩ load)

	Gain	Half-Power Frequency (kHz)
Phototransistor	1	25
Photo IC (single-gain stage)	1	500
Photo IC (double-gain stage)	100	25

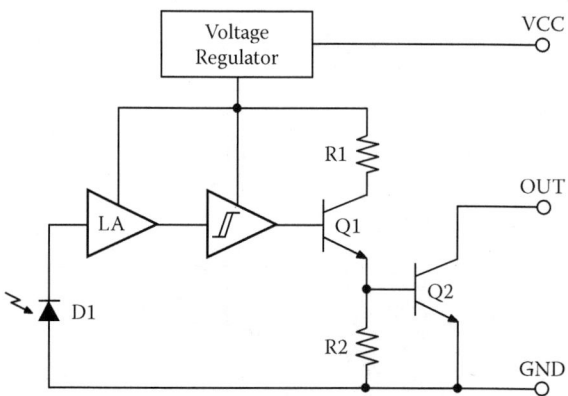

FIGURE 6.17 A more complicated photo IC. This circuit includes a voltage regulator, photodiode, amplifier, Schmitt trigger, and open collector output driver stages. Buffer, inverter, internal pull-up resistor, and totem pole outputs are commonly available.

The current-carrying capabilities of the devices are equal to those of a photosensitive diode occupying approximately 25,000 μm^2, and the chip sizes for all three units are similar. Table 6.1 clearly shows the advantage of the photo IC over the phototransistor. This advantage becomes minor when the device is used as a two-terminal part, where the photodiode becomes common to the collector of the transistor. The capacitance of the equivalent circuitries would then approach the capacitance of the phototransistor, negating the speed advantage.

Popular photo IC designs incorporate a Schmitt trigger, which provides a hysteresis characteristic to improve the performance across varying input signals. Once the input energy level reaches a predetermined point, the circuit will switch "on." The input energy level must drop significantly below this level for the circuit to switch back to the "off" state. The ratio of turn "on" to turn "off" is called the *hysteresis ratio*, and Figure 6.17 shows this circuit. This circuit is shown in the open collector output configuration but is available in the totem pole output configuration and with a 10 kΩ pull-up resistor output. All of the output types are available with a buffer or inverter output option. The photo IC with hysteresis characteristics is also available with and without a built-in voltage regulator. The regulator allows the photo IC to be used with supply voltages up to 16 V. Units are available that will go to higher voltages, but power dissipation may become too excessive for the application.

The unit shown in Figure 6.17 effectively allows amplification and a hysteresis characteristic internal to the sensor, and is widely used in limit switches and encoders, with maximum speeds between 100 to 300 kHz.

7 Special-Function Photointegrated Circuits

7.1 BASIC THEORY

Chapter 6 is devoted to a basic understanding of photo ICs and bipolar technology. The relatively simple circuits offer cost-effective solutions to many applications. The photo IC requires more complex processing for chip fabrication and packaging; therefore, the photo IC is more expensive. For comparable volumes, the photo IC attracts a premium price over simple phototransistors.

The use of photo ICs is increasing because of two strong technological pressures in the marketplace. The first pressure comes from the user's desire to purchase a complete building block or function. This building block can be tested by the supplier to a set of specifications, thus reducing the complexity and cost of the final system. The average application is growing in electronic complexity, mechanical complexity, and volume. As a result, the user is purchasing a more complex part. Photo ICs lead to both improved system performance and reduced system cost.

The second pressure comes from the need for improved performance. The photo IC offers the option to integrate (lower cost/smaller size) a discrete function or perform a function that is extremely difficult, utilizing discrete components. A more thorough analysis of system needs and cost is required to justify this approach in a specific case because a custom photo IC with unique packaging requirements requires a major expenditure in both tooling costs and technical effort. If the photo IC is commercially available, however, the cost comparison becomes very straightforward. Three of these standard function photo ICs that we will be discussing are triac driver, synchronous driver detector, and the color sensor.

7.2 TRIAC DRIVER PHOTOSENSORS

The triac driver photo IC is designed to interface electric controllers that generate a DC output to control AC loads. The triac driver comes in two basic configurations: a standard triac driver and a zero crossing triac driver. Both triacs are available in 400, 600, and 800 V versions. The 400 V version is suitable for use on a 120 V AC line voltage, whereas the 800 V is suitable up to 500 V AC line voltages. It can control AC loads up to 100 mA or provide trigger current of up to 100 mA for controlling a power triac that can control higher-current AC loads. Figure 7.1 shows a schematic of the circuit for controlling a resistive load up to 100 mA. The circuit would be the same for controlling a power triac, except that the photo IC would be connected between the power triac trigger lead (gate) and one side of the AC line.

116 Optoelectronics: Infrared-Visible-Ultraviolet Devices and Applications

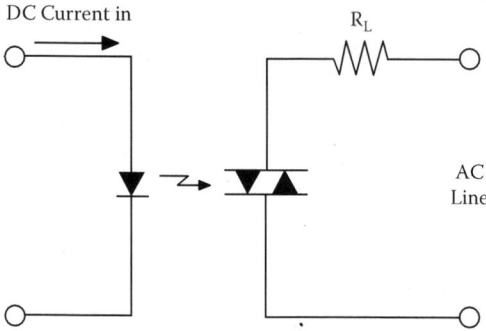

FIGURE 7.1 Electrical schematic of zero current crossing triac photo IC. The symbol for a triac is used for the photo IC because the device functions very much like a triac. At loads up to 100 mA, the device may be used in place of a triac, with IRLED current substituting for gate current.

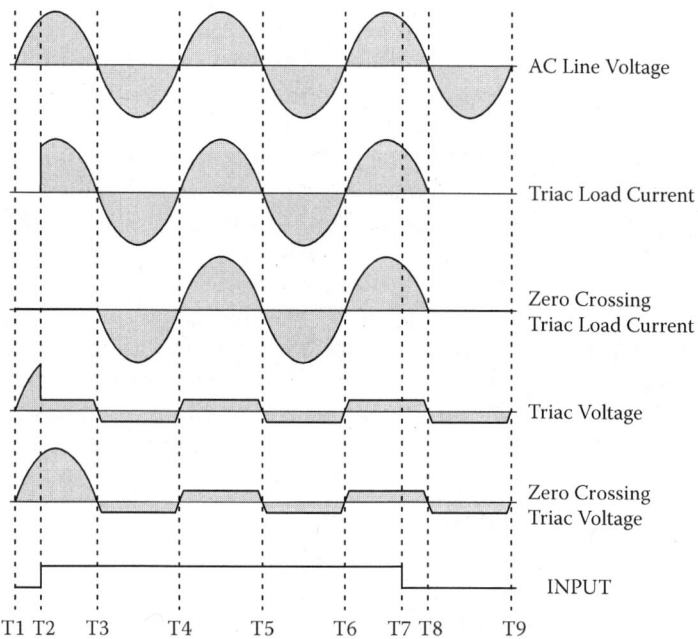

FIGURE 7.2 Principal voltage and current waveforms with resistive load. A resistive load indicates that voltage and current are in phase. The "off" condition is reached when the current waveform reaches zero with both triacs. The standard triac turns on when the LED comes on; the zero crossing triac waits until the input voltage crosses zero to turn on.

Figure 7.2 shows the time relationship of the source or AC line voltage and load currents, the triac voltages, and the control pulse from the IRLED with a resistive load. When the control current for the IRLED turns "on," the source voltage (less the voltage drop across the triac) is applied to the load. When the control current for the IRLED turns "off," the load current and voltage continue their sinusoidal path

Special-Function Photointegrated Circuits

until the load current passes through zero. At that time, the triac turns off and the line voltage is dropped across the triac. Note the differences between the standard triac driver and zero crossing triac driver in the current and voltage waveforms. The turn-off conditions remain the same, but the turn-on condition will change. The zero crossing triac will not turn on until the voltage approaches zero.

Figure 7.3 shows the schematic of the zero crossing triac or triac driver. The schematic for the standard triac driver is the same, once D1, D2, PD1, PD2, M1, and M2 are removed. These devices control the "zero" crossing function. The design is two circuits mirrored for AC operation. One circuit conducts in one polarity, whereas the other circuit conducts with the other polarity. When photons hit the base of Q3 or Q4 (NPN transistors), the collector pulls current from the base of Q1 or Q2 (PNP transistors). Once the current through the resistors R1 or R2 is high enough to reach the V_{BE} of transistors Q3 or Q4, these transistors will turn on and pull more current from the bases of Q1 or Q2, and the triac will latch. It will stay latched until the current is reduced below the threshold current of the device.

The zero current (standard) triac driver photo IC has the inherent disadvantage that the turn-on point is not controlled. Turn-on can occur when the line voltage is "high" across the load, which will cause high current pulses if the load is capacitance. The zero crossing triac overcomes this disadvantage. The waveforms are identical, except that the circuit does not turn on until the line voltage approaches zero. When the photons hit the photodiodes D1 or D2, they will turn on the MOS transistors (M1 or M2). These transistors are used to keep the base of Q3 or Q4 from reaching a V_{BE}

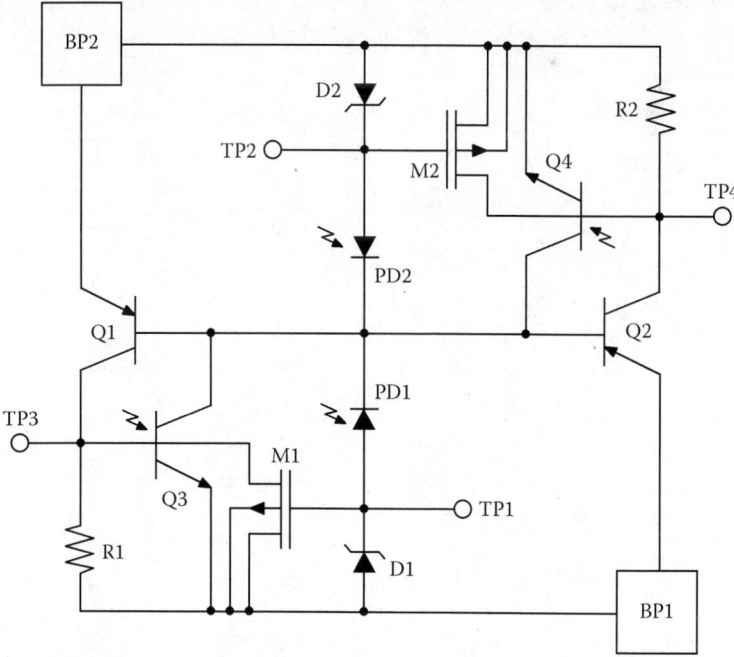

FIGURE 7.3 Electrical schematic of zero crossing triac driver photo IC. Note the symmetry in the circuit. This, of course, is necessary to facilitate the AC functioning of the device.

118 Optoelectronics: Infrared-Visible-Ultraviolet Devices and Applications

and turning on until the voltage differential across the chip is lowered to the V_{TH} of transistors M1 or M2. When they turn off, they let the photocurrent turn on Q3 or Q4. The V_{TH} of these transistors set the zero crossing voltage. The zener diodes are designed to protect the gates on the MOS transistors from being damaged with voltage spikes. Figure 7.4 shows a typical layout for a zero crossing triac driver.

7.3 SYNCHRONOUS DRIVER DETECTOR (SDD)

With the advent of improved bipolar and the newer MOS technologies, more powerful photointegrated circuits are available. The SDD sensor is used when high ambient light is present and other light sources could confuse a standard sensor. Figure 7.5 shows a block diagram for a typical SDD. The circuit operates under the voltage regulator, which allows wider Vcc ranges and improves noise impunity.

The SDD drives a IRLED light source (BP4) with a pulsed constant current drive. An onboard oscillator uses a counter to select every 16 or 32 periods that sets the frequency of the IRLED. The light from the IRLED is projected on the photodiodes (PD1, PD2). PD1 senses the light and amplifies it. PD2 is a photodiode that is 1/8th the size of PD1, and is used as a reference. The capacitor C2 is used to integrate the background light and cancels this unwanted input from the signal.

The amplified signal is passed through C1 to the second stage of amplification. Because of the high gain of the 1st stage of amplification, the capacitor C1 blocks the DC component of the signal, so AMP3 will not be forced into saturation, allowing AMP3 to further increase the signal gain. The SDD has a very high level of sensitivity (<50 nW), which is accomplished by using the reference PD to cancel out ambient

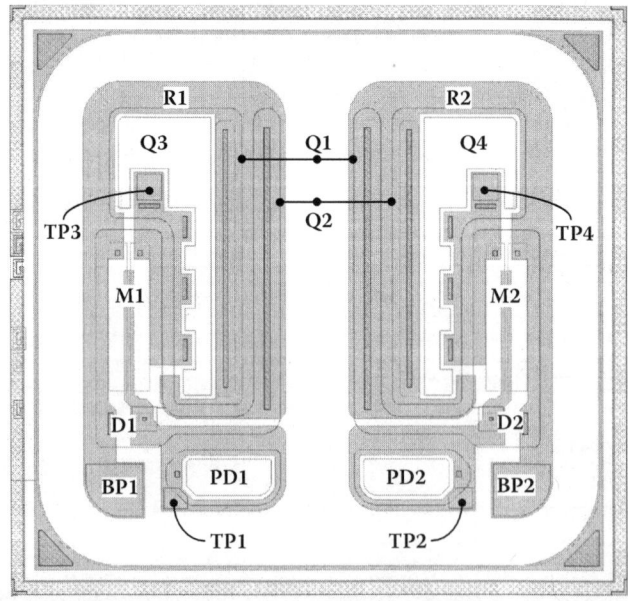

FIGURE 7.4 Typical layout for a zero crossing triac driver.

Special-Function Photointegrated Circuits

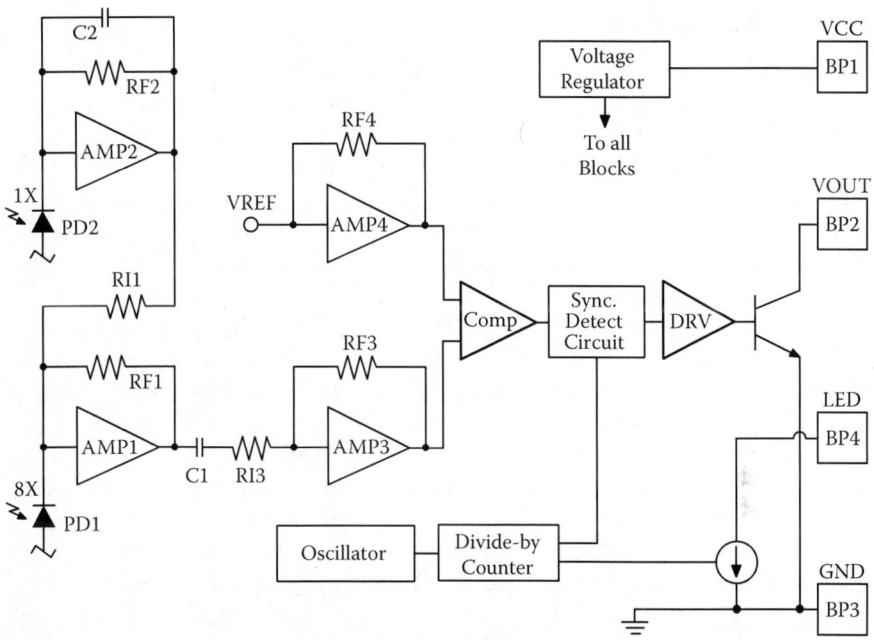

FIGURE 7.5 Block diagram for a synchronous driver detector.

light levels, AC decoupling of the 1st to 2nd stage of amplification, and high gain of the 1st stage of amplification.

The resultant signal is compared to a reference voltage at the comparator. This generates an output pulse from the comparator every time the photodiodes receive an input. This output drives the synchronous detection logic. This logic section allows the part to be unaffected by extraneous signals at the photodiodes. The clock comes out of the divide-by-counter and drives the IRLED as well as the detection circuit. This logic circuit performs several operations:

1. It looks for an input signal.
2. It determines if the input signal corresponds to the IRLED signal.
3. It determines if the rising edge of the signal is within a preset window.
4. It counts the pulses to verify that the signal is in sync with the IRLED pulse.
5. After 4 to 8 pulses (determined by manufacturer), it allows the output to switch on.
6. If any pulse is missed before all of the pulses have latched the logic, the counter resets to 0 and must "see" the minimum number of pulses to set the output on.
7. The logic will hold the output on until a pulse is missed. Any missing pulse will cause the output to switch off.

A typical layout for a SDD chip is shown in Figure 7.6. The SDD tends to operate more slowly than other sensors because it must see the minimum number of good pulses into the photodiode to latch.

7.4 COLOR SENSOR

One of the most cost-effective photo ICs is the color sensor. The sensor uses a known light source (usually a white light source) to determine the color of objects placed in its field of view. The sensor averages the reflected light from the light source to an array of photodiodes with different color filters. An 8 × 8 interdigitated photodiode array is used to minimize the effect of nonuniformity. The photodiodes are enhanced to react to the visible light. Red, blue, and green filter coatings are put on three of the photodiode groups, each group consisting of eight individual photodiodes. One group has no filter (clear).

The current from each group of photodiodes is amplified and sent to a switching network. The red, green, blue, or clear photodiode group is selected using S2 and S3. (Figure 7.7 is a block diagram for a color sensor IC.) This light output current is sent to a current-to-frequency converter. This converter generates a frequency that is proportional to the current out of the amplifier. The frequency determines the magnitude of the input light signal. S0 and S1 set the maximum frequency range for the converter, which runs from a few kilohertz to 500 kHZ, depending on the range selected and the magnitude of the light signal. The slower the run rate, the more the chip will average the light signal. The lower speeds are also used for slower applications. The chip can acquire and convert data in as little as 3 μsec and is configured to drive a TTL or CMOS load. OE Not pin, shown in Figure 7.7, is used to disable the output. Maximum resolution and accuracy may be obtained using frequency-measurement, pulse-accumulation, or integration techniques.

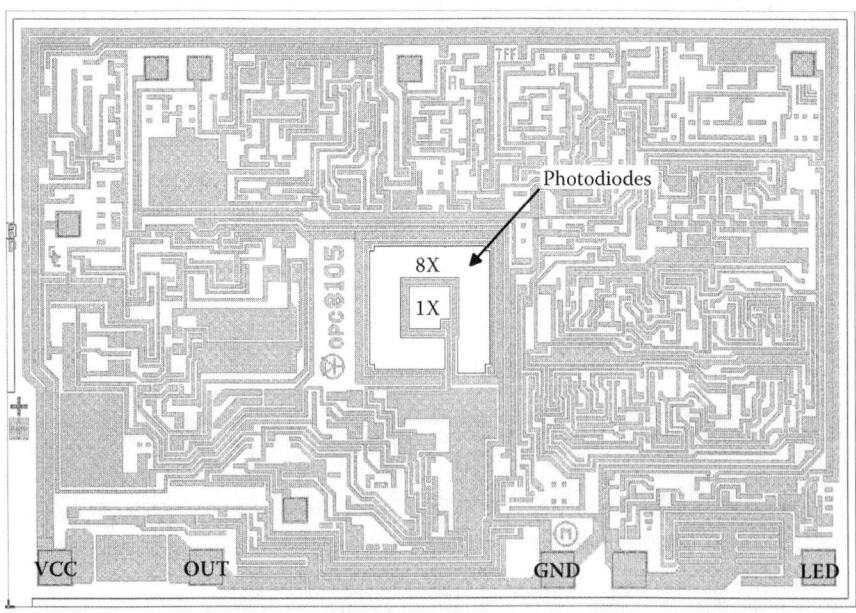

FIGURE 7.6 Typical layout for an SDD.

Special-Function Photointegrated Circuits

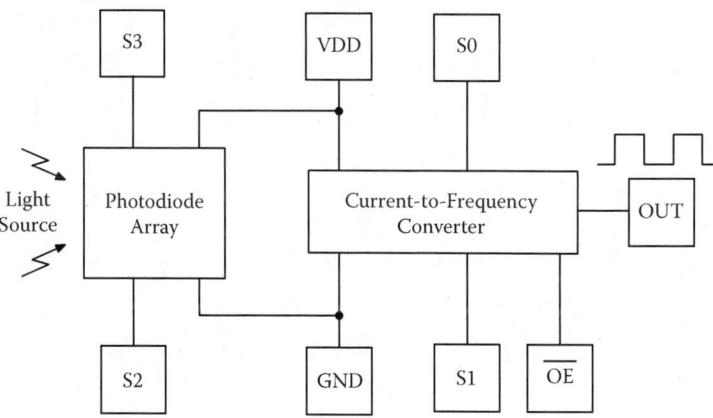

FIGURE 7.7 Block diagram for a color sensor.

Once the frequency is measured for each of the different colors, the signals can be compared to distinguish the hue, saturation, and luminance (HSL) of the light. Hue is the dominant wavelength of the color, saturation is the amount of color, and luminance is the degree of brightness of the light. An algorithm is used with a microprocessor to calculate the HSL.

Using the newer technologies, several other specialized photo ICs are available. They include the following:

Linear sensors
Light-to-voltage converters
Optical rotary encoders
Optical linear encoders
Light-to-frequency converters
Ambient light sensors
Programmable sensors

By combining a photo-sensing photodiode with integrated circuitry onto a single monolithic photo IC, designers are able to customize the IC to special applications that in the past were difficult to develop because of space and/or cost. As the need for sensors increases, custom ICs (ASIC) will be used to solve the space, cost, complexity, and reliability problems in a wide variety of applications.

Part 3

The Coupled Emitter (IRED) Photosensor Pair

8 The Transmissive Optical Switch

8.1 TRANSMISSIVE SLOTTED OPTICAL SWITCH

The transmissive (slotted) switch consists of a light source (usually an IRLED; a visible or UV LED may be used if appropriate), receiver (usually a photosensor or Photologic®), a means of interrupting the flow of the light beam, and a means of fixing these in a mechanical configuration that allows repeatability of the function. Part 1 of the book covered the light source, whereas Part 2 covered the photo receiver. This chapter covers the coupled pair in the slotted mode.

Figure 8.1 shows a simple slotted switch in the conducting (on) and nonconducting (off) conditions. When a photosensor is receiving light (conducting), the slotted switch is in the "on" condition. In order to understand the variables, a simple DC circuit will be used to illustrate the switch states of "on" and "off."

Figure 8.2 shows the typical devices used as a photosensor for slotted switches. A photodiode requires the most additional circuitry to operate properly. The phototransistor is the most versatile, the photodarlington provides the most gain, and the Photologic® is the easiest to interface with additional electrical circuitry.

Slotted switches are designed with different types of emitter devices as well as photosensors. Chapter 15 explains how to interface with an emitter, whereas Chapter 16 explains how to interface with different types of photosensors.

Each slotted switch has either a predefined (hole for the light to go through) or lensed aperture. The predefined aperture is defined using an opaque material with an opening having the width of the aperture and a length typically longer than the diameter of the lens of the component used in the switch. Figure 8.2 shows the typical light patterns for three aperture configurations.

Figure 8.2 shows the effect of an aperture in front of a standard side-looking device. A typical side-looking device has a lens diameter of 0.06" (1.52 mm); a device without an aperture will have an equivalent aperture size of about 0.06" (1.52 mm) diameter. With the addition of an aperture width of 0.05" (1.27 mm) in front of the lens, the apparent sensing area is reduced. By making the aperture even smaller, such as 0.01" (0.25 mm), the area of recognition becomes more critical. The use of apertures provides the ability to modify the effective critical viewing area of the device.

Most apertures are rectangular in shape, allowing the greatest sensitivity in the direction across the component or across the slot in the switch.

Figure 8.3 shows the equivalent light transmission at half-power points for various aperture configurations and a slot width of 0.125" (3.18 mm).

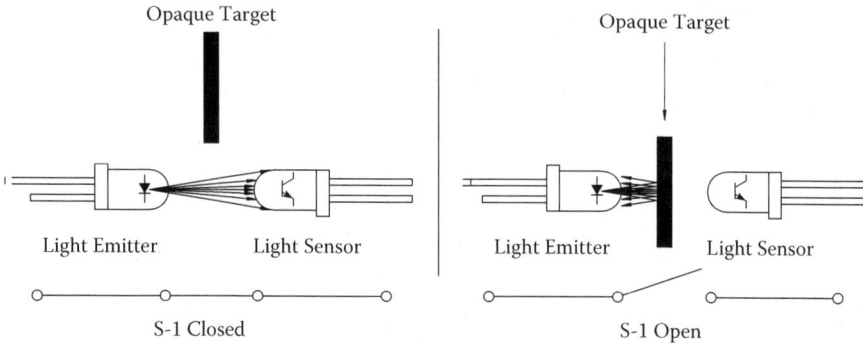

FIGURE 8.1 Slotted (transmissive) switch. Optical motion sensing, in its most basic form, is illustrated by the simple slotted (transmissive) optical switch. The position of the opaque target or "flag" determines whether the switch is open or closed.

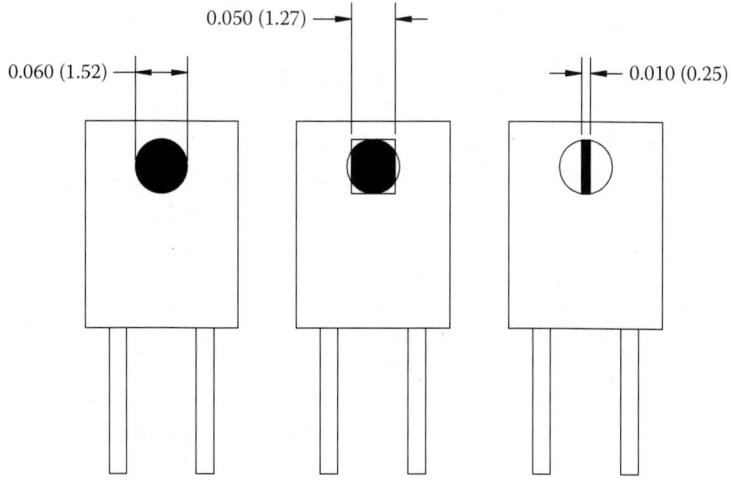

FIGURE 8.2 Equivalent aperture size.

Typical aperture sizes and configurations are given in Table 8.1.

When both devices do not have an aperture in front of them, minimal light is blocked and the area that is illuminated by the emitter and seen by the photosensor is significant. Figure 8.4 shows the transition of current flow as a target passes in front of different aperture sizes. This will be discussed later in the chapter. As an aperture is placed in front of both devices, the equivalent area is changed. The smaller the aperture, the less the light energy that is sent toward the photosensor and the less the light energy that is available to be sensed by the photosensor. When both apertures are equivalent in size (none and none or 0.05" and 0.05" or 0.01" and 0.01"), the equivalent area that can be recognized is minimum in front of both the emitter and photosensor, with the largest area of recognition in the center of the slot. When different aperture sizes are used, it is customary to have the largest aperture on the

The Transmissive Optical Switch

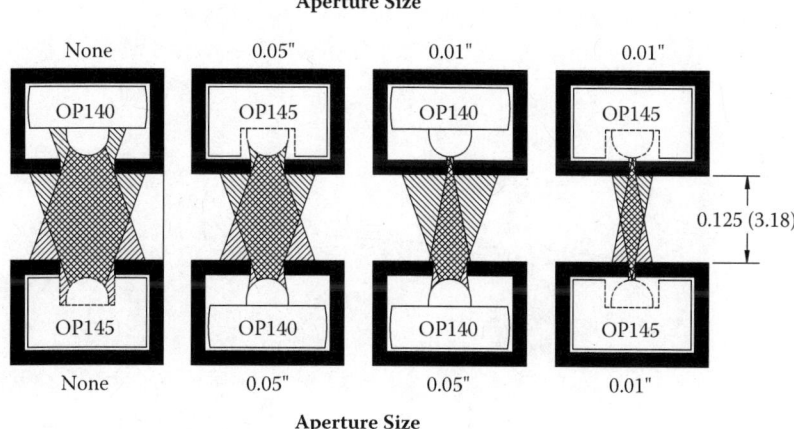

FIGURE 8.3 Aperture interface.

TABLE 8.1
Typical Aperture Configurations and Aperture Sizes (Left)

Emitter	Sensor	Emitter	Sensor
None	None	None	No blockage
0.05" (1.27 mm)	0.05" (1.27 mm)	0.05"	Minimal blockage
0.05" (1.27 mm)	0.01" ((0.25 mm)	0.01"	Significant blockage
0.01" (0.25 mm)	0.01" (0.25 mm)		

emitter side of the slot. This provides a minimum area of recognition of a target in front of the photosensor, with a gradual increase in recognition area until the target is closest to the emitter.

Figure 8.4 shows the switching characteristics of a slotted switch utilizing a phototransistor as the photosensor. As can be seen, when the target passes in front of the light sensor, the amount of output current $I_{C(ON)}$ changes proportional to the amount of light striking the lens of the phototransistor. These graphs show typical transition patterns as the target passes across different locations on the slotted switch.

The graph on the left that shows when an opaque target passes next to the emitter, where the aperture in front of the emitter is 0.05", the transitions from high to low or low to high are typically less than the width of the aperture. This apparent narrowing of the effectiveness of the aperture may be due to the sensitivity of the phototransistor. As long as sufficient light is striking the phototransistor, the output level will be saturated. The two major reasons for this are as follows:

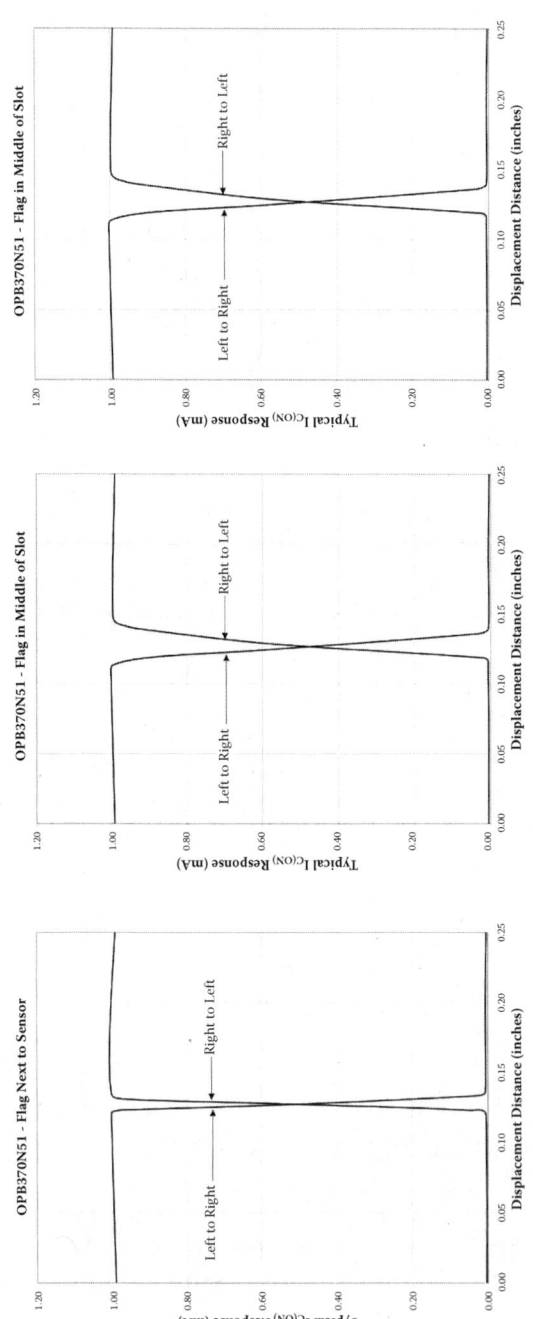

FIGURE 8.4 Slotted switch transition patterns for phototransistor sensor OPB370N51; emitter aperture size is 0.05″, and light sensor aperture size is 0.01″ wide.

The Transmissive Optical Switch

1. As the target passes further in front of the equivalent active area, the output reduces in amplitude.
2. The lens in front of the phototransistor reduces in efficiency as it radiates further from the center.

The actual transition crossing point of the right-to-left and left-to-right transitions are dependent on the alignment of the photosensor and emitter light beams. The graph on the right shows an opaque target passing next to a photosensor with an aperture of 0.01" in front of it, thus showing that the transition area is dependent on the size of the aperture and alignment of the light beam to the phototransistor. Typical side-looking components have a lens diameter of 0.06", thus providing a lens aperture of approximately 0.06" with no aperture present.

Figure 8.5 shows the switching characteristics of a slotted switch utilizing a Photologic device as the photosensor. There are six configurations of Photologic device available: Totem-Pole, Inverted Totem-Pole, Internal Pull-up Resistor, Inverted Internal Pull-up Resistor, Open-Collector, and Inverted Open-Collector. All of the configurations provide similar switching characteristics. When the sensor is radiated with sufficient light energy, the device switches from one state to the other. The Photologic device has a built-in optical hysteresis, thus providing an electrical signal that does not oscillate. As can be seen in Figure 8.5, each switching condition shows a switching location as well as a return switching "Back" position. The absolute locations of each of these switching positions is dependent on the sensitivity of the Photologic as well as the intensity of the light emitter.

Different types of housing material offer unique application configurations. A slotted switch is made with either a transmissive (clear to the sensing wavelength of light) or opaque (blocks the sensing wavelength of light) housing.

"Opaque" housings are typically used in applications where there are ambient light radiating on the photosensor may cause error signals. If sufficient ambient light is striking the photosensor, it may not recognize the target in the slot of the switch. A typical error signal may occur by reflection of light from behind the device, off the target, and back onto the photosensor. Dust may be a concern with these devices because the majority of devices with opaque housings have open apertures. As dust builds up in the aperture slot, less light can reach the photosensor.

"Transmissive" housings are typically used in applications where dusty or other particulate environments. They can be considered optically clear since the devices are optically open to the environment and mechanically closed since they are protected from particulate.

Some switches are available with a base-mounting fixture for the housing, resulting in a device that is "open" to the environment. These devices may be used in areas with minimal particle contamination and minimal ambient light intensity.

Slotted switches are normally a single channel with one emitter and one photosensor and can be used in many switching applications. A single-channel slotted switch can provide several functions as shown in Table 8.2:

Some slotted switches are designed with dual channels (one or two emitters and two photosensors) and are typically used with either linear or rotary encoding. A single-channel slotted switch can provide speed and end of travel, whereas a dual

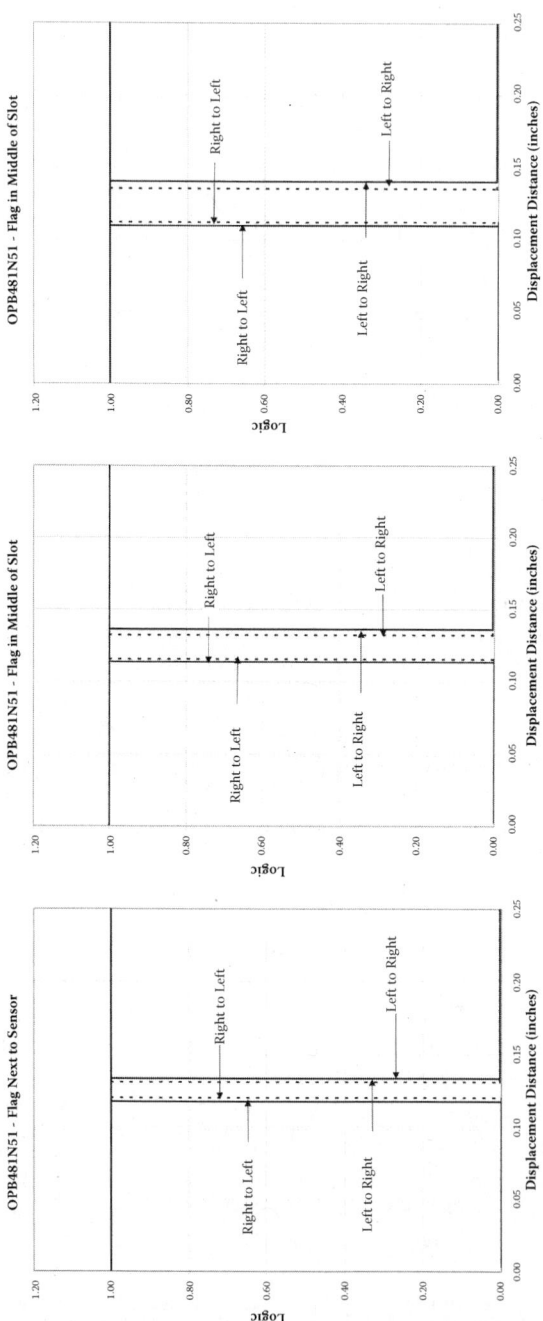

FIGURE 8.5 Slotted switch transition patterns for Photologic® sensor OPB481N51; emitter aperture size is 0.05", and light sensor aperture size is 0.01" wide.

TABLE 8.2
Functions of a Single-Channel Slotted Switch

Function	Description or Application
End of travel	Identify the final location of a moving object.
Object recognition	Identify when an object is between the emitter and sensor blocking the beam. Door closure, paper acknowledgment, coin recognition, currency recognition, tube recognition, etc.
Encoder	Identify transitions on a linear and rotary encoder to provide acknowledgment of travel along a linear tape or on a motor or generator.

channel slotted switch can also provide direction. Dual-channel devices have fixed distances between the channels; thus, encoder tapes or wheels must be designed and built to fit this spacing.

8.2 ENCODER WHEEL DESIGN

The rotational direction of a shaft can be readily determined by utilizing the two channels of a dual optical interrupter, an encoder disc with a number of openings around the circumference, and some simple electronics. The speed and relative shaft location information is available as a byproduct and requires some additional electronics.

Figure 8.6 is a pictorial definition of terms used in this bulletin and should be referred to for clarification. Figures 8.6 and 8.7 show the layout and dimensions for two types of slotted switches. A *period* is defined as a fraction of 360 electrical degrees or the mechanical width of one opening plus one closure at the central point of the slot near the circumference of the encoder disc. In shaft encoding terminology, *quadrature* is the term defining the ability to determine the direction of movement by the phase relationship between the outputs of two channels. System design normally uses 90° for this phase shift. Speed can be determined by accumulating the number of signal pulses for a fixed period of time and dividing by the number of periods per revolution, thus obtaining the revolutions for this time period. Relative location is determined by dividing 360 by the number of periods around the circumference. A pulse is generated for each of these rotational periods. Counters may be used to count the periods or to count the quadrature transitions (4 per period). Next we will describe the method of obtaining the information (rotation direction, speed, and relative location) rather than what is done with the information.

8.3 PERFORMANCE CHARACTERISTICS

First, we should determine the number of periods on the disc. This can be an arbitrary number or referenced to the resolution of the steps on the disc. As an example: if 0.7° steps are required, you would count the 4 quadrature transitions per period, the number of periods per revolution would be 128 (Periods = 360°/[degrees per

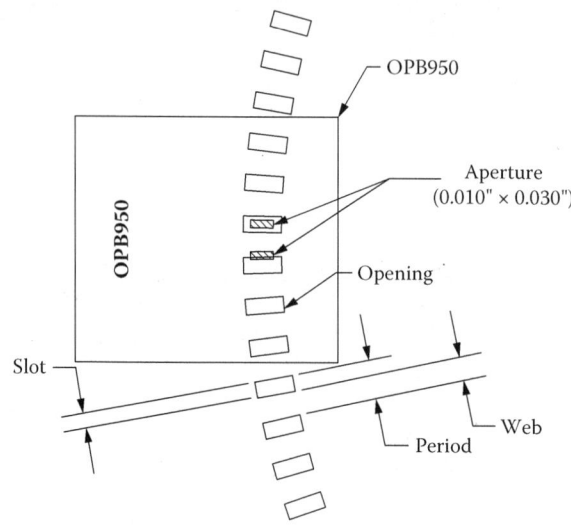

FIGURE 8.6 Wheel spacing.

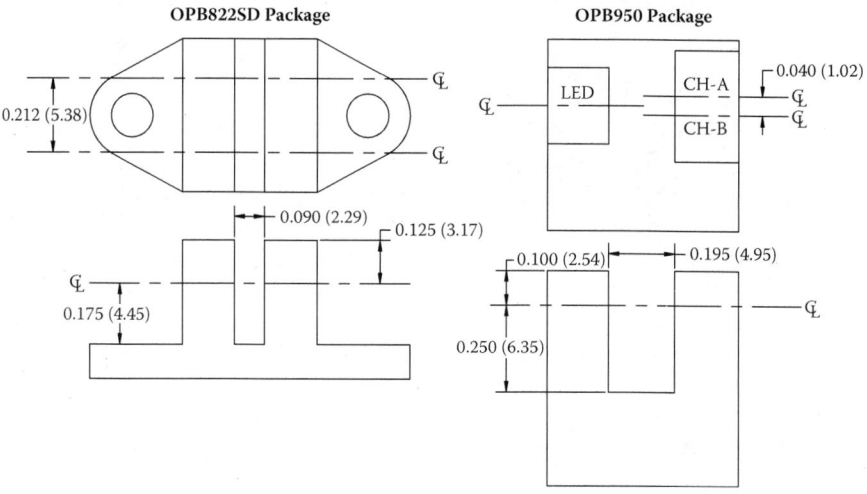

FIGURE 8.7 Package outline.

transition × number of transitions] = 360/[128×4]). We will next calculate the period size, opening and closure width, centerline pitch radius, slot length, and overall disc radius. The logical sequence for different off-multiples showing quadrature transitions will be covered last.

An off-multiple of periods between the center line of the sensor apertures is required for the 90° phase shift. The off-multiple can be ¼, ¾, 1¼, 1¾, 2¼, etc., periods. The period width is calculated by taking the spacing between centerlines of the sensor divided by the off-multiple (Period Width = Distance between Aperture Centers/Off-Multiple).

Period Width=(Distance between Aperture Centers/Off-Multiple)

Example: (0.212"/4.25)=0.050 (Reference OPB822SD)

(0.212"/0.75)=0.283 (Reference OPB822SD)

(0.040"/1.25)=0.032 (Reference OPB950)

(0.040"/0.75)=0.053 (Reference OPB950)

The opening in the disc should be greater than the width of the aperture (0.010" for both the OPB822SD and OPB950), otherwise this would decrease the guaranteed output signal. The period width will be divided into an opening for the light (slot) as well as an area for closure (Web), blocking the light. In order to have a 50% duty cycle, large period widths can be separated into two equal sections, whereas small period widths may require the slot to be smaller then the Web.

For a small period width with respect to the device aperture size, the aperture width should be taken into consideration. The opening (slot) width for the period width is calculated by taking 50% of the period width and subtracting 50% of the device aperture width (Opening Width=[0.5×Period Width]−[0.5×Aperture Width]). The closure (Web) width is calculated by taking the period width and subtracting the opening width (Slot Width=Period Width−Web Width).

Slot Width=([0.5×Period Width]−[0.5×Aperture Width])

Example: ([0.5×0.050"]−[0.5×0.010"])=0.020" (Reference OPB822SD)

([0.5×0.283"]−[0.5×0.010"])=0.137" (Reference OPB822SD)

([0.5×0.032"]−[0.5×0.010"])=0.011" (Reference OPB950)

([0.5×0.053"]−[0.5×0.010"])=0.022" (Reference OPB950)

Web Width=(Period Width−Opening Width)

Example: (0.050"−0.020")=0.030" (Reference OPB822SD)

(0.283"−0.137")=0.146" (Reference OPB822SD)

(0.032"−0.011")=0.021" (Reference OPB950)

(0.053"−0.021")=0.031" (Reference OPB950)

The centerline pitch radius, or the optical centerline of the slots, of the encoder disc is determined by multiplying the period width by the number of periods per revolution and dividing by two times pi (Pitch Radius=[Period Width×Periods per Revolution]/[2π]).

Centerline Pitch Radius = ([Period Width × Periods per Revolution]/2π)

Example: ([0.050" × 128]/6.283) = 1.016" (Reference OPB822SD)

([0.283" × 32]/6.283) = 1.440" (Reference OPB822SD)

([0.032" × 128]/6.283) = 0.652" (Reference OPB950)

([0.053" × 128]/6.283) = 1.087" (Reference OPB950)

The slot length should be larger than the sensor aperture opening length. As a good rule of thumb, you can use two times the disc eccentricity of the wheel plus the sensor aperture length. The maximum slot length would depend on the mechanical design of the disc. Remember, the slot should totally fit inside housing of the encoder.

Minimum Slot Length = (Aperture Length + [2×Eccentricity]) (assume 0.010" eccentricity for these calculations)

Example: (0.080" + [2×0.010"]) = 0.100" (Reference OPB822SD)

(0.030" + [2×0.010"]) = 0.050" (Reference OPB950)

Disc material and thickness will vary depending on the application. The minimum width is determined by the material stiffness, opacity to near infrared, and eccentricity expected. The maximum width is determined by the width of the slot and eccentricity expected.

The minimum disc radius is equal to the centerline pitch radius plus half the minimum slot length plus mechanically suitable support distance. The maximum disc radius should take centerline pitch radius plus optical centerline of the aperture to the bottom of the encoder slot minus two times the disc eccentricity.

Minimum Disc Radius = (Centerline Pitch Radius + [0.5×minimum slot length] + support distance)

Example: (1.016" + [0.5×0.100"] + X" = 1.066" + X" (Reference OPB822SD)

(1.440" + [0.5×0.100"] + X" = 1.490" + X" (Reference OPB822SD)

(0.652" + [0.5×0.050"] + X" = 0.702" + X" (Reference OPB950)

(1.087" + [0.5×0.050"] + X" = 1.137" + X" (Reference OPB950)

Maximum Disc Radius = (Centerline Pitch Radius + Optical Center Line of aperture to Bottom of the encoder slot − [2 * Disc Eccentricity])

The Transmissive Optical Switch

Example: $(1.019" + 0.155" - [2 \times 0.010"]) = 1.149"$ (Reference OPB822SD)

$(1.441" + 0.155" - [2 \times 0.010"]) = 1.571"$ (Reference OPB822SD)

$(0.652" + 0.235 - [2 \times 0.010"]) = 0.867"$ (Reference OPB950)

$(1.080" + 0.235 - [2 \times 0.010"]) = 1.295"$ (Reference OPB950)

The disc should be made from a material that is compatible with the application. The types of material could be plastic, metal, or glass with openings or a pattern applied to the disc. Encoder applications requiring higher precision usually select glass as the disc material. The encoder wheel should have the pattern as close to the sensor side of the encoder as possible, taking into account end play of the encoder.

The digital sequence expected from each configuration is dependent on the position of the apertures of the device in reference to the slots in the encoder wheel. As can be seen, the sequence of where channel A and channel B cross the opening is dependent upon the off-multiple and the spacing of the channels as shown in Figure 8.8.

As the channel A of the OPB822SD is in an opening for the 0.212/4.25 configuration with the pulse width of 0.050", channel B is in the middle of an opening. On the other hand, when channel A of the OPB950 is in an opening for the 0.040/0.75 configuration, channel B is in the middle of a closed part of the encoder wheel. This changes the logical sequence, as shown in Figure 8.9. By knowing the expected logical sequence, the direction as well as the speed of the wheel can be identified. As in the OPB822SD example, the expected sequence is 1-1, 1-0, 0-0, 0-1, where the expected sequence for the OPB950 is 1-0, 1-1, 0-1, 0-0. As can be seen, the 1-0 and 0-1 logical level sequences are reversed, therefore, you need to know the expected sequence to identify the direction of the encoder wheel. Off-multiples ending in "0.25" will output a quadrature direction sequence opposite to the sequence with off-multiples ending in "0.75".

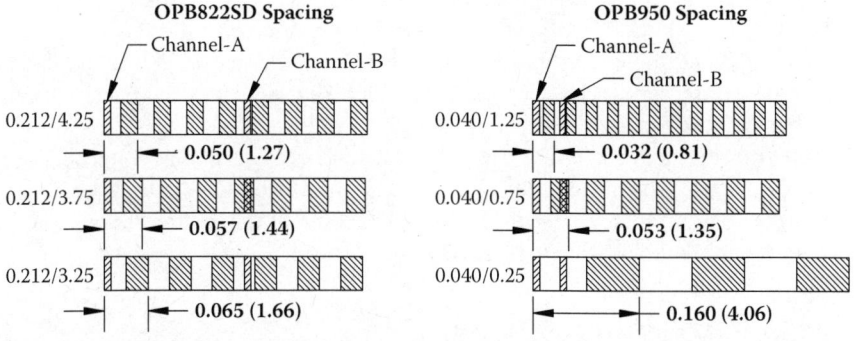

FIGURE 8.8 Timing diagram.

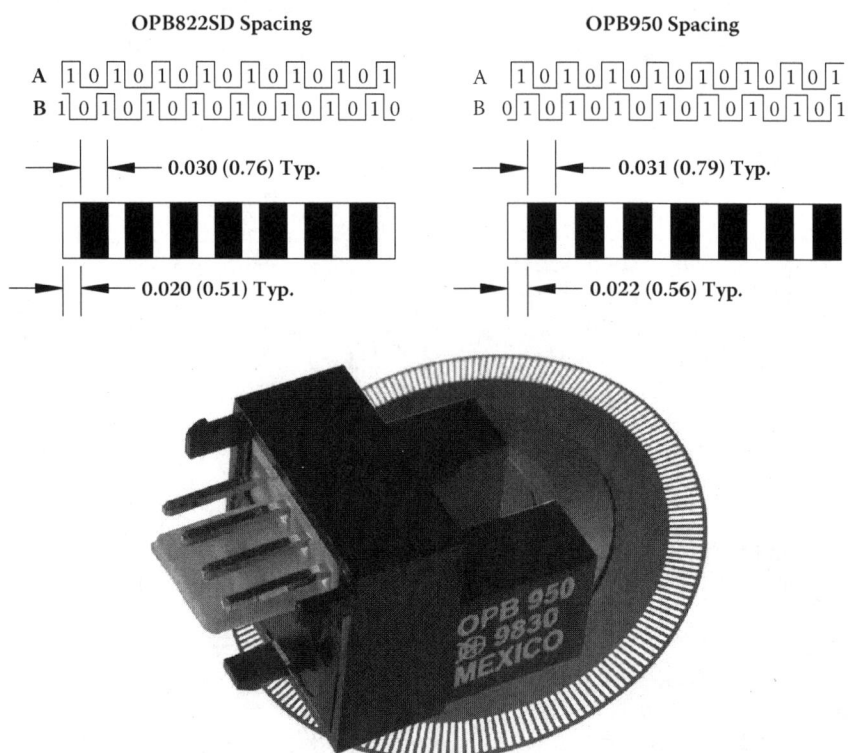

FIGURE 8.9 Logical output.

8.4 FLAG SWITCHES

Flag switches are a special type of slotted switch in which the target that breaks the energy beam is preassembled on the assembly. A flag switch consists of a standard slotted switch with either a phototransistor or Photologic photosensor.

Figure 8.10 shows how a typical slotted switch works with the flag in the rest position, allowing the energy beam to irradiate the photosensor. When the flag is moved using a piece of paper or other target, the flag interrupts the energy beam, thus changing the electrical state of the device.

Slotted switches come in a variety of sizes and configurations. Table 8.3 and Figure 8.11 show some of the typical configurations of slotted switches. Additional versions will be discussed later in this chapter. Some of the configurations have a name or nomenclature as shown in Table 8.3.

8.5 SLOTTED SWITCH SUMMARY

- Emitters provide a light source to directly illuminate the photosensor.
- Emitters can be

The Transmissive Optical Switch

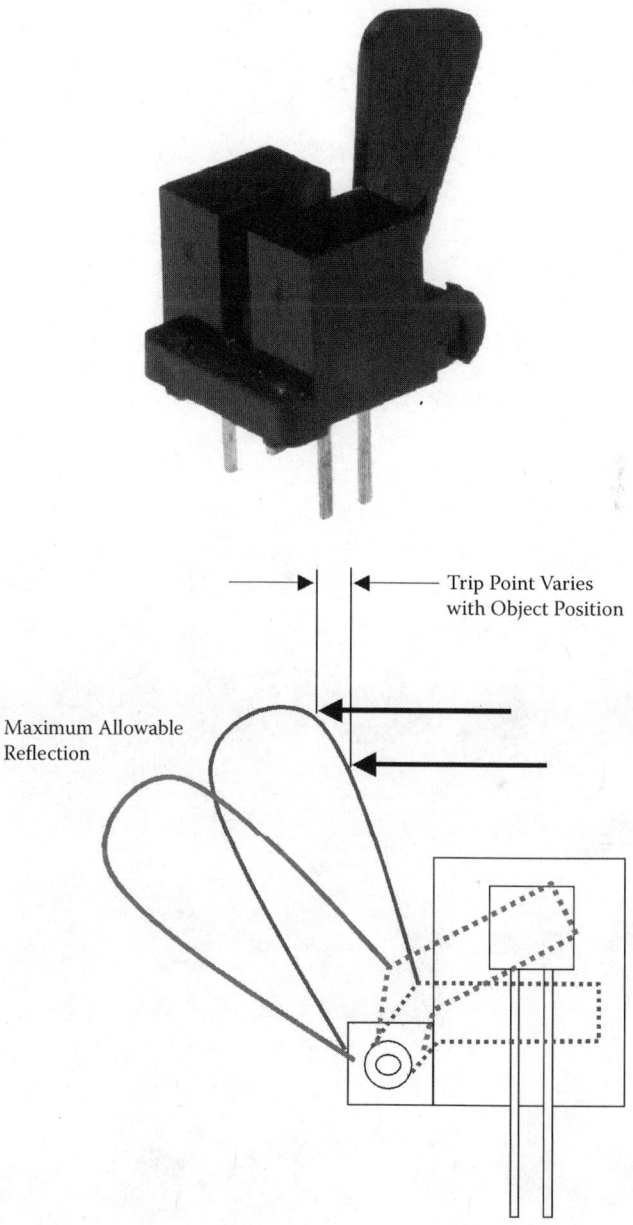

FIGURE 8.10 Flag switch configuration.

- Near-infrared LEDs
- Visible LEDs
- UV LEDs
- Vertical cavity surface-emitting lasers (VCSELs)

TABLE 8.3
Names of Typical Configurations of Slotted Switches

Configuration	Emitter	Sensor	Mounting Tabs	Mounting Configuration
N	LED	Transistor or logic	None	PCBoard or remote
L	LED	Transistor or logic	Next to emitter	PCBoard or remote
P	LED	Transistor or logic	Next to photosensor	PCBoard or remote
T	LED	Transistor or logic	Both sides	PCBoard or remote
TS	LED	Transistor or logic	Two on one tower	PCBoard or remote
Flag	LED	Transistor or logic	Varies	PCBoard or remote
Side	LED	Transistor or logic	Varies	PCBoard or remote
Wide	LED	Transistor or logic	Varies	PCBoard or remote
Dual	LED	Transistor or logic	Varies	PCBoard
Flow	LED	Transistor or logic	Varies	PCBoard or remote

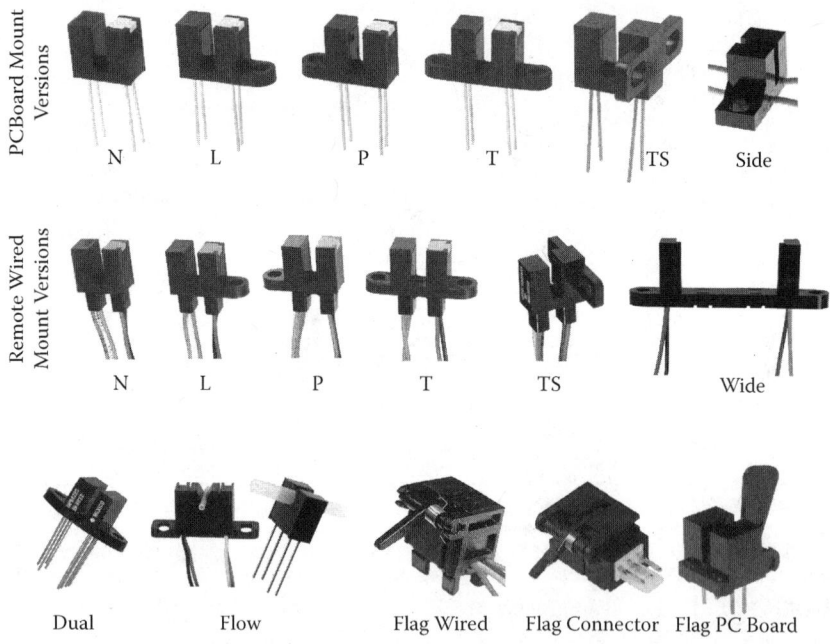

FIGURE 8.11   Typical slotted switch configurations.

The Transmissive Optical Switch

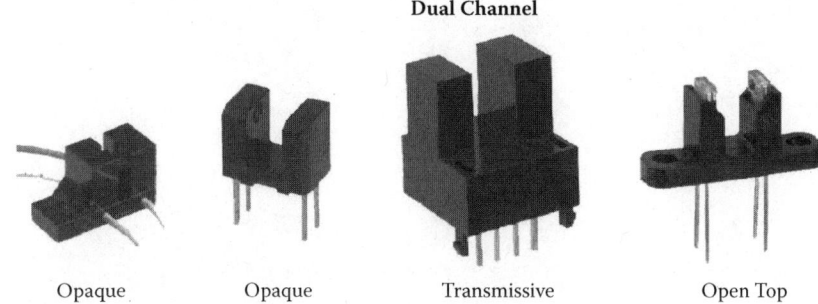

FIGURE 8.12 Opaque and transmissive slotted switches.

- Photosensors can be
  - Photodiodes
  - Phototransistors
  - Photodarlington transistors
  - Photologic®
- Apertures are used to optimize the area being looked at (narrow the critical area)
- Housings (see Figure 8.12):
  - Opaque material—blocks light at critical wavelengths
  - Transmissive material—clear to light at critical wavelengths
  - PCBoard mounting
  - Remote wire or cable interface
- Slotted switches come in a variety of configurations:
  - Single channel
  - Dual channel
  - Narrow-to-wide slot width
  - Short or long device towers
  - None, one, two, or special mounting features
- Encoders are designed with
  - Single-channel switches
  - Dual-channel switches
  - Wheels are designed utilizing the following properties of the switch:
    - Single channel
    - Aperture size
    - Slot depth
    - Slot width
- Custom versions may be designed if standard product does not meet the mechanical or electrical design criteria.

9 The Reflective Optical Switch

9.1 ELECTRICAL CONSIDERATIONS

Reflective switches may be divided into separate categories: *surface detectors* and *mark or line detectors*. The output may be analog or Photologic® (digital). Their sensing range can be from a few hundredths of an inch to several inches or feet. The basic reflective switch has both the transmitter and the receiver mounted together on the same side of the surface or line to be detected. Most units operate on a combination of diffused reflectance and specular reflectance. Figure 9.1 shows these two principles.

A highly polished surface (such as a mirror) will yield a high specular reflectance of energy, whereas a surface with diffused reflectance (such as white bond paper) will have a high diffused reflectance component of energy. In order to take advantage of these components, the design of the reflective switches is different for each case. The unit optimized for specular reflectance usually has the emitting and receiving elements mounted along the legs of an isosceles triangle with the reflective surface placed at their extended intersection. The unit optimized for diffused reflectance usually has the sensing and emitting elements mounted parallel to each other with the reflectance surface perpendicular to their extension. Part 1 discussed the dispersion of energy from the IRLED as the distance increased from the unit. Figure 9.2 shows this graphically for a convex lens device. Note that the reflective surface maintains the dispersion of energy such that the actual beam appears to follow the inverse square law relationship. Thus, the actual energy the sensor sees in the reflective mode is a factor of four rather than two lower than the energy received at the reflective surface. Because the decrease of energy per unit area versus distance actually deviates from the inverse square law relationship (the energy source at close distances), the basic inverse square law assumption is technically false but is good as a first-order approximation.

This is brought about by the reflecting surface further dispersing the energy. As the surface becomes less polished, the specular reflected component will decrease and the diffused component will increase. Figure 9.3 shows a unit designed for specular reflection to intersect at 0.200 in. from the front. Due to the diverging beam pattern from the IRLED, the actual peak response occurs at 0.150 in. This is because the diverging pattern of the energy from the IRLED and the inside portion of the light beam has a shorter optical path, causing higher outputs before the actual mechanical peak is reached.

A unit designed for diffused reflectance usually is fabricated with devices having plano lenses. This reduces the efficiency of energy transfer but allows a flat

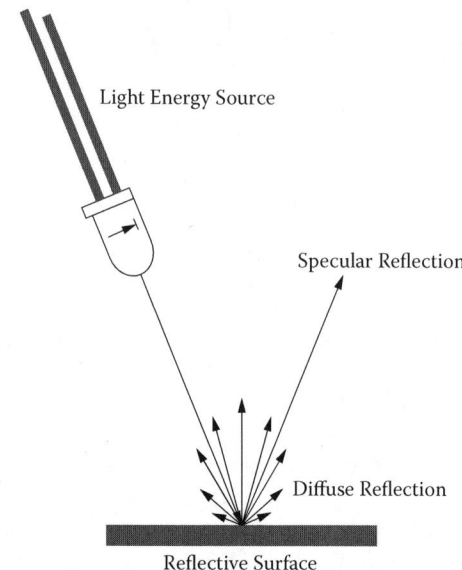

FIGURE 9.1 Diffused and specular reflection. The angle of the specular reflection component is equal to the angle of incidence, whereas the diffused component could be directed anywhere above the surface.

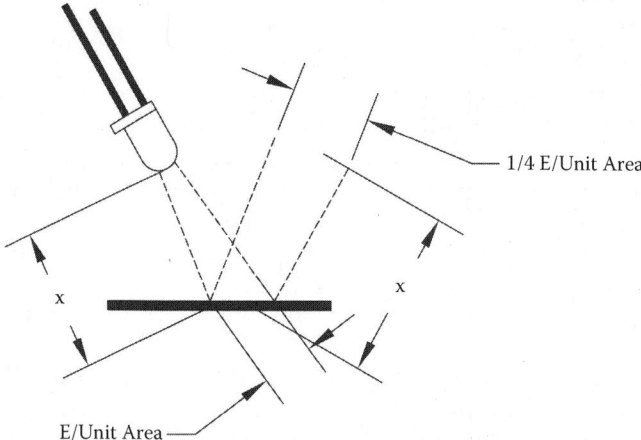

FIGURE 9.2 Representation of reflected energy versus distance. A good approximation of the reflected energy can be obtained by using the inverse square law. However, the diffusion and beam irregularities are not accounted for by this mode.

surface (non-dust-collecting) on the face of the unit. The peak response normally occurs close to the face of the unit (approximately 0.050 in.). This is shown in Figure 9.4.

In both cases, the effective photo-sensitive area detects the energy reflected back from the reflecting surface. The amount of energy the sensor receives is a function of

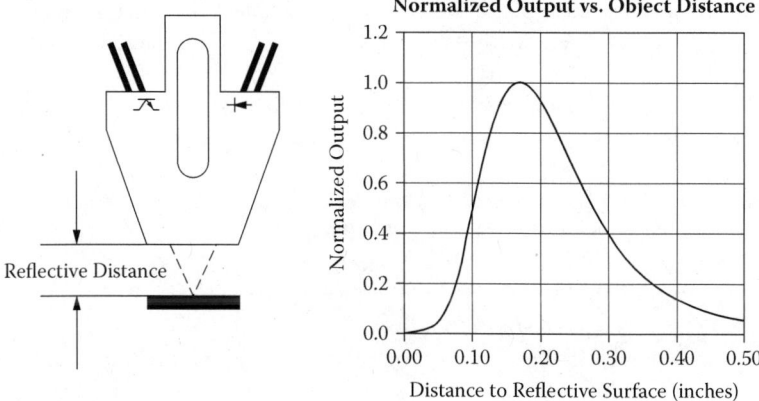

FIGURE 9.3 This curve is critical to the design engineer because it shows the sensing distance of the sensor. Optimum response is obtained by designing the specular reflective surface to pass about 0.150 in. in front of the focused reflective sensor.

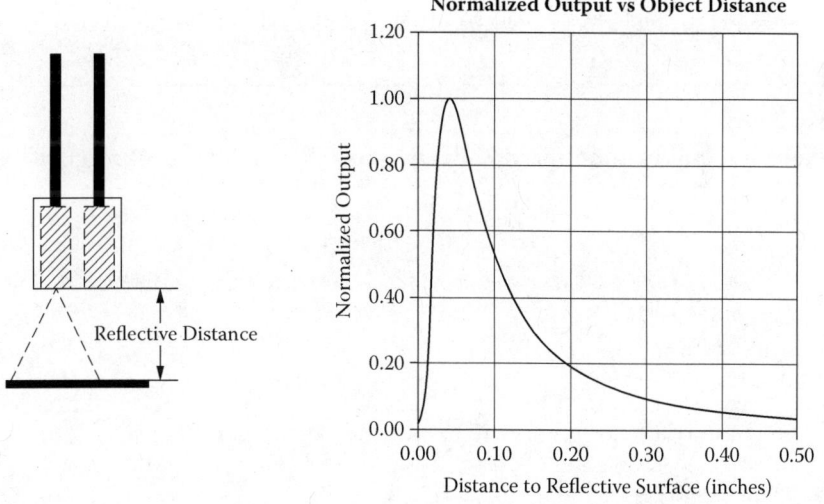

FIGURE 9.4 As in Figure 9.3, the output versus distance curve represents the sensing range of the sensor. For this curve, an unfocused reflective switch is used to detect a diffuse surface (reflective, but not specular).

the reflectance of the reflective surface with respect to the projected sensor-receiving area, the distance from the sensor to the reflective area, and the perpendicularity of the reflective surface to the plane through the transmitter and receiver. The unfocused reflective switches are less susceptible to perpendicularity due to the wide irradiance angle of their devices with plano lenses.

Table 9.1 shows relative output from an infrared reflective switch with respect to different types of surfaces. Analysis of the data shows several interesting points:

1. The black velvet paint in surface 4 gives excellent contrast for either unit.
2. The graphite pencil mark in surface 7 provides relatively small change to the reflected energy, because it both improves the reflectivity (smeared shiny graphite) and disrupts the pore fibers of the paper.
3. Color makes very little difference (9, 10, and 11).
4. The 3M tape #476 is an excellent nonreflecting surface (12).
5. Black anodized aluminum may look dark to the human eye but reflects a considerable amount of infrared light (6).
6. Dye-based black ink reflects more infrared light than carbon-based black ink.

The reflective sensor will normally work better than the transmissive sensor in three types of applications. The first of these is obviously when there is access to only one side of the object to be detected. The second of these is in liquid level detection.

Figure 9.5 shows an outline of the principle of operation for a fluid level detector. This is more precise than a transmissive sensor and also allows the electronics to be remote with respect to the liquid.

The principle of operation is quite simple. The IRLED energy normally is deflected by the angle at the bottom of the rod and returns to be detected by the

TABLE 9.1
Relative Output of an Infrared Reflective Switch with Various Reflective Materials

	Surface Type	
1	Reference reflective surface Kodak 90% diffuse card Eastman Kodak Cat. # E152-7795	100%
2	Gray side	20%
3	Newspaper with ink	60%
4	Aluminum foil tape	120%
5	Aluminum foil tape painted with flat black velvet (3M #101-C10 Black)	0.2%
6	Black anodized aluminum	50%
7	White bond paper	94%
8	No. 3 graphite on white bond paper	45%
9	White smooth plastic surface	90%
10	Blue smooth plastic surface	90%
11	Red smooth plastic surface	90%
12	Fiber-tip pen, black (Stabilo)	10%
13	3M tape #476 (a dull black surface)	0.2%
14	Black carbon-based ink	8%
15	Black dye-based ink	20%

The Reflective Optical Switch

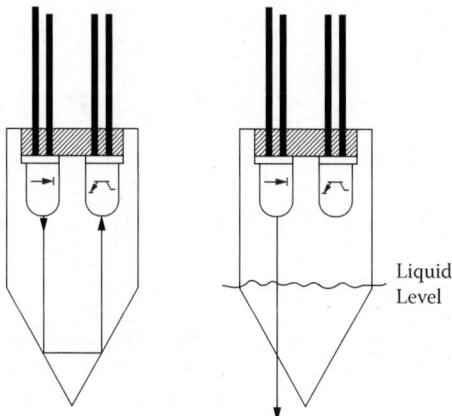

FIGURE 9.5 Liquid level sensor. The basic design of a fluid level sensor may be expanded to include multiple reflective steps to detect different fluid levels.

photosensor. When a liquid such as gasoline covers the reflecting surface, the energy is not reflected but refracted out into the fluid (Snell's law when the index of refraction changes from air to a higher index of refraction).

The third application involves precise height measurement, where the width across the material is very large compared to the distance from the detector to the material. Figure 9.6 illustrates this principle. The application might require a tray or holder adjustment to banknotes or other light, small objects being automatically processed. The distance they dropped could be held to a minimum to prevent misorientation. As the number of checks in the holder is increased, the distance between the checks and reflective sensor would decrease. The sensing electronics would get an increasing signal. Once the checks get close enough to the sensor that the signal starts decreasing (see Figure 9.3), the electronics triggers the check holder to drop or unload, increasing the distance, and starting the cycle over.

Mark sense or line detectors are optically much more complicated. Special attention should be taken to determine the difference between reflected light signal strength from the background material and from the mark or line. This difference is often expressed as the ratio of the two signal levels, and it is often called the *contrast ratio*:

$$Contrast\ Ratio = \frac{Background\ Signal\ Strength}{Mark\ Signal\ Strength} \tag{9.1}$$

Contrast ratio is one of the most important parameters needed to achieve a reliable sensing condition. It is affected by the mark or line, background color as well as material, ambient light conditions, and the wavelength of the sensor. A good rule of thumb for a good contrast ratio is 4:1 or more. A low contrast ratio, such as 2:1, will not take into account production lot variations on the sensors as well as dust built

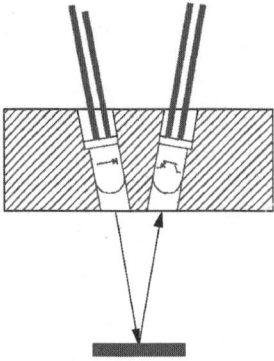 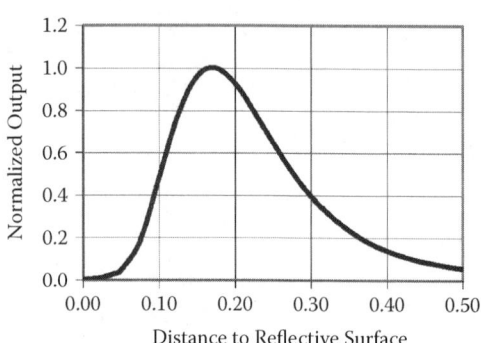

FIGURE 9.6 Optical technique used to reject reflections from close objects. At a certain distance, the output signal becomes large enough to trigger secondary activity, such as unloading a tray moving part.

up in the optical path, ambient light changing conditions, and variations in sensing distance due to sensor-mounting tolerances.

In many instances, simply increasing or decreasing the IRLED, light output can increase a low contrast ratio. For example, when a mark or line is overflooded with light, the extra reflections will increase the mark signal strength, decreasing the overall contrast ratio.

Contrast ratio is not only affected by the reflectivity of the two mediums, the mark or line and the background, but also from the size of the mark relative to the size of the effective viewing area of the sensor. Figure 9.7 shows how an effective viewing area larger than the mark would decrease the overall contrast ratio by failing to absorbed part of the light energy. For example, if the background-to-mark contrast ratio reflectivity is 12:1, it may not be sufficient if the mark area is 60% of the effective viewing area of the sensor:

$$Contrast\ Ratio = \frac{12}{(0.40 \times 12) + (0.60 \times 1)} = 2.22 \qquad (9.2)$$

This illustrates how an excellent contrast ratio can be severely affected if the mark or line to be detected is smaller than the effective viewing area of the reflective sensor.

Normally, both the collimation of the energy source and limitation of the effective viewing area are added to the mark or line detector. Four techniques can be utilized to better collimate the energy source. First, changing the light source from a wide beam divergence, noncoherent IRLED to a LASER, or a form of low-beam divergence, coherent light source such as a VCSEL (vertical cavity surface-emitting laser) will reduce the viewing angle by creating a small light spot on the mark. The second technique to improve the contrast ratio is to add a converging lens to reduce the dispersion of energy from the IRLED (see Figure 9.2) to focus the light energy. Third, adding apertures, or windows, to reduce the effective viewing area of

The Reflective Optical Switch

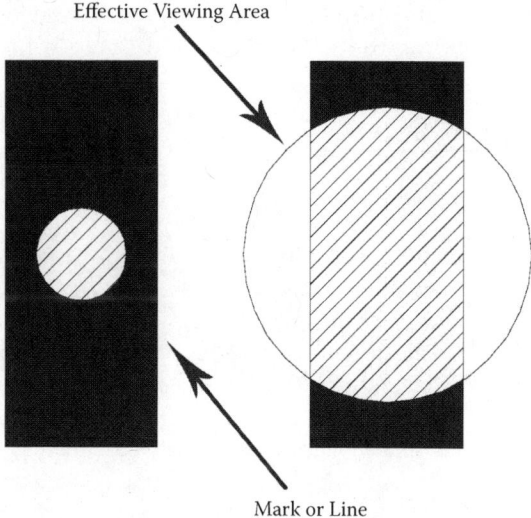

FIGURE 9.7 Contrast ratio is reduced when the object is smaller than the size of the effective-viewing area of the sensor.

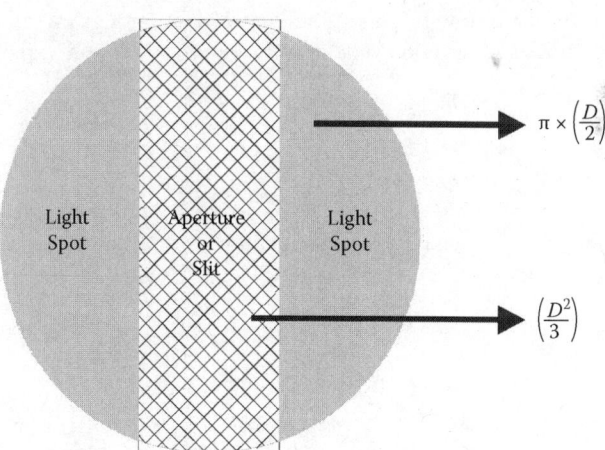

FIGURE 9.8 Adding an aperture or a slit to the emitter or detector decreases the reflected signal but increases the resolution of the reflective sensor.

the sensor can be an efficient and low-cost approach to improve the resolution and, therefore, the contrast ratio of the sensor. This may be done on both the emitter and detector side. The trade-off is that the signal output due to the reflected energy will be reduced, but the benefit of the trade-off can be significant. For example, adding an aperture width equal to one-third of the light spot diameter, as shown in Figure 9.8

reduces the potential reflected energy by about 60%, but the resolution of the reflective sensor triples.

$$I_{C(ON)} = I_{REFLECTED}\left(\frac{SLIT\ AREA}{LIGHT\ SPOT\ AREA}\right) = \frac{D^2/3}{\pi \times (D/2)^2} = I_{REFLECTED} \times 0.42 \qquad (9.3)$$

Finally, the fourth technique is to utilize a light pipe, such as a fiber-optic pipe, to decrease the energy loss and reduce the effective viewing area of the sensor.

To determine the resolution of the reflective sensor, it is necessary to understand the switching curve due to an abrupt change in reflectivity with distance. The output of the reflective switch is not abrupt; it changes as different amounts of energy are reflected back to the sensor. Figure 9.9 pictorially represents the general wave shape that will be seen as two abrupt changes in reflectivity pass by the effective viewing area of the sensor.

As the mark (nonreflective target) starts absorbing the energy from the IRLED, the sensing element will start to turn "off." If the system has enough resolution, the sensor will turn "off" prior to the end of the mark, the next abrupt change in reflectivity. The reflected light energy will start increasing as the trailing edge of the mark goes by, increasing the sensor output. The switching curve of the reflective sensor due to an abrupt change in reflectivity is analogous to the switching curve of the slotted switch due to an abrupt change in light transmission.

The interval from 1 to 6 will remain fairly consistent for a given setup. As different units from various production runs are introduced, the main variations viewed are as follows:

a. Variation in slope from 1 to 3
b. Variation in slope from 4 to 6

As the resolution of the reflective switch increases, the turnoff delay will decrease, and the flat portion between 3 and 4 will get wider. In other words, 3 will move to

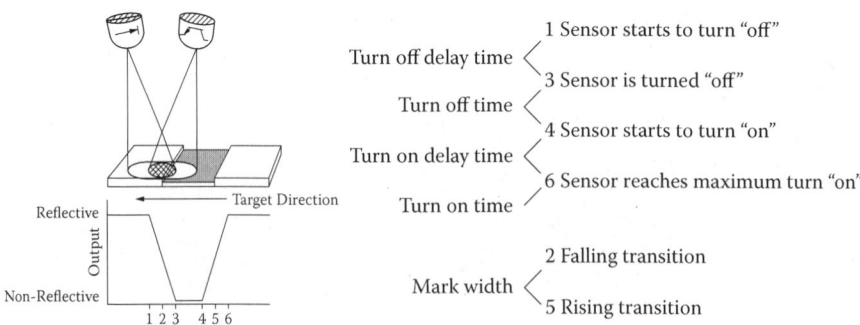

FIGURE 9.9 Pictorial representation of a reflective sensor signal output due to two abrupt changes in reflectivity. The gray area represents a mark.

the left, and 4 will move to the right. As the resolution of the sensor decreases, the opposite will occur. Points 3 and 4 will become one point and start to increase the "nonreflective" output value. As a result, the overall contrast ratio will decrease. The waveform of a sensor with poor resolution will resemble the mark as the low peak of a triangular wave.

Photologic® reflective sensors utilize similar mechanical configurations as the previous analog reflective sensors: they can be focused or nonfocused. Focused reflective sensors are best utilized when the target detection is in close proximity and background reflection is a concern. Photologic nonfocused reflective sensors are best utilized for medium-to-long sensing distance. The main difference between the analog and Photologic sensor is the output. The Photologic sensor output is not continuous but digital; it only has two states: digital low and digital high. As a result, minimal addition of electrical components is required to interface the sensor output with microcontrollers. Photologic sensors incorporate a photointegrated circuit as their sensing element. Some advantages and disadvantages of the photointegrated circuit (photo IC) over the analog photosensor are described in Chapters 6 and 7. Figure 9.10 shows a typical output of a focused Photologic reflective sensor.

The curve represents the sensing distance of a digital reflective sensor when an object is brought toward or moved away from the face of the sensor. The difference between the trip points toward and away is called *hysteresis*. Hysteresis is needed on a digital sensor to prevent oscillation of the output if the light signal level changes close to the trip point. The sensing range of the digital sensors will depend on the target size and reflectivity. As the target reflectivity changes, the trip points toward and away move, changing the hysteresis.

As with analog sensors, contrast ratio is the key to a successful sensing system. Contrast ratio is not only important for mark detection but is also important for other types of reflective sensing methods. For example, long-distance reflective

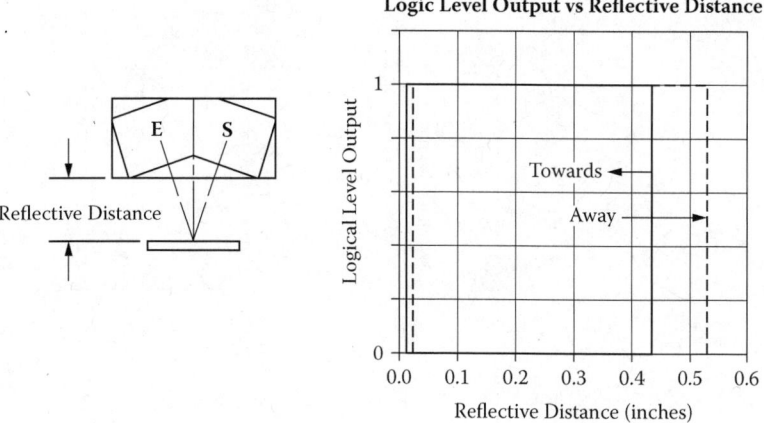

FIGURE 9.10 This curve shows the sensing distance of the sensor. Optimum response is obtained by designing the specular reflective surface to pass about 0.225 in. in front of the focused reflective sensor.

sensors should have enough contrast ratio to distinguish a target from the background. Background can be other surfaces of different reflectivity or just ambient light interference.

To achieve a long sensing distance, it is often helpful to pulse or modulate the signal from the IRLED. Modulation of the IRLED is accomplished by adding an oscillator circuit or other means of turning the IRLED on and off at a given frequency. Modulation of the light signal allows the detector to distinguish between the signal emitted by the IRLED from other light signals such as sunlight and interior artificial lightning. The detector is able to differentiate between the IRLED-modulated signal and other signals because most light sources do not produce periodic signals witht high frequency. The signal rejection is often achieved by addition of a receiver circuit or by utilizing a photo IC designed specifically to detect the modulated signal of the IRLED. In more advanced photo ICs, the IRLED oscillator and receiver circuit are integrated in the same chip. This sensing scheme, often called *Synchronized Driver Detection* or *SDD*, allows a reflective sensor not to be affected by most ambient light conditions while also achieving long sensing distance. Figure 9.11 shows the sensing range of a nonfocused reflective sensor utilizing an SDD scheme. The sensor is able to greatly improve the sensing range compared to a reflective sensor with a similar mechanical configuration, such as the one shown previously in Figure 9.4. The advantages of sensors utilizing SDD scheme are the strong rejection of ambient light noise and improved sensing range. The limitation is usually the sensing speed, which can be typically in the milliseconds range. On the other hand, analog sensors typically have a sensing speed in the low microseconds to nanoseconds range.

The field of view of long-range reflective sensors depends on the overlap of the viewing angle of the emitter and detector. Close to the face of the sensor, the overlap area or the field of view of the sensor is small, which allows the reflective sensor to

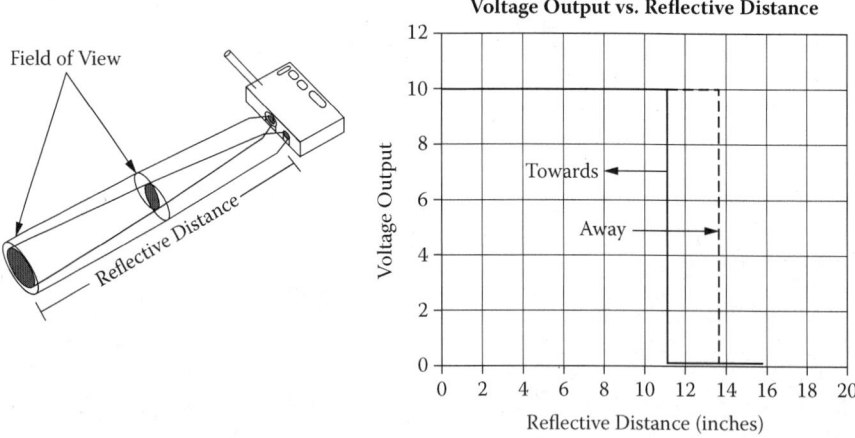

FIGURE 9.11 This curve shows the sensing distance of a long-range reflective sensor. Resolution depends on the target size and field of view of the sensor. Sensing occurs when a target is within the field of view. The field of view is the area in which the light source and photosensor's viewing angle overlap.

The Reflective Optical Switch

detect small objects. At a long distance from the face of the sensor, the field of view area increases, allowing the sensor to have maximum sensitivity for a given reflected signal. It is important to take advantage of this feature by utilizing a reflective target with an area as large as possible in order to maximize the contrast ratio.

9.2 MECHANICAL CONSIDERATIONS

The housings of reflective optical switches serve the same dual role as transmissive optical switches. They not only serve the mechanical function of locating the transmitter and receiver but also incorporate a fastening system to the system or subsystem.

The focused reflective assemblies normally have the shape of an arrowhead. A mounting slot is included to provide adjustment of focus for optimum signal. Figure 9.12 shows a photograph of three different types of reflective units. These are usually fabricated with axial discrete emitters and sensors (egress leads in same plane as the lens) with convex lenses to improve signal levels.

The short-sensing-distance unfocused reflective sensor is fabricated from either discrete plano lens devices or an individual housing or holder with the discrete IRLED and phototransistor chips mounted in the housing. Figure 9.13 shows a photograph of three such sensors.

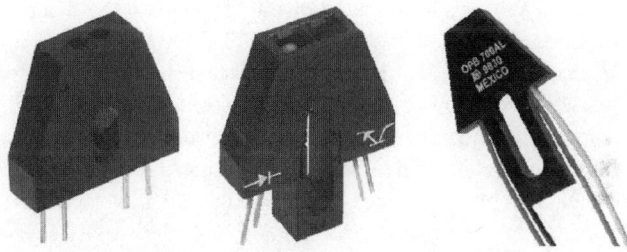

FIGURE 9.12 Photograph of focused reflective optical switches. As was the case with transmissive switches, a variety of models are commercially available as standard products. More complex applications often require customized designs.

FIGURE 9.13 Photograph of unfocused reflective optical switches. The square shapes and parallel leads of these devices are due to the flat-lensed discrete emitters and sensors (or chips) mounted on parallel optical axes.

FIGURE 9.14 Photograph of medium- and long-distance unfocused reflective optical switches. Some models are available at fixed sensing distance, whereas others may have an adjustment feature for different sensing distances.

The two sensors on the left are made with discrete transmitters and receivers, whereas the two on the right are fabricated from chips mounted on a header or lead frame with a molded plastic housing, which forms a container for the epoxy protecting the chips as well as an energy-blocking media between the chips.

Because these are made with plastic material, a dye can be added during fabrication. This dye effectively blocks all wavelengths shorter than 700 nm and passes the longer wavelengths of IR energy. The flat surface prevents the buildup of dust. The major drawback of this type of assembly is the difficulty in mounting. They may be mounted in a molded or machined cavity or also mounted by their leads in either a socket or on printed circuit board.

Medium- and long-range sensing unfocused reflective sensors are typically fabricated utilizing discrete components with convex lenses. Owing to the additional complexities of these sensors, they are relatively bigger than the two previous types of sensors discussed. Figure 9.14 shows a picture of three of these sensors.

The analog sensor on the left utilizes two discrete axial components with convex lenses and a dust-proof window. The sensors in the middle and right utilize two discrete components with convex lenses or external lenses to improve the optical performance of the sensor. In addition to that, the housings of most of these sensors hold additional electrical components necessary to add optical and electrical flexibility to the sensor. For example, the sensor in the middle has a sensitivity adjustment point to increase or decrease the sensing distance.

Part 4

The Optical Isolator and Solid-State Relay

10 Electrical Considerations

10.1 BACKGROUND

There are many situations in which signals and data need to be transferred from one system to another within a piece of electronic equipment, without making a direct electrical connection. Often, this is due to the different components operating at significantly different voltages. One part that may be operating at 5 V DC could be used to control a triac that is switching 240 V AC. In order to protect the low-voltage side, the link must be electrically isolated. A very effective method of achieving this is to implement it optically.

Optically coupled isolators, also called *optocouplers*, are used to isolate one electrical system from another in an electronic circuit. They allow direct circuit control with complete electrical isolation of input from output. These isolators are considered the best, most cost-effective devices to eliminate associated differential ground, ground loop, and EMI/RFI problems.

The optical isolator or optical coupler uses a beam of light to send a signal to an optical receiver. It is similar to a transformer in that the input is electrically isolated from the output. This is because a transparent barrier separates the two. An IRLED is connected to the input circuit, optically coupled to a silicon photosensor at the output circuit. Both the IRLED and photosensor are housed in a single package with a light-conducting medium between them.

The optical isolator or coupler is normally driven from a DC source, and the output drives an electronic or DC circuit. Couplers are available that operate from an AC source or drive an SCR or triac to control an AC load. In principle, they are very similar to the transmissive switch discussed in Part 3. A transparent material that provides coupling of the IRLED energy to the photosensor replaces the air gap. Most applications utilize the units in the digital mode, taking advantage of the isolation of signal from input to output. A portion of Chapter 10 will discuss utilization of the coupler in a linear mode and how to take advantage of this type of operation.

Figure 10.1 shows the construction of a coupler utilizing a discrete-packaged IRLED and photosensor mounted in a housing and separated by a light pipe. The physical separation of the source and receiver provides the electrical isolation of input to output. This obviously will vary as a function of the distance and the material separating the units electrically.

Table 10.1 lists currently available couplers by package, isolation voltage, and output function. As the coupler resembles a transformer in function (providing electrical isolation), various agencies control some of the electrical and mechanical specifications. If the coupler were to short input to output, there would be a possibility of damage to equipment or personnel operating the equipment.

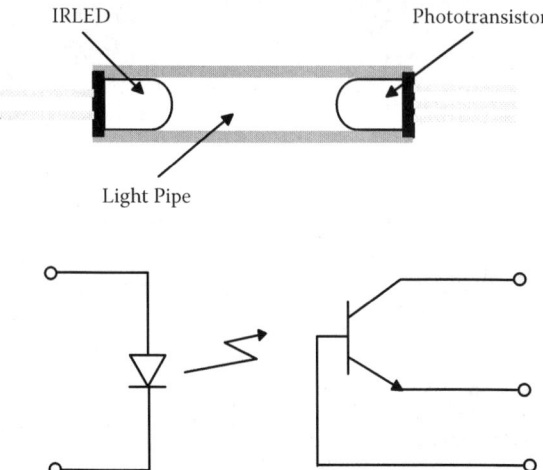

FIGURE 10.1 Optical isolator utilizing discrete transmitter and receiver. Perhaps the simplest design for an optically coupled isolator is to place discrete axial components at opposite ends of a hollow tube. The early designs, and some of the most popular designs today, are based on this method of manufacture.

TABLE 10.1
Listing of Currently Available Types of Couplers

Package	Isolation Voltage	Output
Coaxial discrete	10–50 kV	Photodiode, Transistor, Darlington, and IC
TO-18, TO-5	1 kV	Photodiode, Transistor, Darlington
6–8 pin DIP	1.5–7 kV	Photodiode, Transistor, Darlington, IC, SCR, triac driver
PC board mounted	6 kV	Phototransistor
Rectangular case		Photodarlington

This possibility of damage obviously varies from country to country, that one of the major causes of overload is voltage spikes or surges. The quality of the line voltage varies throughout the world. The amount of overload protection must then vary in accordance with electrical service provided and by the degree of conservation that exists within the controlling agency. Table 10.2 lists the major agencies.

10.2 FUNCTION

In order to understand the function of the coupled pair, the input must be examined first, and then the output, so that the two can be put together in one element. Figure 10.2 shows the energy output waveform from an IRLED with a trapezoidal input waveform. The assumption is made that the turn-on and turn-off times of the IRLED are negligible.

If the input waveform was sinusoidal, then the output waveform would be similar, with the IRLED not turning on until approximately 0.9 V and turning off when the

Electrical Considerations

TABLE 10.2
Certification Agencies

Agency

UL (U.S.)	Underwriters Lab
VDE (Germany)	Verband Deutscher Elektrotechniker
TUV (Germany) independent agency	Technischer Uberwachungs-Verein
CSA (Canada)	Canadian Standards Association
BSI (Britain)	British Standards
SGS FIMKO (Finland)	Nordic Certification Service
NEMKO (Norway)	Norges Elektriske Materiellkntroll
DEMKO (Denmark)	Subsidiary of the UL
SEMKO (Sweden)	A division of Intertek Group

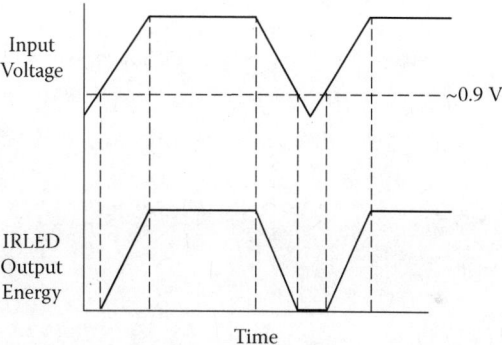

FIGURE 10.2 IRLED output waveform versus trapezoidal input waveform. The infrared output of the diode tracks the voltage waveform above 0.9 V. Below 0.9 V, no infrared is emitted.

input voltage dropped below 0.9 V. Figure 10.3 shows the same type of plot when two back-to-back IRLEDs are used at the input.

There are several inconsistencies that should be considered when examining the effect of Figures 10.2 and 10.3. The energy output is not linear with increasing voltage or current. The energy output is inconsistent from one IRLED to another. The energy output must be offset by the turn-on and turn-off times of the IRLED.

Now apply the output from Figure 10.3 to a photodiode and to a phototransistor. Figure 10.4 shows this relative relationship. R_L has been adjusted in magnitude so that output voltage is identical.

Without the adjustment in R_L, the voltage across the load for the photodiode would be much lower. It would approximate the voltage across the transistor R_L divided by h_{FE}. The curves must also be offset by the turn-on and turn-off times of the photosensor. The important point is the nonlinearity of gain versus current on the phototransistor. Figure 10.5 shows the electrical schematic of a coupler with phototransistor output.

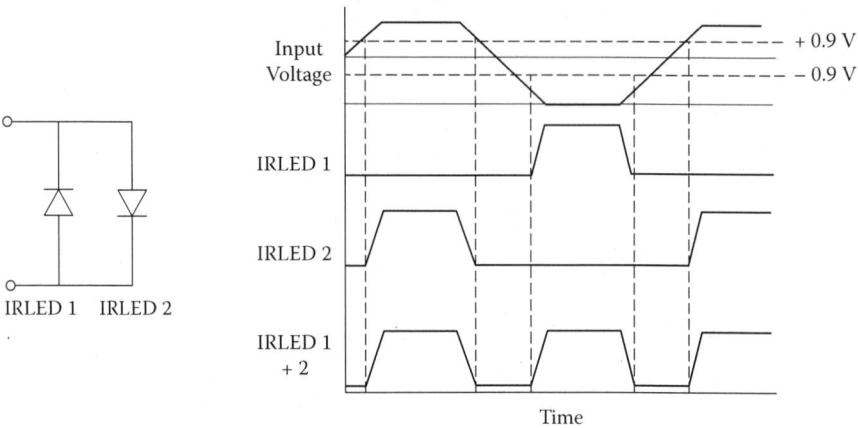

FIGURE 10.3 IRLED output waveform versus trapezoidal input waveform. Two IRLEDs in parallel, at opposite polarities, are used in AC input optocouplers. A typical application would be a telephone ring detector for an answering machine or computer.

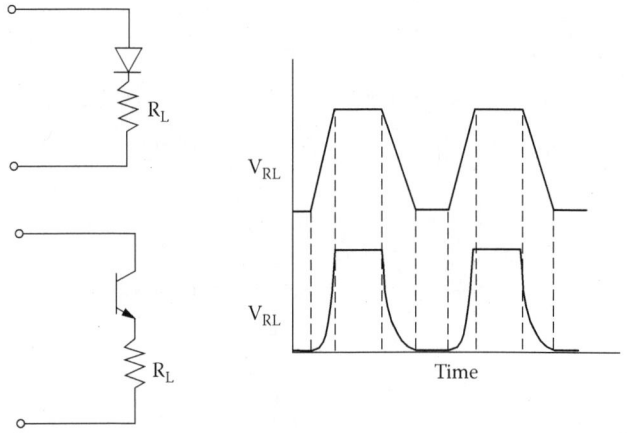

FIGURE 10.4 Output waveform from photodiode and phototransistor with input waveform from Figure 10.3. The linear response of the photodiode (top waveform) results in a waveform identical to the IRLED voltage waveform. The nonlinearity of the phototransistor (lower waveform) introduces some distortion.

Assume that the output of the IRLED is linear from 2 to 20 mA. Table 10.3 shows the output of the phototransistor versus the step input from the IRLED and the subsequent calculation of h_{FE}. The equivalent I_b shows a linear input to the phototransistor. Typically, a phototransistor is designed to be reasonably linear in the range of 100 μA to 10 mA. The h_{FE} becomes more nonlinear above 10 mA due to current crowding. The nonlinearity in h_{FE} below 100 μA is simply due to nonlinearity in h_{FE} efficiency.

Electrical Considerations

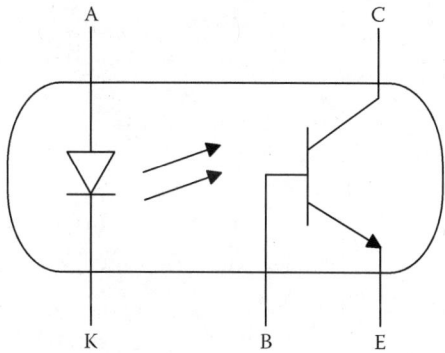

FIGURE 10.5 Electrical schematic of coupler with phototransistor output. The vast majority of couplers produced follow this basic schematic. Most are packaged in simple six-pin plastic dual-in-line packages (PDIPs).

TABLE 10.3
Phototransistor Output and h_{FE} versus Input

IR LED Relative Output	$I_{TRANSISTOR}$ (mA)	h_{FE}(XSTR)	Equivalent I_b (µA)
2	0.83	335	2.5
4	1.75	348	5.0
6	2.73	364	7.5
8	3.8	380	10.0
10	5.0	400	12.5

The four-pin plastic dual-in-line package (PDIP) is widely used for its low cost, dual sourcing, and performance characteristics. The most popular type is the phototransistor output with an electrical schematic, as shown in Figure 10.5. Many design engineers add a resistor (base to emitter) in order to improve the fall time. Figure 10.6 shows the switching time of a typical unit versus the value of base resistance.

Assume a base resistance of 200 kΩ. This would improve the fall time from approximately 8 µs to 4.4 µs. The normal base-to-emitter bias level would be approximately 0.7 V. This means that 3.5 µA of base current would be required for the 200 Ω resistor. This current would have to be subtracted from the photocurrent of the phototransistor.

The performance of the coupler versus temperature generally will track the performance of a reflective or transmissive switch versus temperature. Normally, a coupler has the transmitter and receiver located in closer proximity to each other and, as a result, the photosensor and the IRLED operate at closer to the same junction temperature. Figure 10.7 shows the relative change in output for an IRLED and a phototransistor. Figure 10.8 shows their combined characteristics.

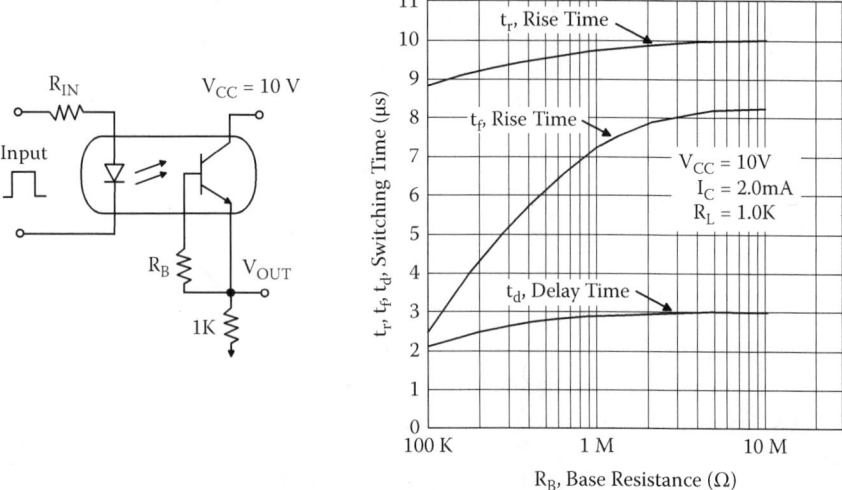

FIGURE 10.6 Delay, rise, and fall time versus base resistance. The base resistor facilitates the trade off possible between collector current and switching time. Any current that passes through the base resistor must be subtracted from the photocurrent.

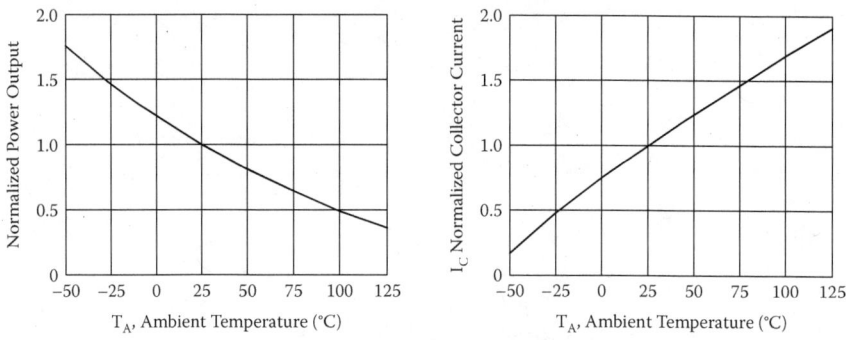

FIGURE 10.7 Normalized output versus ambient temperature for an IRLED and phototransistor. The IRLED output and collector current changes with respect to temperature are opposite functions. Increasing temperature results in less infrared energy, but simultaneously, the phototransistor becomes more sensitive.

10.3 DIFFERENT TYPES

Optocouplers are available in many different types. Besides the PDIP package discussed earlier, they are also available in several small form factors (small outline package [SOP]). There are also these types of optocouplers: AC couplers, specialty, hermetic or military, and high-technology couplers. The standard type includes the six-pin PDIP. A drawing is shown in Figure 10.9.

Electrical Considerations

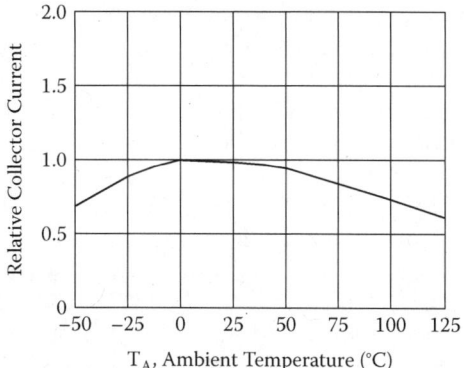

FIGURE 10.8 Relative collector current versus ambient temperature for a typical optocoupler. At temperature extremes, one effect always dominates the other. Below 25°C, the low transistor gain takes over. Above 25°C, the low IRLED output dominates. Optimal operation takes place in the 0 to 50°C range.

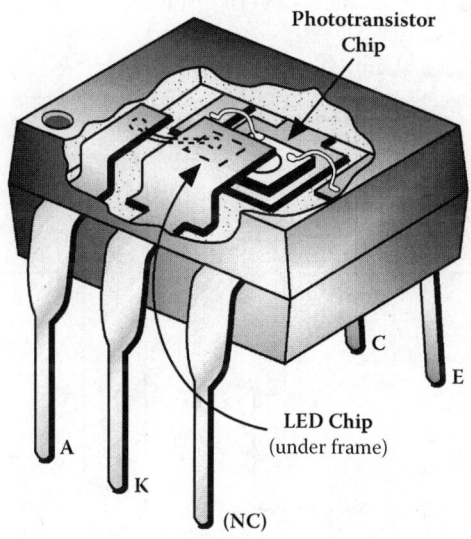

FIGURE 10.9 Drawing of six-pin PDIP. This standard design is still one of industry's most popular optocoupler packages. Many output types are possible, including complex IC outputs.

Each manufacturer has a variety of standard couplers with varying specifications listed under their own unique part numbers. They are available with either phototransistor or photodarlington output, current transfer ratios (CTR) as high as 500%, and input current as low as 0.5 mA. The fabrication techniques for these couplers are covered in Chapter 11, Section 11.1.

The AC couplers are of several different types. One of the major volume parts consists of an IRLED input coupled to an SCR or triac driver output. These devices are very useful in controlling AC loads by DC electronics. This type of part uses a high-efficiency IRLED that provides high light output at lower drive currents. These devices sense threshold levels over a wide range of input voltages by utilizing a single external resistor. The hysteresis of the input buffer provides extra noise and switching immunity. The diode bridge allows for easy use with AC input signals. The internal clamping diodes protect the buffer and IRLED from a wide range of overvoltage and overcurrent transients. Variations in optical coupling from the IRLED to the detector will have no effect on the threshold levels, because the threshold sensing is done prior to driving the IRLED. A functional diagram of this type of device is shown in Figure 10.10. Another popular type of AC coupler that consists of two back-to-back IRLEDs along with a phototransistor is shown in Figure 10.11.

Specialty couplers are also available from a variety of manufacturers. They generally differ significantly in shape. They feature high isolation voltages up to 50,000 V DC with guaranteed CTRs up to 700%, and with input currents as low as 5 mA. Figure 10.12 shows a coaxial part with 10 kV DC isolation. Figure 10.13 shows another coaxial part, with 15 kV DC isolation, and Figure 10.14 shows one with 50 kV DC isolation.

Figure 10.15 shows a slightly different packaging approach, which uses a side-viewing discrete sensor and side-emitting discrete IRLED mounted in a molded package.

Hermetic or high-reliability couplers are designed such that they can be processed and tested to a military-type specification such as Mil-S-19500/486A or Mil-S-883

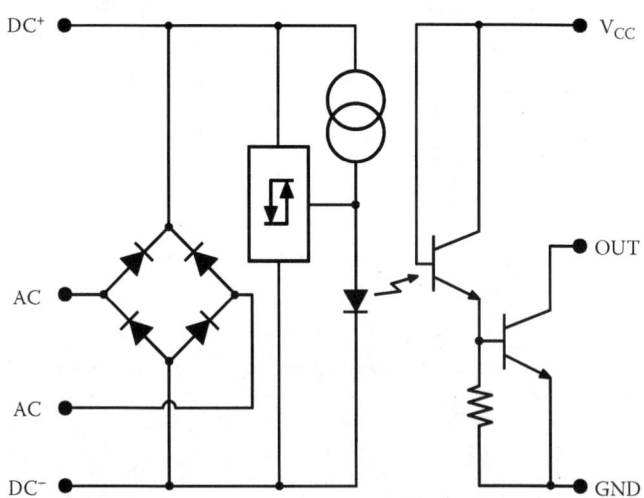

FIGURE 10.10 Optocouplers operating from an AC input to a DC output. In many applications, an AC signal requires monitoring. The bridge-type design is an effective method of AC/DC conversion.

Electrical Considerations

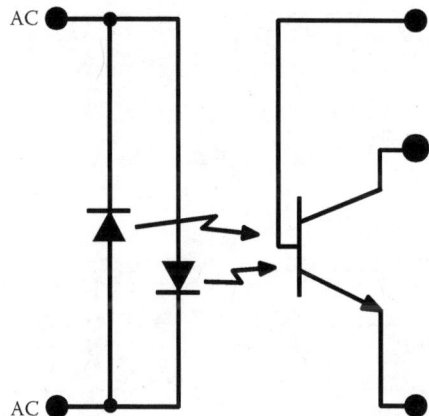

FIGURE 10.11 Back-to-back IR LEDs coupled with a phototransistor.

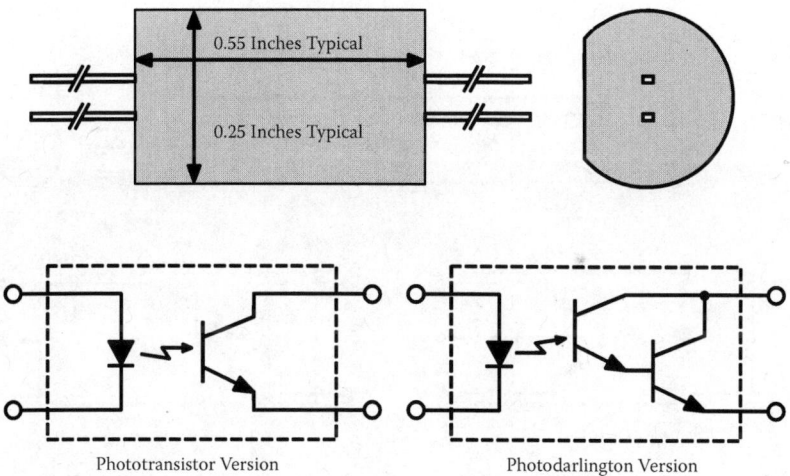

FIGURE 10.12 10 kV DC optocoupler. The physical separation of the emitter and sensor makes high isolation voltage possible. The design is commercially available with either a phototransistor or photodarlington output.

Class B. These are available in the TO-18 outline, the TO-5 outline, and the hermetic DIP outline packages. Some manufacturers have in-house high-reliability processing to augment the military-approved parts. This allows the customer to generate a military-type specification on a coupler that has not had a specification released by the military.

Some optocouplers are more complicated than the types of couplers previously discussed. They are IRLED or DC input combined with a photosensitive integrated circuit output. The 6N135 and 6N136 are graded on current transfer ratio and are the simplest form of a photo IC. The electrical schematic is shown in Figure 10.16.

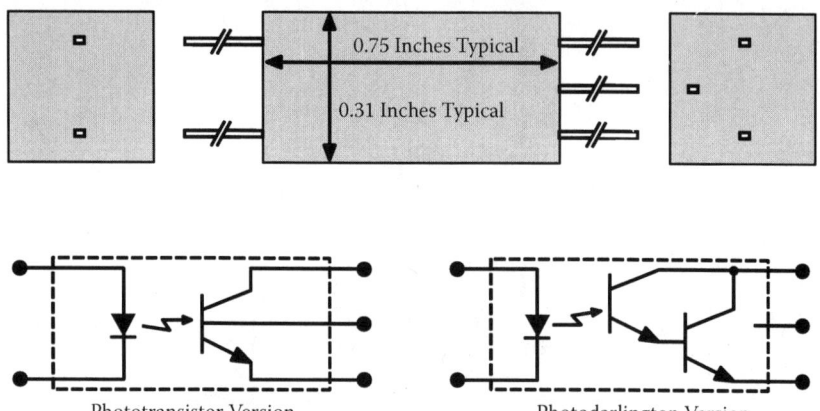

FIGURE 10.13 15 kV DC optocoupler. This particular design incorporates hermetic TO-46 and TO-18 discrete components. The military and high-reliability electronics (e.g., aircraft) market is served by these products.

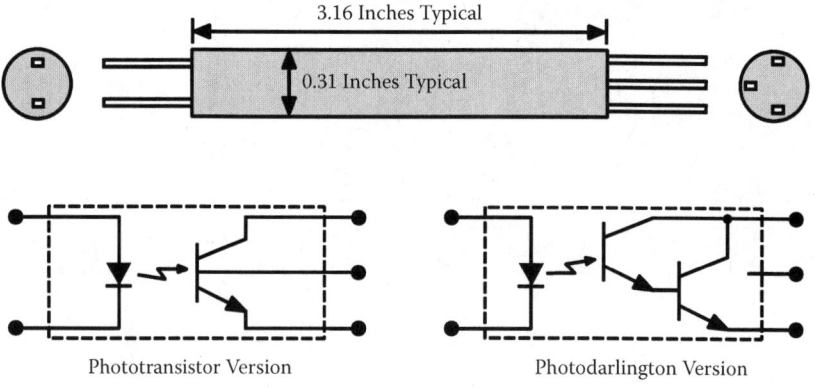

FIGURE 10.14 50 kV DC optocoupler. Increasing the physical separation between IRLED and phototransistor makes very high voltage isolation possible. An internal high-dielectric constant light pipe is necessary to transmit the infrared energy from IRLED to sensor.

Optocouplers with TTL-compatible output have isolation voltages up to 15,000 V DC and speeds up to 5 MHz. This compares to 25 kHz in a conventional phototransistor coupler.

A similar unit with higher gain characteristics but lower frequency response is the 6N138 and 6N139. The gain is increased from 7 to 300%, whereas the data rate is reduced from 1 Mb/s to 300 kb/s. The electrical schematic of this optocoupler is shown in Figure 10.17.

Another series of optocouplers couple the output of an IRLED to an integrated circuit that incorporates a photodiode, a linear amplifier, and a Schmitt trigger on a single silicon chip. The devices feature TTL/LSTTL-compatible logic level output that can drive up to 8TTL loads directly without additional circuitry. Figure 10.18

Electrical Considerations 165

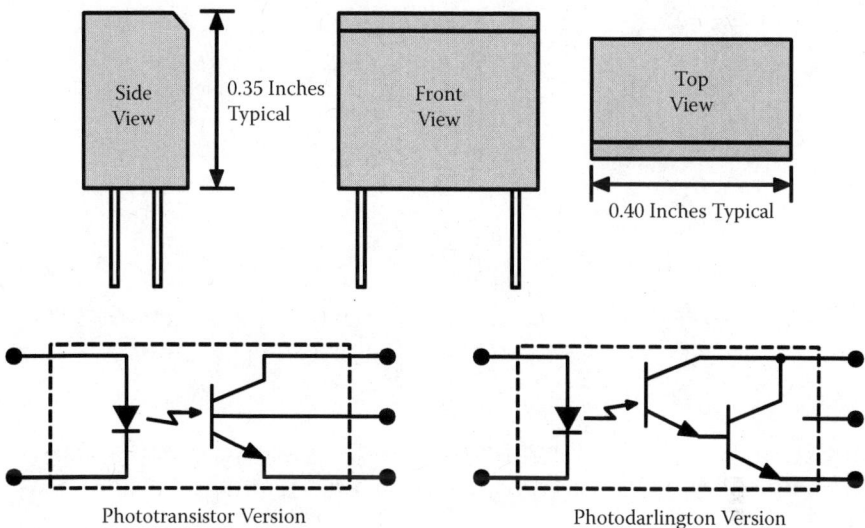

FIGURE 10.15 Side-emitting discrete IRLED and side-viewing discrete sensor in a molded package. The feature of easy PC board mounting is provided, whereas relatively high isolation voltage is maintained.

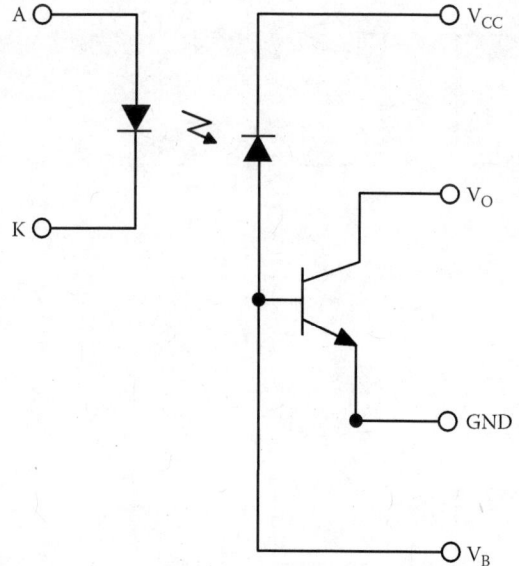

FIGURE 10.16 A schematic of a simple optocoupler utilizing photo IC. High DC current transfer ratios are possible using a photodiode coupled into an output transistor.

shows the electrical schematic for one option with a Totem-Pole Output buffer. Another version of this with a Totem-Pole Output inverter is shown in Figure 10.19.

Figure 10.20 shows the schematic of an Open-Collector Output buffer, and the Open-Collector Output inverter is shown in Figure 10.21.

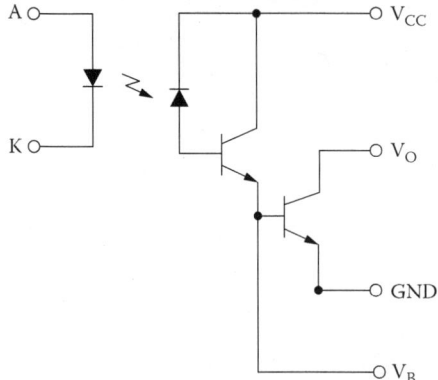

FIGURE 10.17 A schematic of a Darlington output optocoupler utilizing photo ICs. A Darlington pair amplifies the photodiode output in order to increase CTR, at the cost of operating speed.

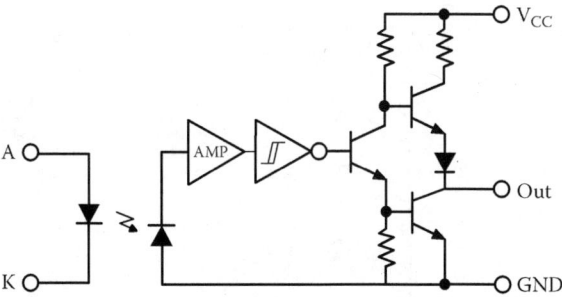

FIGURE 10.18 Electrical schematic of a photo IC coupler with Totem-Pole Output buffer.

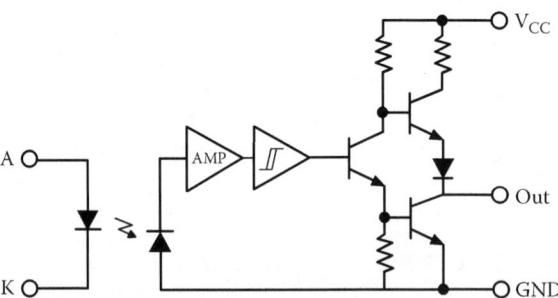

FIGURE 10.19 Electrical schematic of a photo IC coupler with Totem-Pole Output inverter.

These units are also available with other options such as package types, speedup resistors, duals, etc. New types of these high-technology couplers will become available as operating speed is improved, photosensor functionality is increased, input sensitivities are enhanced, and innovative packages are designed.

Electrical Considerations

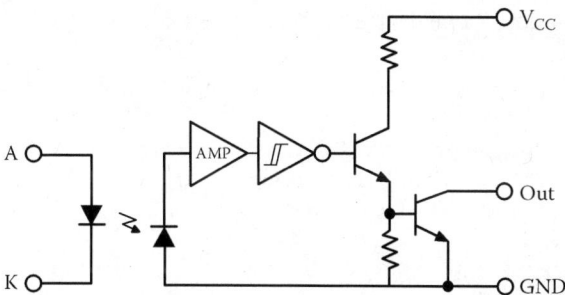

FIGURE 10.20 Electrical schematic of a photo IC coupler with Open-Collector Output buffer.

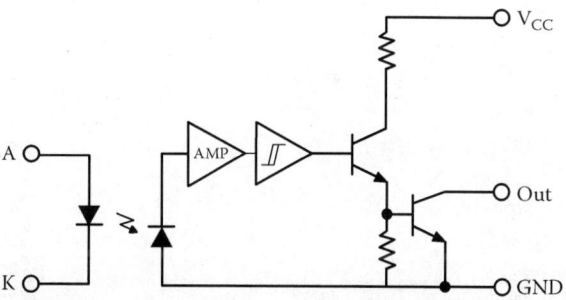

FIGURE 10.21 Electrical schematic of a photo IC coupler with Open-Collector Output inverter.

10.4 INTRODUCTION TO SOLID-STATE RELAYS (SSR)

This portion of the chapter will cover the basic operation of the solid-state relay (SSR) utilizing zero-crossover internal silicon-controlled rectifiers for the AC output and a high current drive transistor with an internal protection diode for the DC output version.

All SSR devices are electronic control devices, meaning that all switching is done electronically. Unlike mechanical relays, SSR devices have no moving parts, thus assuring users longer and more reliable operation. Because SSR devices have no moving parts, contact wear is eliminated as well as the need for antibounce circuitry.

The ideal continual cycle time for an SSR is between 200 and 2000 ms. The cycle time should be a half or full cycle on the output for AC versions to minimize noise generation while allowing startup to begin at zero, optimizing the operation of the relay.

Typical configurations of the SSR devices are

- DC input–DC output
- DC input–AC output
- AC input–AC output

The DC input–DC output device has a 4 to 24 V DC input range with load current of 3 A and load voltage of 100 V DC.

The DC input–AC output series devices have input ranges from 4 to 32 V DC with load current up to 15 A and load voltage up to 480 V AC.

The AC input–AC output series devices have input ranges of 5 to 12 V AC, 4 to 32 V AC, and 100 to 240 V AC with load currents up to 40 A and load voltage up to 480 V AC.

10.4.1 Applications

- Alarm systems sensor security
- Appliances
- Agricultural
- Electric vehicles
- Food processing equipment
- General-purpose medium-current switching applications
- HVAC systems
- Industrial control of resistive heaters
- Industrial equipment
- Molding equipment
- Motor controls
- Motor starters
- Office equipment
- On/off lighting applications
- Packaging equipment
- PCBoard-based control applications
- Robotics
- Temperature-controlled systems

10.5 THEORY OF OPERATION

The SSR device controls the output device with the use of an infrared light-emitting diode (LED). As the input signal turns on the LED, it illuminates a photosensor that controls the output device. The three basic configurations of SSR devices that meet most industry applications utilize either DC or AC input and load conditions.

Basic configurations are the following:

- DC to DC
- DC to AC
- AC to AC

DC load devices typically control a lower load current than AC devices. Overvoltage and overcurrent surge conditions are taken into consideration with most SSR devices sold today. Thermal management is a must for most high-current SSR devices and should always be calculated and resolved while evaluating the initial system design.

10.6 DC INPUT–DC OUTPUT SSR OPERATION

The SSR that utilized both DC input and DC output is similar to an optocoupler. The LED is a very fast switching device, whereas the output waveform may vary from a square wave to a sawtooth. This change in output waveform is dependent on the load conditions of the SSR as well as the switching speed. The lower the load resistance and the faster the switching pulse rate, the slower the response on the output (see Figure 10.22).

To properly operate a DC-to-DC SSR, a current pulse should be provided to the LED using the +V and −V nodes. Figures 10.23 and 10.24 show typical configurations used to pulse the input LED for DC-to-DC operation.

The node +V is connected to the positive node of the current source for the LED, whereas −V is connected to the switching transistor that turns the LED on and off.

R_D can be calculated with the following equation:

$$R_D = \frac{(V_{CC} - V_{LED} - V_{CE})}{I_{LED}} = \frac{(5 - 1.25 - 0.3)}{0.01} = \frac{3.45}{0.01} \approx 345$$

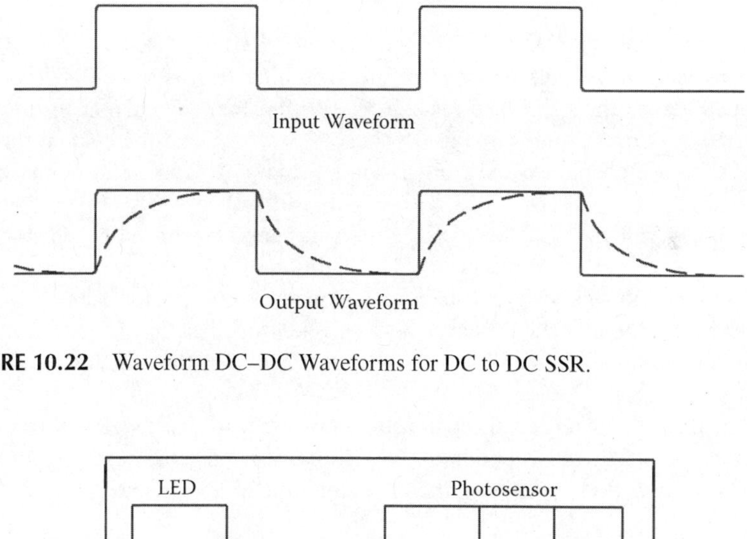

Input Waveform

Output Waveform

FIGURE 10.22 Waveform DC–DC Waveforms for DC to DC SSR.

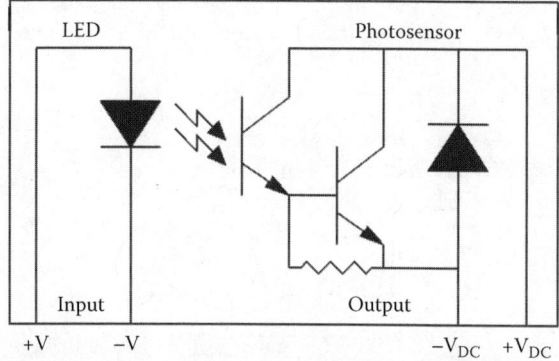

FIGURE 10.23 DC–DC-Hook-up hook-Up for a DC-to-DC SSR.

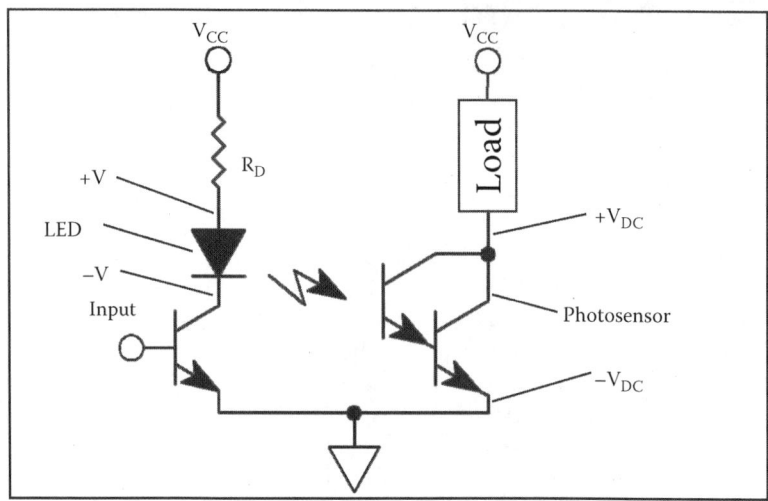

FIGURE 10.24 DC–DC-Hook-up schematic of hook-up for a DC-to-DC SSR.

With a supply voltage (V_{CC}) of 5 V and a forward voltage (V_F) for the LED at 1.25 V and saturation voltage (V_{CE}) of the switching transistor of 0.3 V; 3.45 V is dropped across the LED load resistor R_D. With a forward current through the LED of 10 mA, R_D becomes 345 Ω. Because V_{FLED}, V_{CE}, and the current through the LED are not absolute values, R_D is an approximate value, and about 350 Ω should work.

The supply voltage (V_{CC}) for the drive circuit can vary depending on the characteristics of the system requirements. The addition of a current regulator in place of R_D can assist for higher V_{CC} voltages. Remember, the V_{CC}'s for each side of the device are separate and should never be connected together. In order to minimize the noise between the input and output sections of the SSR, the grounds should also be separated.

Figure 10.25 show a typical relationship between load resistance and transition time for a typical DC transistor output. As the load resistance increases, the RC time constant increases, thus increasing the transition time of the device.

$$\tau \cong RC$$

The capacitance "C" is constant and the load resistance "R" is changed, causing the transition time "t" to change.

10.7 DC INPUT–AC OUTPUT SSR OPERATION

An SSR that utilizes both DC input and AC output can control a variety of high-current devices. Most of these devices have a zero voltage crossover (ZC) circuitry (see Figures 10.26 and 10.27).

Electrical Considerations

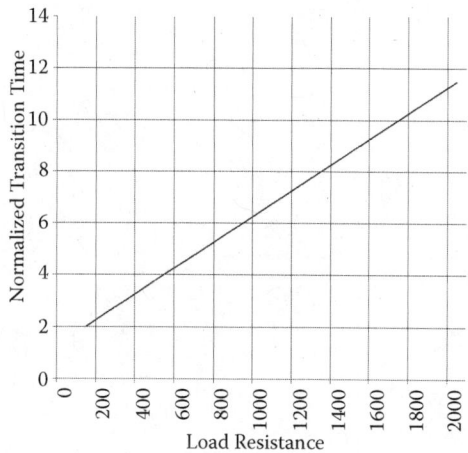

FIGURE 10.25 Lead resistance versus transition time.ai.

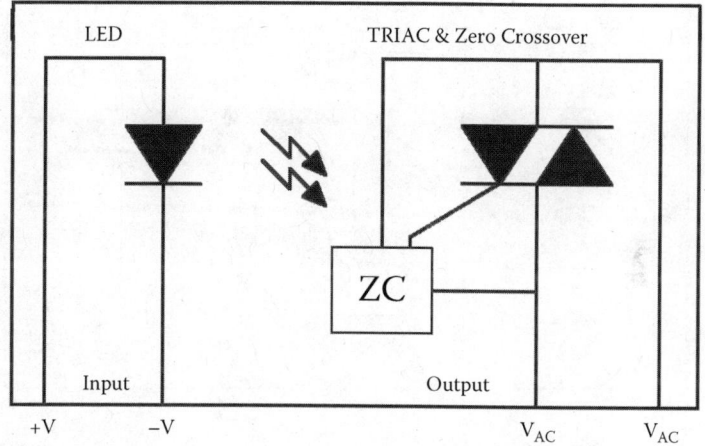

FIGURE 10.26 Hook-up-DC–AC hook-up for a DC-to-AC SSR.

The input operates similar to the "DC input–DC output SSR series" noted earlier, whereas the output operates differently. The output is a triac device with ZC circuitry and allows the device to operate AC output voltages.

When the photosensor element detects light from the input LED, the ZC circuit is activated and begins to activate the triac section of the SSR.

10.8 AC INPUT–AC OUTPUT SSR OPERATION

An SSR that utilizes both AC input and AC output can control a variety of high-current systems (see Figures 10.28 and 10.29). If the output section has dual triac operation, the amount of current that can be controlled is increased dramatically.

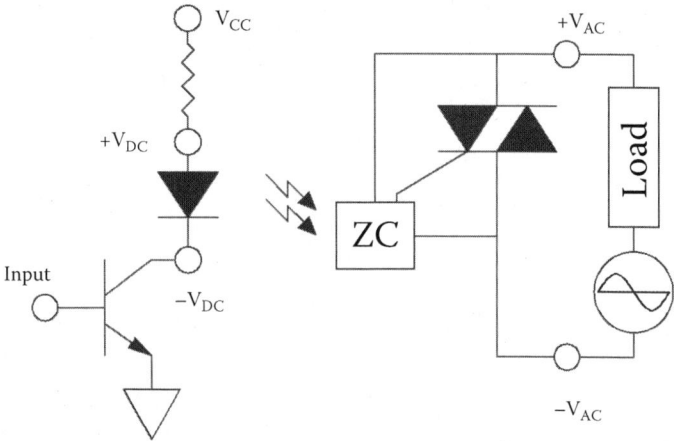

FIGURE 10.27 Hook-up-DC–AC Schematic of hook-up for a DC–AC SSR.

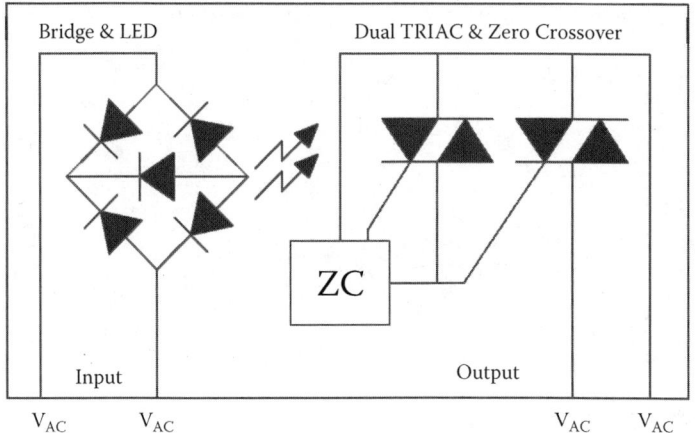

FIGURE 10.28 Hook-up-AC–AC hook-up for AC-to-AC SSR.

The AC input section of the device is a full-wave bridge rectifier, allowing the internal LED to radiate continually while an AC signal is presented to it.

Just as in the other version, the LED illuminates the photosensor, turning on the zero crossover and dual triac section of the device.

10.9 TEMPERATURE CONSIDERATIONS

Power devices such as SSRs control high-power applications. When power is dropped within the device, thermal management may be required. A panel mount type of SSR device is designed to easily be mounted on a heat sink, allowing continual operation over wide temperature and power ranges. The SSR thermal management

Electrical Considerations

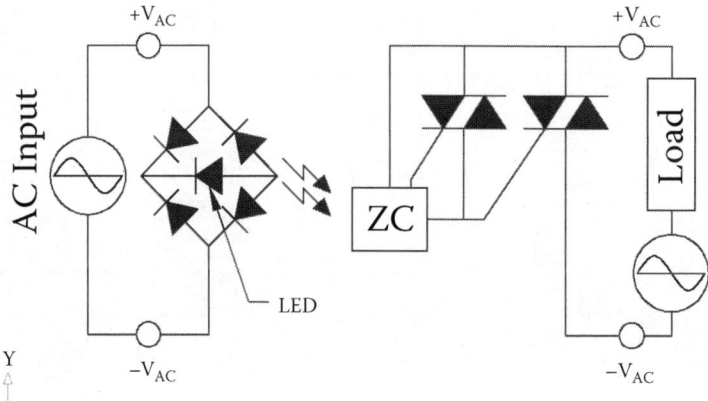

FIGURE 10.29 Hook-up-AC–AC schematic of hook-up for AC-to-AC SSR.

is easier to control with the input circuitry physically and electrically isolated from the output circuitry.

The input section consists of an LED or full-wave rectifier driver for the LED. As long as the input LED has sufficient current and voltage, it transmits light to the photosensor circuitry. As temperature increases, the amount of light radiated from the LED decreases, whereas the gain of the photosensor circuit increases. This offset allows the SSR to operate over a large temperature range with minimal current required to drive the input LED.

10.10 OVERVOLTAGE AND OVERCURRENT SURGE PROTECTION

Most of today's SSR devices have been designed to take into consideration short pulses of higher voltages and higher currents than the specified typical operating conditions. Surges are pulses that are out of the ordinary in having a higher potential than the average operating signal conditions. These surges should be sporadic, thus not requiring additional design considerations for thermal management.

10.11 ZERO VOLTAGE CROSSOVER

Zero voltage crossover (ZC) allows the system being controlled by the SSR to switch at the proper time during the AC load cycle. The positive-going cycle and negative-going cycle are similar but opposite in operation. For this discussion, we will discuss the positive-going cycle sequence.

1. A signal is applied to the LED, turning it on and radiating the photosensor.
2. The photosensor activates the ZC cycle of operation. As long as the photosensor is activated prior to the output cycle reaching the "positive threshold voltage," the triac will start operation when the positive-cycle signal reaches the "positive threshold voltage" (Figure 10.30).

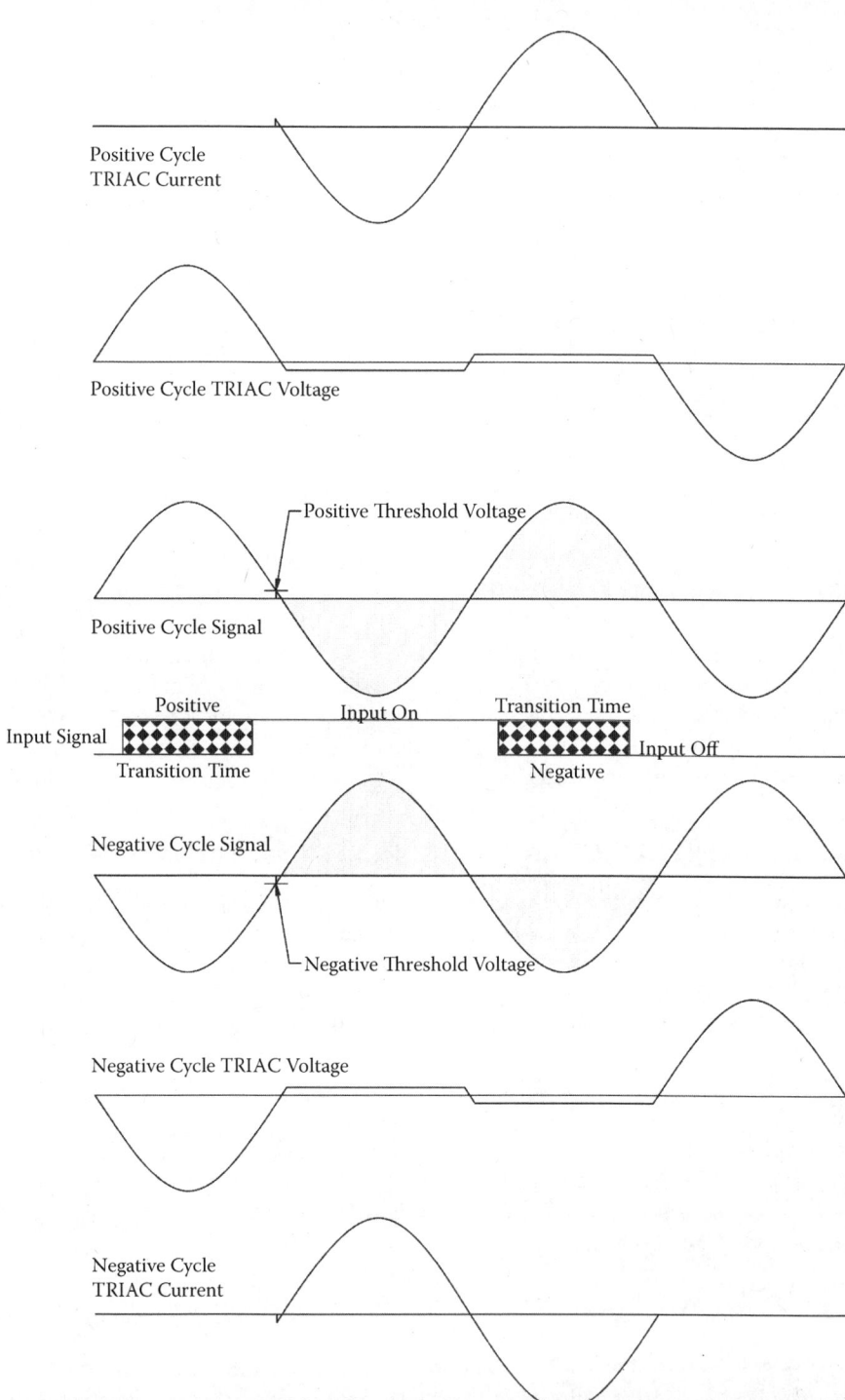

FIGURE 10.30 Triac-operation triac operation.

Electrical Considerations

3. The positive side of the triac is initiated, allowing current to flow. As the signal reached the zero crossover, the negative side of the triac will start conducting current. As long as the cycle is in the negative half of the positive-cycle signal, the negative side of the triac remains on. When the positive-cycle signal goes through zero, the positive side of the triac will begin operation and continue until the zero crossover is reached and the negative side of the triac takes over again. This cycle continues until the LED is turned off.
4. When the LED is turned off, the photosensor starts the shutdown cycle. When the signal reaches the zero crossover, the conducting triac turns off.

10.12 CONCLUSION

- SSR devices are provided to the industry in three typical basis configurations:
  DC to DC
  DC to AC
  AC to AC
- Overvoltage and overcurrent surge protections are designed into most SSR devices.
- Thermal management should be part of the initial system design.
- SSR devices are much more reliable than mechanical relays.
- SSR devices isolate the input electronics from the output electronics utilizing an optical signal. Both sides of the SSR should have different power systems, including ground returns.

11 Mechanical and Thermal Considerations

11.1 MECHANICAL CONSIDERATIONS

11.1.1 DISCRETE COMPONENTS

The two broad classifications of discrete component package types are hermetic and plastic. Each offers its own distinct advantages. In many environments, hermetic packaging may be mandatory. It has excellent resistance to water and other solvents, while offering the broadest operating temperature range and resistance to thermal shocks. Plastic packaging, in addition to offering a cost advantage over the hermetic packaging, exhibits excellent optical properties. Overall emission efficiency is also superior because optical interfaces are minimized. Finally, resistance to mechanical shock and vibration is excellent because both chip and bond wire are fully encased in supportive material. Figure 11.1 shows a picture of some common LED and photosensor packages in the industry.

11.1.2 SURFACE MOUNT CHIP CARRIERS (SMCCs)

Surface mount chip carriers (SMCCs) offer solutions to many applications that require more than standard or conventional discrete components or assemblies. These devices are used when the desired function cannot be accomplished with conventional through-hole, leaded, or individual surface mount components. SMCC devices allow multiple LEDs and sensors to be packaged together to create compact sensing solutions. Space savings can be as much as 80% compared to discrete-packaged components. The polyimide chip carrier typically has four main parts: substrate, frame, components, and encapsulation. The substrate is fabricated from high-temperature copper-clad laminate. Standard PC board processing provides the plated and nonplated holes, circuit patterns, and chip-mounting features. To make the substrate compatible with die attach and wire bonding techniques, the copper surface is plated with a nickel barrier and gold. The frame layer is made from the same polyimide laminate as the substrate and is used to protect the die and contain the encapsulation material. After the chip components are mounted and bonded, the frame is screened, printed with a pattern of nonconductive epoxy, aligned with the matching substrate cells, and laminated to the substrate under elevated temperature and pressure. The encapsulant is a conformal coating that is applied to fill the component cavities. After curing the encapsulant, the array is sawn into individual product elements and is ready for final test and packaging. Polyimide is an excellent substrate because of its strength, high processing temperature, and close match with

178 Optoelectronics: Infrared-Visible-Ultraviolet Devices and Applications

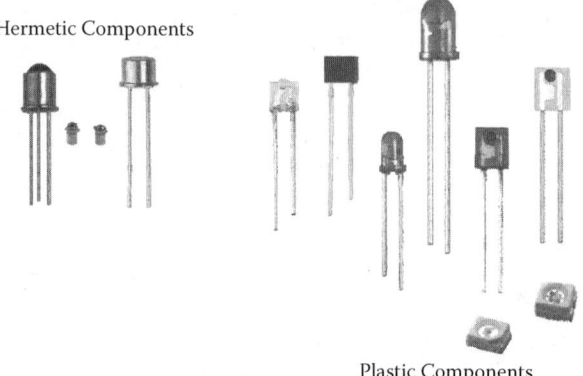

FIGURE 11.1 Photograph of common discrete component packages for a variety of requirements.

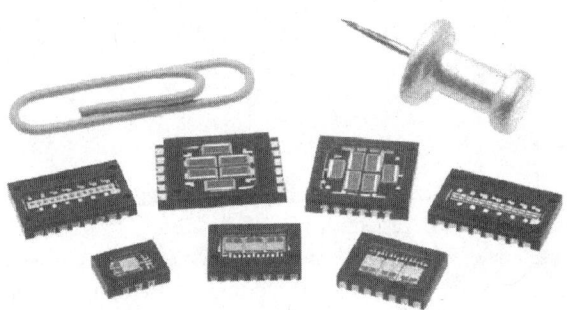

FIGURE 11.2 Photograph of SMCC. These devices combine a variety of different technologies including LEDs, VCSEL, or photosensors. A photosensor can be a photodiode, phototransistor, photodarlington, and simple and complex photo ICs.

the expansion coefficient of silicon devices. Figure 11.2 shows a picture of some surface mount chip carriers.

11.1.3 SMCC Advantages

- SMCC devices offer unique LED and sensor configurations to suit a special set of application requirements.
- The standard materials and processes result in packages that accommodate extended temperatures beyond the range of many commercial components.
- Chip carriers withstand the challenges of low-cost automated handling, placement, and reflow soldering.

- In comparison to a custom IC solution, the development cost of a typical SMCC circuit is far less and modifications are quicker and easier.
- Array processing in SMCC fabrication minimizes cost and optimizes quality.

11.1.4 THE EMITTER AND PHOTOSENSOR ASSEMBLIES

The assemblies category consists of a variety of products that can be used in many noncontact sensing situations. Optoelectronic assemblies typically consist of an optical transmitting device and a sensing device. The optical transmitting device can be either an UV, visible LED, IRED, or vertical cavity surface emitting laser (VCSEL). The sensing device can be either a photodiode, or phototransistor, or photodarlington, or Photologic® device. The optical pair (transmitter and sensor) is mounted in either a reflective, slotted, or transmissive switch inside a housing.

Housings for transmissive and reflective switches serve a dual role. They hold the discrete transmitter and receiver in a fixed location and also contain mounting holes or some other means of attaching the assembly to the machine needing the switch function. The housing materials most widely used are a 10% glass-filled polycarbonate or polysulfone. These are used in an injection-molding process to form the housing shape. Polysulfone has the unique property of transmitting wavelengths longer than 700 nm while blocking the shorter wavelengths. It is useful in housings that are used in a high-dust ambience because there is no mechanical opening to clog up and thus block the optical path. The 10% glass-filled polycarbonate is used in most applications because it has higher solvent resistance and also blocks ambient energy in the long wavelengths. Other plastics are used for more specific applications. Figure 11.3 shows a picture of some commercially available optical switches.

11.1.5 OPTOISOLATORS

Optoisolators are devices designed to isolate electrical energy between two circuits. They are, typically, sealed packages consisting of an optical transmitter (IRLED

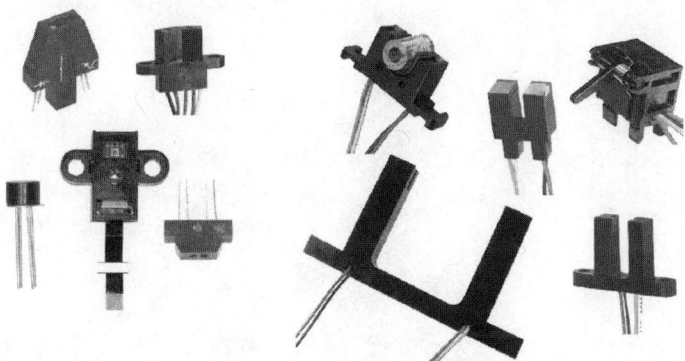

FIGURE 11.3 Photograph of commercially available reflective and transmissive assemblies.

or VCSEL) on one side and an optical sensor (photodiode, phototransistor, photodarlington, Photologic) on the other side. Optoisolators are designed for applications that typically require voltage isolations from voltages of 1000 V or greater. In principle, optoisolators are similar to the transmissive switch discussed previously. The air gap is replaced by a transparent material, which provides coupling of the IRED energy to the photosensor. The physical separation of the source and receiver provides the electrical isolation of input to output. This obviously will vary as a function of the distance and the material separating the units electrically. Table 11.1 lists some currently available couplers by package, isolation voltage, and output function. As material technology advances, optoisolators may decrease in size and increase voltage isolation.

Optocouplers may be divided into different types: standard DIP and SMD, through-hole, AC couplers, high-speed analog or digital, and high reliability. Figure 11.4 shows a photograph of different optoisolators.

TABLE 11.1
Listing of Some Optoisolators

Package	Isolation Voltage	Output
Discrete components	10–50 kV	Photodiode transistor, darlington, and photo IC
TO-18, TO-5	1 kV	Photodiode, transistor, darlington
DIP	1.5–7 kV	Photodiode, transistor, darlington, photo IC SCR, triac driver
Surface mount	6 kV	Phototransistor, photo IC
Rectangular case	16 kV	Photodiode, transistor, darlington, and photo IC
Fiber optics	15 kV or more	Photodiode transistor, darlington, and photo IC

FIGURE 11.4 Photograph of various optoisolators available in the market.

Mechanical and Thermal Considerations

11.2 MANAGING JUNCTION TEMPERATURE OF HIGH-POWER LIGHT-EMITTING DIODES

High-power LEDs fulfill a growing number of applications. They can have any typical peak wavelength, such as UV LEDs, visible LEDs, or infrared LEDs. Visible LEDs have a high luminous efficacy when compared to low-medium-high pressure UV tubes, fluorescent, and traditional incandescent technology. Most of the common technology has already reached near-maximum luminous efficacy, but the efficacy of LEDs is forecasted to continue increasing in the future owing to improved light extraction and thermal management.

Typically, LEDs have been driven at a low power with a minimum power dissipation of 150 mW; therefore, most lighting applications required numerous low-power LEDs. The new high-power LEDs have power dissipations ranging from 500 mW to as much as 10 W in a single package. With improving power output and luminous efficacy, these high-power LED components can and will replace other lighting technologies in most applications. When current flows through the LED chip layers, heat and light are generated at the P-N junction. The heating effect is very rapid; therefore, the specifications for light performance, V_F and dominant wavelength are measured within milliseconds with a junction temperature close to 25°C. Seconds later, because of heat buildup, these parameters will have drifted closer to the actual operating temperature conditions of the application.

When using high-power LEDs in applications, many design considerations must be considered. These include the following:

How much power or luminous flux is required?
What is the desired dominant wavelength or color temperature (in case of visible LEDs)?
What is the required mean time to failure MTTF?
How much power or flux degradation is tolerable?

The temperature of the LED's P-N junction impacts these issues. Junction temperature directly alters the performance and reliability of LEDs in the following ways:

Reduced output power: At constant operating current, the power or luminous efficacy decreases by about 5% for every 10°C rise in junction temperature.
Reduced forward voltage: At constant operating current, forward voltage decreases by about 20 mV for every 10°C rise in junction temperature.
Shifted dominant wavelength: Dominant wavelengths shift by about 1 to 2 nm for every 10°C change in junction temperature.
Shifted color temperature: White LEDs are more sensitive to changes due to junction temperature because the color temperature can change significantly. LEDs emit white light by combining standard blue emission with a phosphor overcoat that absorbs the blue flux and reemits a wide range of wavelengths throughout the visible range. Reemission efficiency is highly dependent on the wavelength of the blue flux, which shifts as junction

temperature changes. If the dominant wavelength of the blue LED shifts out of the efficient range of the phosphor coat, more blue flux escapes the package, which increases the color temperature. Although infrared and UV LEDs cannot be perceived by human vision, the shifting of their spectrum can have a negative impact in their application.

11.2.1 Reduced MTTF and Accelerated Degradation

Catastrophic failure and LED degradation are mechanical and chemical processes that occur at rates described by the Arrhenius model. Their rates are inversely proportional to the exponent of the inverse of junction temperature. The impact of junction temperature cannot be overstated. Successful thermal management is vital to successful design.

11.2.2 Generating Heat

Junction temperature depends on three factors:
Power dissipation
Thermal resistances of the substrate and assembly
Ambient conditions

Power dissipation determines how much heat is generated, whereas thermal resistances and ambient conditions dictate how efficiently heat is removed. All of photons and most of the heat (phonons) produced by an LED are generated at the P-N junction. Because the junction is very small, the heat generation rate per unit area is very large. A 1 W LED chip (1 mm × 1 mm) generates 100 W/cm^2. This rate is higher than many of today's high-power microprocessors.

11.2.3 Removing Heat

To maintain a low junction temperature, all methods of removing heat from LEDs should be considered. The three means of heat transference are conduction, convection, and radiation.

Thermal conduction is the transmission of heat across matter. Thermal conductivity within and between materials is proportional to the temperature gradient and the cross-sectional area of the conductive path. Conversely, conductivity is inversely proportional to the length of the conductive path. LEDs are typically encapsulated in a light-transmissive plastic, which is a very poor thermal conductor. Nearly all heat produced is conducted through the back side of the chip. For an interface with area A and thickness l, the rate of heat conduction has the following proportion:

$$Q°C\, A \cdot \Delta T/l \qquad (11.1)$$

Convection is the transfer of heat by currents in a liquid or gas. Convection rate is proportional to surface area and the temperature gradient between the surface and

Mechanical and Thermal Considerations

the fluid. LEDs do not benefit from convection at the component level, because their surface area is too small. Convective technologies include fans, heat pipes, and liquid cooling. For a surface with area A_S and temperature T_S, convection has the following proportion:

$$Q \mu A_S \cdot [T_S - T_A] \qquad (11.2)$$

Thermal radiation is electromagnetic radiation from an object's surface due to the object's temperature. Radiation is proportional to the object's absolute temperature raised to the fourth power and its surface area. Heat sinks with large surface area are effective at radiating heat. For a surface with area A_S and temperature T_S, the following proportion applies:

$$Q \mu A_S \cdot [T_S - T_A]^4 \qquad (11.3)$$

11.2.4 Thermal Equilibrium

Heat transference is an equilibrium condition. All three types of heat transference become more efficient as temperature gradients increase. The junction temperature will rise until the rate of heat transference out of the system is equal to the rate of heat generation at the junction.

11.2.5 Analogy to Electrical Circuits

Thermal systems are analogous to electrical circuits, with similarities between the following:

Power dissipated (Q) and current
Thermal resistance ($R\theta$) and electrical resistance
Temperature difference (ΔT) and voltage

The Ohm's law equivalent is

$$\Delta T = Q \cdot R\theta \qquad (11.4)$$

Heat input is calculated:

$$Q = I_F \cdot V_F \qquad (11.5)$$

where I_F is the operating current and V_F is the measured forward voltage of the LED.

Thermal resistance is usually unknown and should be calculated using Equation 11.1 and measured ΔT and Q. For thermal interface materials (TIM), the thermal resistance of the material, $R\theta_{TIM}$, depends on its thermal conductivity, K, expressed in W/m·K. Thermal resistance is calculated:

$$R\theta_{TIM} = k_{TIM} \cdot [l/A] \qquad (11.6)$$

where l is the length of the thermal path and A is the cross-sectional area of the thermal path. To minimize thermal resistance, the cross-sectional area should be maximized and the thickness of the interface should be minimized.

Temperatures within the thermal system can usually be measured directly. Junction temperature is the exception because the junction is inaccessible.

Fortunately, the forward voltage of an LED has distinct temperature dependence that makes the junction its own thermometer once it has been calibrated.

11.2.6 Determining Junction Temperature from Forward Voltage

The forward voltages of nearly all LEDs manufactured with III to V compounds decrease between 1 and 3 mV per 1°C increase in temperature. The following test can be conducted on single components or on large assemblies with multiple LEDs. The temperature–forward voltage curve is empirically generated as follows:

1. Connect the LED to a constant-current power supply, and install the device in a controlled oven with the power off. Set the operating current, I_F, to the expected application condition.
2. Set the temperature to 25°C, and allow sufficient time for the oven and assembly to stabilize. Turn the power on for a short period, preferably less than 10 µs, and record the forward voltage, V_F. When possible, use sense cables to measure V_F. As the LED is on for a very short period, it does not significantly heat itself, and $T_J \sim T_A$.
3. Repeat step 2 at 50, 75, 100, and 125°C. This test may be destructive. Plot T_J as a function of V_F, and derive a best-fit line. The temperature dependence is not linear, but within the operating range a best-fit line is quite accurate. Figure 11.5 shows an example of such a curve for an infrared LED.
4. Drive the assembly at the application I_F. The V_F will decrease until thermal equilibrium is reached. Cross the stabilized V_F with the plot generated in step 3 to derive the junction temperature.
5. Repeat the procedure for multiple current loads to fully characterize the system across all power dissipations.

11.2.7 Passive Thermal Management

Passive thermal management systems have no moving parts and do not consume additional energy. They rely primarily on conduction and radiation to remove heat from the junction. The typical method is to attach LEDs to a thermally conductive substrate, such as a metal-core IMS substrate or ceramic substrate, and then attach the substrate to a heat sink. Novel technologies such as Optotherm® make it possible to attach the LEDs directly to the heat sink. Heat is conducted to the heat sink and radiated from its surface. Thermal performance is enhanced by reducing the length and thermal resistances along the path to the heat sink and by increasing its surface area.

Mechanical and Thermal Considerations

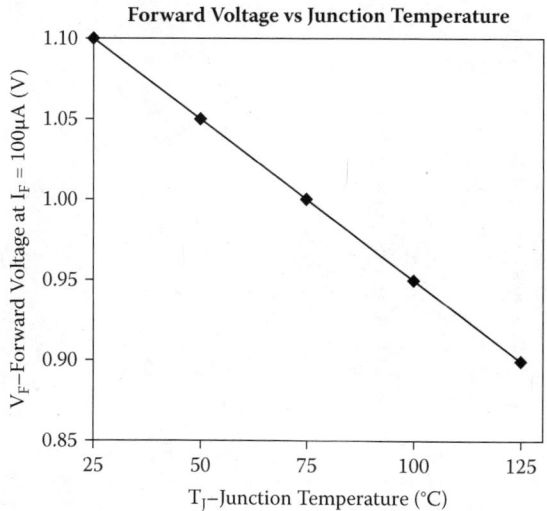

FIGURE 11.5 Voltage drop versus junction temperature for an IR LED. This curve offers the key to accurate measurement of junction temperature at various power levels, which is necessary to determine proper thermal deratings of LEDs. The initial and final V_F values are different for other wavelengths.

11.2.8 Active Thermal Management

Active thermal management systems involve convection by incorporating fans, heat pipes, and liquid cooling. These technologies enable significantly better thermal management and should be considered for ultrahot applications. In most cases, they are more complex and require better design to avoid decreasing the reliability of the system (Figure 11.6). These trade-offs are manageable if extreme thermal management is required.

The assembly's thermal characteristics are expressed by the following equations:

$$\Delta T_{J-A} = Q \cdot R\theta_{J-A} \tag{11.7}$$

$$\Delta T_{J-A} = Q \cdot [R\theta_{J-C} + R\theta_{C-A}] \tag{11.8}$$

$$\Delta T_{J-A} = Q \cdot [R\theta_{J-C} + R\theta_{C-S} + R\theta_{TIM} + R\theta_{H-A}] \tag{11.9}$$

ΔT_{J-A} and Q must be measured, and $R\theta_{J-C}$ is provided by the LED vendor. $R\theta_{C-A}$ is the combined thermal resistance of the rest of the assembly. Equation 11.9 can be used to calculate ΔT_{J-A}, if sufficient data is supplied by the substrate, thermal interface material, and heat sink vendors; however, it is recommended to calculate $R\theta_{C-A}$ by rearranging Equation 11.8 to:

$$R\theta_{C-A} = [\Delta T_{J-A}] / Q - R\theta_{J-C} \tag{11.10}$$

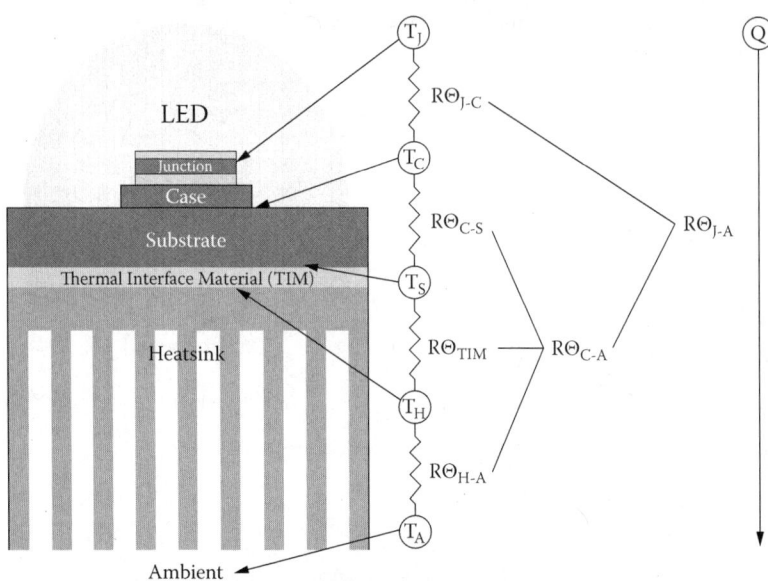

FIGURE 11.6 Thermal model for single-component assembly.

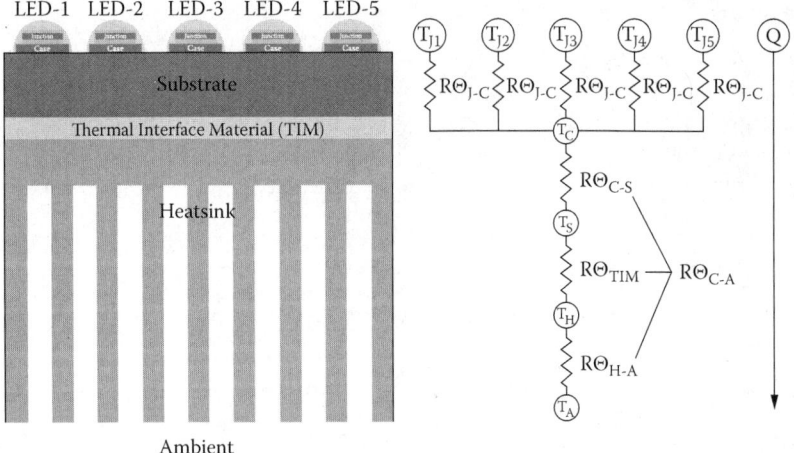

FIGURE 11.7 Thermal model for multiple-component assembly.

The multiple-component assembly's thermal characteristics are described by equations that are similar to those for single-component assemblies (Figure 11.7):

$$\Delta T_{J-An} = Q_n \cdot R\theta_{J-An} \qquad (11.11)$$

$$\Delta T_{J-An} = Q_n \cdot R\theta_{J-Cn} + Q_{Total} \cdot R\theta_{C-A} \quad (11.12)$$

$$\Delta T_{J-An} = Q_n \cdot R\theta_{J-Cn} + Q_{Total} \cdot [R\theta_{C-S} + R\theta_{TIM} + R\theta_{H-A}] \quad (11.13)$$

$$R\theta C - A = [\Delta TJ - An]/Qn - R\theta J - Cn \quad (11.14)$$

For single-component assemblies, the equation for $R\theta_{C-A}$ (Equation 11.10) is derived from Equation 11.8. Note that Equation 11.14 was not derived from Equation 11.12 in a similar manner. T_C is the same for all components on the multiple-component assembly, and $R\theta_{C-A}$ can be derived based on one component's ΔT_{J-A}, Q, and $R\theta_{J-C}$.

For the same component and power dissipation, ΔT_{J-C} will be the same whether the LED is alone or is part of an array. In an array, however, the heat input of all LEDs must be transferred through the substrate, TIM, and heat sink. ΔT_{C-A} and ΔT_{J-A} increase considerably over single-component assemblies.

When making a choice between 1 p-watt component and p 1 W components, the p-watt component must have p times lower thermal resistance than the 1 W component for the junction temperatures of both designs to be the same. The reality is that most package technologies for high-power LEDs have similar thermal resistances. Spreading the heat input to multiple components is recommended because less thermal management is required.

Recommendations for reducing junction temperature without compromising luminous flux are the following:

Use components with better luminous efficacy to reduce I_F and Q.
Increase the number of components at the same total power dissipation to reduce $R\theta_{J-C}$.
Change to better-packaged components to reduce $R\theta_{J-C}$.
Use Anotherm or similar substrates to eliminate $R\theta_{C-S}$.
Use Anotherm or similar substrate heat sinks to eliminate $R\theta_{C-S}$ and $R\theta_{TIM}$.
Increase the heat sink's surface area to reduce $R\theta_{H-A}$.
Add a fan, heat pipe, or liquid cooling to reduce $R\theta_{H-A}$.

Part 5

Open Air and Fiber-Optic Communication

12 Fiber-Optic Communication

12.1 BASICS

The ability to transmit information, and the media we use to do it, has been responsible for an important technical evolution. Progressing from the copper wire of a century ago to today's fiber-optic cable has made it possible to transmit more information, more quickly, and over longer distances.

The transmitter converts an electrical analog or digital signal into a corresponding optical signal. The source of the optical signal can be either a light-emitting diode, or a solid-state laser diode. Figure 12.1 shows a diagram of a fiber-optic transmission system. The most popular wavelengths of operation for optical transmitters are 850, 1300, or 1550 nm.

In a fiber-optic system, these devices are mounted in a package that enables an optical fiber to be placed in very close proximity to the light-emitting region to couple as much light as possible into the fiber. In some cases, the emitter is even fitted with a tiny spherical lens to collect and focus light into the fiber. An example of such an emitter is shown in Figure 12.2.

IRLEDs have relatively large emitting areas and are widely used for short-to-moderate transmission distances because they are much more economical, quite linear in terms of light output versus electrical current input, and stable in terms of light output versus ambient operating temperature. Laser diodes (LDs), on the other hand, have very small light-emitting surfaces and can couple many times more power to the fiber than IRLEDs. LDs are also linear in terms of light output versus electrical current input, but unlike IRLEDs, they are not stable over wide operating temperature ranges and require more elaborate circuitry to achieve acceptable stability. In addition, their added cost makes them primarily useful for applications that require the transmission of signals over long distances.

IRLEDs and LDs operate in the infrared portion of the electromagnetic spectrum, so their light output is usually invisible to the human eye. Their operating wavelengths are chosen to be compatible with the lowest transmission loss wavelengths of glass fibers and highest sensitivity ranges of photodiodes. The most common wavelengths in use today are 850, 1300, and 1550 nm. Both IRLEDs and LDs are available in all three wavelengths.

Figure 12.3 shows the variation of attenuation with wavelength over the three principal windows of operation. These correspond to wavelength regions where attenuation is low, and the transmitters are able to efficiently generate light emission. The increases in attenuation in these regions are due to the presence of hydroxyl radicals in the glass fiber material. The radicals are the result of the presence of water, which enters the fiber material during the manufacturing process.

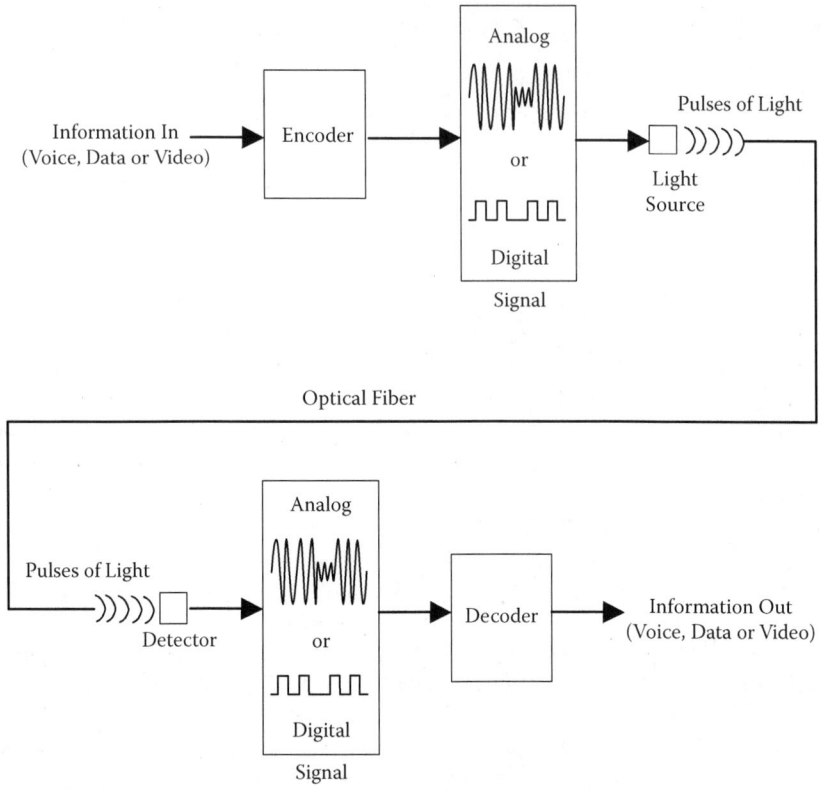

FIGURE 12.1 Fiber-optic information transmission.

The cable consists of one or more glass fibers, which act as waveguides for the optical signal. Figure 12.4 shows a cross section of a typical fiber-optic cable. The hair-thin glass core is surrounded by a cladding material, which helps to confine the transmitted light within the core because the cladding material's lower index of refraction. Light rays will stay confined to the core if they strike the cladding at a shallow angle; however, a ray that exceeds a certain "critical" angle will escape the fiber. This is illustrated in Figure 12.5.

Fiber-optic cable is similar to electrical cable in its construction but provides special protection for the optical fiber within. Figure 12.6 illustrates one type of construction.

Single-mode fibers have small cores that are about 10 μm in diameter and transmit laser light that has a wavelength of 1300 to 1550 nm. These fibers carry higher bandwidth than multimode fibers, and they require a light source with a more narrow spectral width. They also have a higher transmission rate and operate up to 50 times the distance of multimode fibers. The small core along with the single light wave will virtually eliminate any distortion, providing the least signal attenuation and the highest transmission speeds of any fiber type. However, they cost more than multimode fiber systems.

Multimode fibers have larger cores that are 50 to 125 μm in diameter, with 62.5 μm being the most common. They transmit infrared light at wavelengths of 850 to 1300 nm from light-emitting diodes. Multimode fiber gives a high bandwidth at speeds

Fiber-Optic Communication

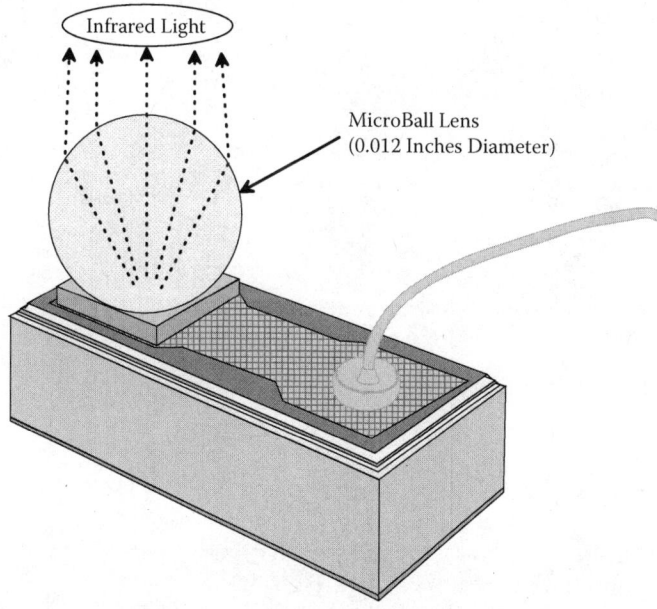

FIGURE 12.2 Fiber-optic LED. Infrared light at 850 nm is emitted from a 2-mil-diameter spot and is transmitted through the microball lens and into a glass fiber that is positioned a close distance from it.

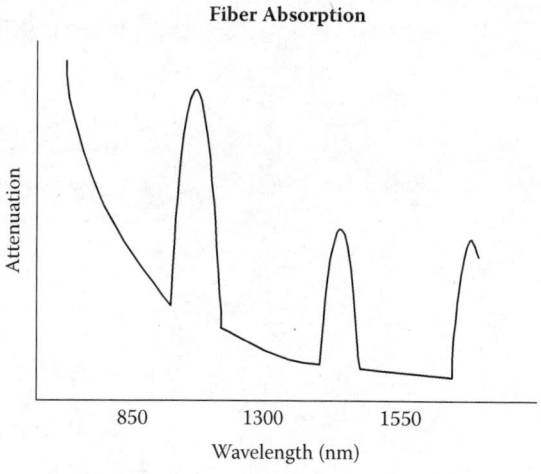

FIGURE 12.3 Optical fiber absorption curve.

of 10 to 125 megabaud over medium distances. Figure 12.7 illustrates the difference between single-mode and multimode fibers.

Once the transmitter has converted the electrical input signal into whatever form of modulated light that is desired, the light must be coupled into the optical fiber. The

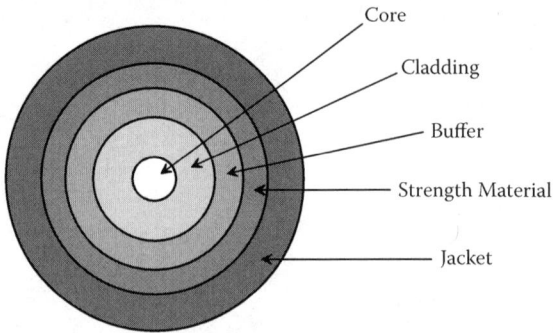

FIGURE 12.4 Fiber-optic cable cross section.

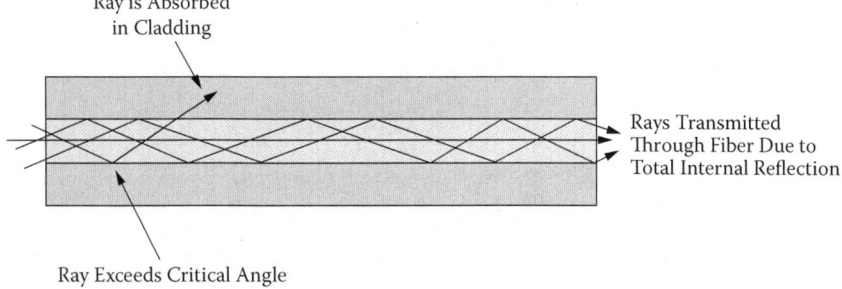

FIGURE 12.5 Light transmission through the fiber due to total internal reflection.

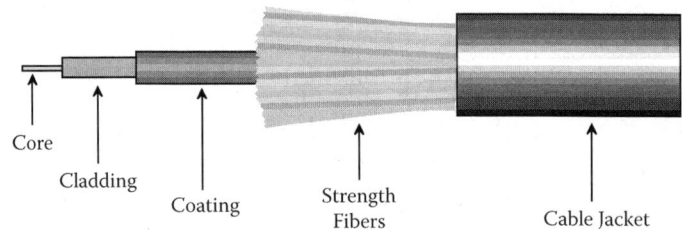

FIGURE 12.6 Typical fiber-optic cable construction.

amount of light that will enter the fiber is a function of several factors. The intensity of an IRLED or LD is a function of its design and is usually specified in terms of total power output at a particular drive current. Sometimes, this figure is given as actual power that is delivered into a particular type of fiber. All other factors being equal, more power provided by an IRLED or LD translates to more power coupled into the fiber.

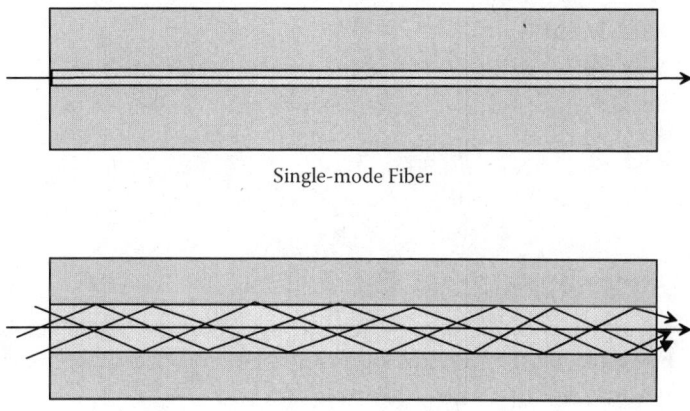

FIGURE 12.7 Single-mode and multimode fibers.

The light that gets into a fiber is a function of the area of the light-emitting surface compared to the area of the light-accepting core of the fiber. The smaller this ratio is, the more the light that gets into the fiber. The acceptance angle of a fiber is expressed in terms of numeric aperture. The numerical aperture (NA) is defined as the sine of one-half of the acceptance angle of the fiber. Typical NA values are 0.1 to 0.4, which correspond to acceptance angles of 11° to 46°. Optical fibers will only transmit light that enters at an angle that is less than or equal to the acceptance angle for the particular fiber.

There is always a loss due to reflection from the entrance and exit surface of any fiber. This loss is called the *Fresnel loss* and is equal to about 4% for each transition between air and glass. There are special coupling gels that can be applied between glass surfaces to reduce this loss when necessary.

The optical receiver is usually a PIN or avalanche photodiode that is combined with the necessary circuitry to change the optical signal back into a replica of the original electrical signal.

12.2 ADVANTAGES

There are several reasons that fiber-optic systems are revolutionizing telecommunications. The advantages are described in the following subsections.

12.2.1 Less Expensive

Long, continuous lengths of optical cable can be made cheaper than equivalent lengths of copper wire. Small diameters make it feasible to manufacture and install much longer lengths than for metal cables. Continuous lengths of 12 km of optical fiber are common.

12.2.2 Small Diameter and Light Weight

The amount of information that is required for many of today's applications is increasing at a rapid rate. Because optical fibers are thinner than copper wires, more fibers can be bundled into cable of a given diameter than copper wires. It is commonplace to install new cabling within existing duct systems. The smaller diameter and light weight make installations easy and less expensive.

12.2.3 Higher Capacity

Optical fibers are thinner than copper wires, so more fibers can be bundled into cable of a given diameter than copper wires. More information such as phone calls or cable channels can be delivered using less space.

12.2.4 Nonconductivity

Another advantage of fiber cable is that it does not conduct electricity. Because it has no metal components, it can be installed in areas with electromagnetic interference, including radio frequency interference. Also, because no electricity passes through the fiber, there is no fire hazard.

12.2.5 Low Power

Lower-power transmitters can be used instead of the high-voltage electrical transmitters needed for copper wires. This saves money and makes installations easier.

12.2.6 Less Signal Loss

The loss of signal in optical fiber is less than that in copper wires. Unlike electrical signals in copper wire, the light signals in one fiber do not interfere with those in an adjacent fiber. This results in more dependable data transmission, clearer phone calls, and better television signals.

12.2.7 Digital Signals

Optical fibers are ideally suited for digital transmission, and this makes them especially effective for data transmission in computer networks.

12.2.8 Security

The dielectric nature of optical fibers makes it impossible to remotely access the signal that is being transmitted in a fiber cable. The fiber would have to be physically accessed, and this would be easily detected by security surveillance. This makes optical fiber ideal for transmitting data by banks, military, and other government bodies that have security concerns.

13 Wireless Communication

13.1 BASIC THEORY

Wireless communication can provide many advantages over traditional wired network systems. Some of the main advantages of wireless communication are ease of use and flexibility, short amount of time to implement, optimized inventory control (RFID), and substantial cost savings. There are several different wireless technologies that employ different segments of the electromagnetic spectrum, which we will discuss in this chapter. We will broadly cover the spectrums of electromagnetic radiation used for wireless communications and then discuss the basics of using infrared LEDs as the groundwork for Chapter 14.

The first optical wireless telephone transmission occurred in 1880, when Alexander Graham Bell demonstrated his photophone (U.S. Patent nos. 235,496, 235,497, and 235,616). The photophone transmitted human sound waves via a mouthpiece onto a beam of sunlight using a silvered reflecting mirror. The mirror reflector when set in vibration by the corresponding sound waves imparted the variations onto the beam of sunlight. A lens was used to focus the beam of sunlight to some distant point. The receiver Bell developed used photosensitive selenium cells and a parabolic condensing mirror to better focus the sunlight onto the selenium cells. The output voltage then varied in proportion to the sound waves being transmitted. The photophone was operated at a distance of 600 ft between two buildings and in the process of developing the photophone; Bell also devised a method of annealing selenium to enhance its photovoltaic properties.

The wireless technologies employ different segments of the electromagnetic spectrum, including radio frequencies (RF), microwave, infrared, and visible electromagnetic radiation. These wireless technologies are characterized as nonionizing radiation because the energy level is not high enough to strip away electrons from atoms; higher-energy x-rays and gamma rays, however, would be considered ionizing radiation. These different wireless technologies can be used for very short distances (≤ 1 m) to very long distances up to millions of kilometers for deep space communications. Wireless derived its name from early radio efforts, and the term now refers to all spectrums of electromagnetic radiation spectrum that do not need wires for operation. The Federal Communications Commission (FCC) regulates and licenses most of the RF and microwave telecommunication industry in the United States.

RF frequencies range from the extremely low frequency (ELF) of 3 Hz to the ultrahigh frequency band (UHF) 300 to 3,000 MHz. Some typical RF wireless applications include cellular telecommunication, radio and television broadcasting, data networks, remote controls, and police and fire department radio communications. Microwave electromagnetic radiation ranges from 1 GHz, as recognized by the IEC and IEEE, up to the extremely high-frequency (EHF) range of 300 GHz. Microwaves are also used in telecommunications for point-to-point communications, for ground-stations-to-orbiting-satellites telecommunication applications, and wireless LAN protocols such as 802.11 (2.4 GHz). Radar, in a noncommunication application, utilizes microwave radiation to identify the range, speed, and characteristics of distant objects. Some of the wireless standards available are IEEE 802.15.1 (Bluetooth—2.4 GHz and 2.5 GHz, version 1.2), IEEE 802.15.4 (Zigbee—2.4 GHz), and IEEE 802.11a/b/g (WiFi—2.4 GHz b/g and 5 GHz/a).

In the infrared region, there are two broad operating ranges: those at a wavelength of 850 nm and those at 1550 nm. Unlike RF and microwave, the optical infrared region is not required to be licensed by the FCC. These two ranges are also used for infrared fiber-optic communication as well, though these IR wireless systems must operate on a clear line of sight. Wireless IRDC (infrared data communication) operates at a wavelength of 850 nm to 900 nm and at relatively short distances of approximately 1 m up to 10 m (16 Mbps). This range became popular because of readily available low-cost semiconductor LEDs and laser diodes. The 1550 nm wavelength range allows for higher data rates (2.5 Gbps) and longer transmission ranging up to several kilometers using Class 1M (IEC—International Electrotechnical Commission) infrared lasers and receivers with telescopic lenses to collect the infrared beam. IR communication offers higher-security transmission for both the 850 nm and 1550 nm wavelengths than RF. IR is harder to detect and intercept and can also be encrypted, thus making it a highly secure method of communication.

The International Organization for Standardization developed the basic networking systems model known as OSI (Open Systems Interconnect). The seven-layer OSI model contains the protocols for each layer, allowing for communication back and forth through these layers. The basic description of the OSI model is as follows:

Layer 1 is the physical layer and defines the standards for the physical connection of equipment and standards the medium (IRDC or RF) must meet.
Layer 2 is the data link layer and describes how to address data over the physical layer.
Layer 3 is the network layer and performs the network routing function.
Layer 4 is the transport layer and describes the transfer of data between end users; it is also used for error correction.
Layer 5 is the session layer, which controls the connections between computers and the communication operation process, termination, and restart.
Layer 6 is the presentation layer, which encodes and decodes the data for the application layer and also includes any data encryption.
Layer 7 is the application layer and connects an application or program to a communication protocol. The end user, however, must initiate the application process to invoke a file transfer.

The OSI model was developed in the late 1970s and, since its inception, newer protocols have emerged from the IEEE and Internet Engineering Task Force (IETF) with fewer layers.

Wireless IR communication using an IRLED is pulsed at a low duty cycle and a high current to provide the transmitting energy. The photosensor having a high-gain amplifier is used to detect the transmitted data. The pulsing of the IRLED and the receiving photosensor will be discussed in additional detail in Chapter 14.

One of the earliest applications using IRLEDs was in the remote control for TV sets. At the time, three GaAs IRLEDs with an emission angle of 50° between half-power points were placed in series and used as a transmitter. A photodiode and amplifier were located in the TV set. The hand-held battery-operated transmitter would send coded data to the receiving photodiode. Some of the energy would be directly received by the photodiode. More of the energy would simply illuminate the room by bouncing from walls and ceiling to the receiving photodiode. This concept had significant advantages over the earlier ultrasonic systems used in the first television remote control. The transmitted data was much more impervious to spurious noise and was capable of transmitting and receiving more complex data. Figure 13.1 shows some of the typical encoding schemes used to transmit information.

Longer-distance transmission utilizes a focusing lens but requires more precise alignment between the transmitter and the receiver. Molded acrylic lenses can be fabricated that will project a 10 to 15 ft circle at 500 ft. A GaAlAs IRLED in the T-1 3/4 package has a typical power output of 120 mW/cm² in an area located 1.130 in. from the lens tip when it is driven at 3 A. If the focusing lens captured this power and dispersed it into a circle that is 10 ft in diameter at 500 ft, the intensity would drop to approximately 240 μW/cm². A 1.5 in. convex lens would illuminate a 0.100 in.×0.100 in. photodiode with approximately 171 μW of IR energy. The photodiode would sink approximately 5 μA of photocurrent. A communication link for office building to building or a perimeter surveillance system could be easily built using this approach. A laser could be substituted for the IRLED and lens system. The benefit of the lens system can be quickly seen if the same calculation is made, where distance is the variable for the 50° wide angle T-1 3/4 and the 16° narrow angle T-1 3/4. Table 13.1 shows a comparison of a system without a magnifying lens for 50°

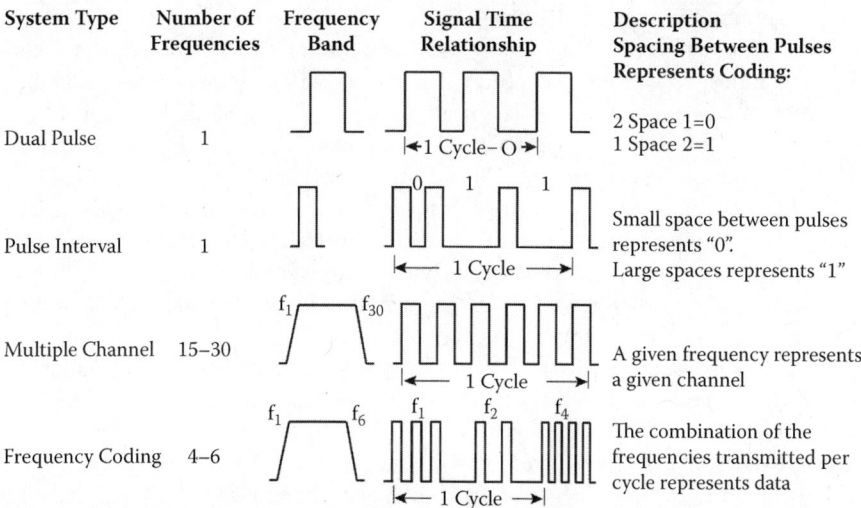

FIGURE 13.1 Typical IRLED encoding methods.

TABLE 13.1
Comparison of System without a Magnifying Lens
16° T-1 ¾ @ 1.4 A—approximately 30 in. separation
50° T-1 ¾ @ 1.5 A—approximately 6 in. separation

and 16°. A 4X magnifying lens and photodiode is used with a 0.100 in.×0.100 in. photosensitive area.

A much more sensitive receiver system is required with the broadband broadcasting system such as is used in remote control of TV systems. The receiving system is discussed in more detail in Chapter 14.

13.2 BACKGROUND

The principles outlined in Section 13.1 offer exciting opportunities for infrared data communication systems. These may be divided into three basic categories and will be discussed with the following assumptions:

1. The IRLED is driven at a peak current of approximately 3 A.
2. The receiver will detect 1×10^{-9} A pulses.

The first category is a hand-held transmitter that will operate effectively up to 30 ft. The target area or accuracy of pointing would be an area 28 ft in diameter. A standard T-1 3/4 GaAlAs IRLED with a beam angle of 50° between half-power points is the transmitter. The receiver is a photodiode measuring 108 sq. mils mounted in a metal can TO-5 package or comparable package. The transmitter and receiver schematic diagrams are shown in Figure 13.2.

The emitter driver, IRLED, photodiode, and amplifier could be built in 1000-piece quantities for approximately $1.00. Chapter 14 will go into more detail on this. This type of IR communication scheme could be used for a variety of applications such as communication with peripheral devices such as printers, PDAs and mobile phones, and in remote controls for TV and appliances.

The second category is an aimed hand-held transmitter and/or receiver. The aiming would require hitting a 100-ft-diameter target. The distance would be controlled by the focusing lens, which would be added in front of both the transmitter and receiver. Table 13.2 shows various angular divergences of lens versus distance, where a 100-ft-diameter circle is the consistent target.

The only changes to the system would be the focusing lens added in front of the IRLED and photodiode. The lens could be molded of clear acrylic plastic, and the cost would be lower than if a glass lens was used. Applications include garage door openers, security systems, utility meter reading, and remote control of mechanical systems.

The third category is quite similar to the hand-aimed transmitter and receiver. However, the target size would be reduced, to either improve the accuracy and the range or to reduce the possibility of detection by outside systems. A decrease to

Wireless Communication

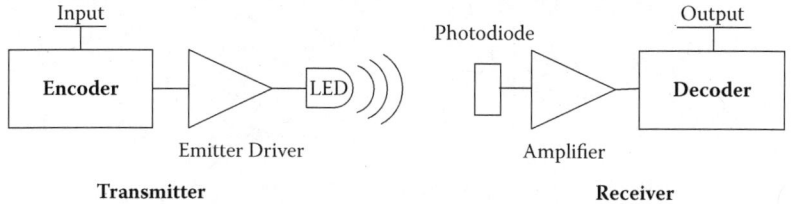

FIGURE 13.2 Schematic of an infrared data communication system.

TABLE 13.2
Angular Divergence of Focusing Lens versus Distance

Diverging Angle	Distance to Target (ft)
27°	100
14°	200
7°	400
3.5°	800

0.25° in the diverging angle of the focusing system would allow the development of a vehicle-speed-measuring system that would detect speeds at up to a 1/2 mile range in the reflective mode. Using this same type system in the transmissive mode, data could be transmitted up to 2 mi.

A low-cost communication link thus becomes very practical, provided that system limitations are not exceeded. The line of sight between the transmitter and the receiver must also be free of any IR-blocking material.

The data rates must be low enough to stay within the average current limitations of the IRLED. Higher data rates may be obtained by using an IR laser as the transmitter at a proportionately higher cost. Applications would include situations where high-speed point-to-point communications would require gaining legal access and right of way, such as in high-density urban areas or in cases where the communication link is needed on a temporary basis.

Part 6

IR Applications

14 Pulse Operation

14.1 THE IRLED

The advantage of running the IRLED at high currents becomes obvious when the output versus IRLED current is examined. Figure 14.1 shows this relationship on both a broad- and a narrow-beam T-1 3/4 unit. The power shown is apertured power measured on a photovoltaic detector 0.250 in. in diameter and located 0.500 in. from the lens side of the mounting flange on the broad-beam unit and 1.429 in. from the lens side of the mounting flange on the narrow-beam unit.

At 3.0 A forward current, the broad-beam unit will typically emit 475 mW/cm^2, whereas the narrow-beam unit will typically emit 114 mW/cm^2. In using these units in the pulse mode, there are three variables to take into consideration: the height of the pulse, the width of the pulse, and the repetition rate of the pulse. The average current (the pulse height times the duty cycle) is related to the maximum DC current allowed. However, the unit is receiving a cycling stress (pulse current causing internal heating and cooling) in addition to the average current. For conservative designs, the maximum average current (pulsed) is limited to two-thirds the maximum DC current. These results indicate that the degradation rates of the IRLEDs under these two conditions are approximately the same (again, a conservative approach). Figure 14.2 shows the degradation rates at an average current of 1 mA (I_{peak} = 1 A, PW = 100 µs, 10PPS) and a DC current of 20 mA. The degradation rate at 10,000 hr for the 1 mA average current is 2.5%, whereas that for 20 mA DC is 7.5%. This would equate to the one-third derating factor. Refer back to the section on reliability in Chapter 3 for the complete use of these curves.

The curve shown in Figure 14.3 shows the maximum peak pulse current versus pulse width. Note that the slope changes after pulse widths exceed 25 µs. This is due to the internal heating caused by the power pulse. Note also that at 300 ms the curve becomes the maximum DC rating for the part.

The maximum peak current is obtained from two factors. Current crowding within the chip causes the nonlinearity in output versus current density. The physical limitation in the bond wire's ability to carry higher currents results in an overall maximum current higher than and independent of the average current.

The method of pulsing the IRLED is rather straightforward. Typically, the battery or power supply charges a capacitor during the off cycle. Figure 14.4 shows a simple schematic utilizing a 9 V battery and assuming a 1 V drop across the switching transistor in the "on" mode.

The average I_F of the IRLEDs at a forward voltage of 4 V is 3.4 A. As the battery ages to 7.5 V, the average I_F will drop to 2.25 A. If the selected transistor switch has an "on" voltage less than 1.0 V (the usual case), then a current-limiting resistor is used to control the current. Under the previous conditions, the peak current pulse

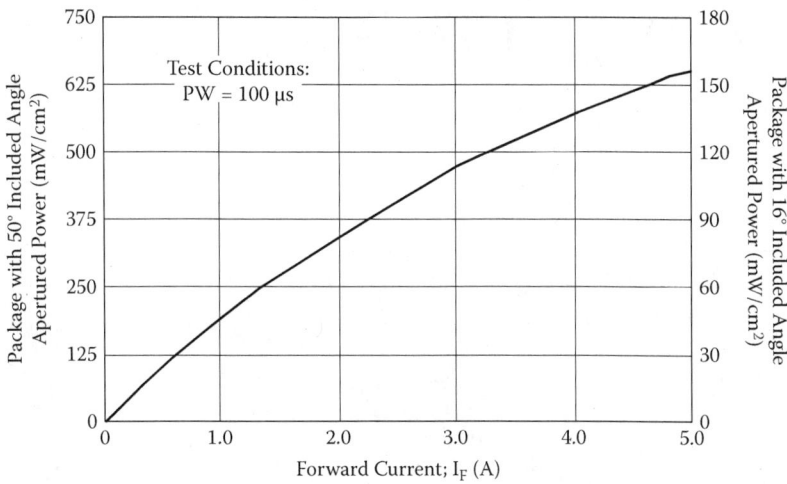

FIGURE 14.1 Apertured power versus forward current for T-1 3/4 GaAlAs IRLEDs. For high forward pulse currents, IRLED output rises dramatically.

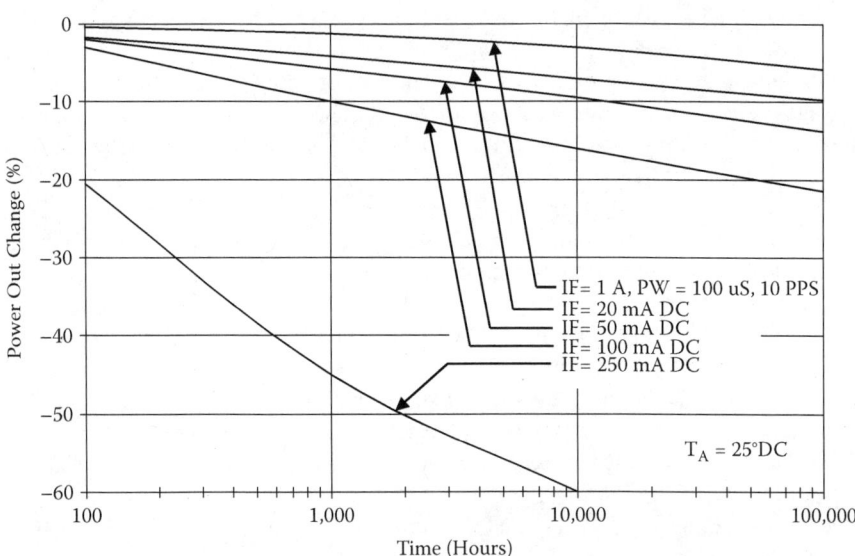

FIGURE 14.2 Percentage change in power output versus time for T-1 3/4 GaAlAs IRLEDs. Pulsing adds the additional stress of the current cycle; however, empirical data demonstrates that carefully controlled high-current pulse operation is no more harmful than low-current continuous operation.

Pulse Operation

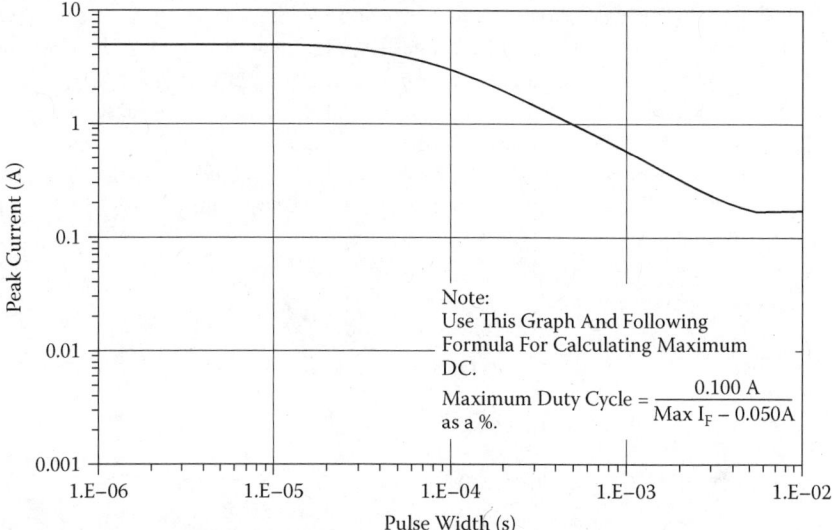

FIGURE 14.3 Maximum peak pulse current versus pulse width for T-1 3/4 GaAlAs IRLEDs. The "envelope" for safe operation is shown. Safe operation is defined as having an extremely low chance of catastrophic failure and acceptably low output degradation.

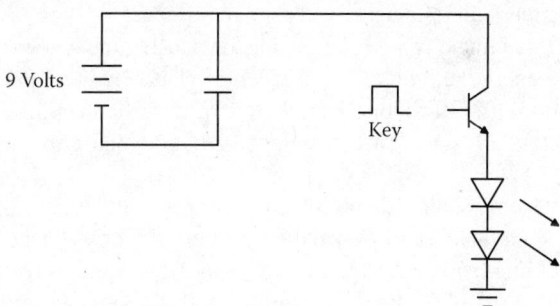

FIGURE 14.4 Simple schematic of pulse circuit for high-current pulsing of two T-1 3/4 GaAlAs IRLEDs. Just as one might logically expect, the high-current is provided by a charged capacitor with switching controlled by a transistor.

width would be a maximum of 80 µs and the maximum duty cycle would be 3.3% (see Figure 14.3).

$$\text{Maximum duty cycle is a \%} = \frac{0.100\,A}{3.040 - 0.050\,A}$$

$$\text{Duty cycle} = 3.3\%$$

(14.1)

In order to demonstrate how to determine the usable power at the receiver, two calculations will be used as examples. The first system will utilize a focusing lens designed to focus the IRLED energy within a 25-ft-diameter circle. The assumption is made that the focusing lens will capture all of the energy within a 30° included angle from the IRLED. The system's worst case of $I_F = 2.25$ A will be used. The effective power (from Figure 4.1) is 375 mW/cm² (50° included angle). The total energy within the 25-ft-diameter circle is equal to the amount of energy within the 0.250-in.-diameter circle.

$$\text{Ratio} = \frac{r_1^2}{r_2^2} = \frac{(0.125/12)^2}{(12.5)^2} = \frac{0.0013}{156.25} = 8 \times 10^{-6}$$

$$\text{Energy} = 375 \text{ mW/cm}^2 \times 8 \times 10^{-6} \tag{14.2}$$

$$= 3000 \times 10^{-9} \text{ W/cm}^2$$

$$= 3 \text{ μW/cm}^2$$

This number will be used in Section 14.2 for a sample calculation. The distance between the transmitter and the receiver is now a variable to be considered in lens design. Table 14.1 shows some varying distances between the transmitter and receiver for the diverging angle of the lens.

A larger diverging angle can be obtained by decreasing the target size. This would also increase the energy/unit area. The 3 μW/cm² would be quadrupled by lowering the target area from a 25-ft-diameter circle to a 12.5 ft diameter. A good two-stage amplifier can detect 20×10^{-9} W/cm², which is 1/150th the 3 mW/cm² used in the example.

The second system used for demonstrating sample calculations will utilize a narrow-beam T-1 3/4 GaAlAs IRLED without an external focusing lens. It will assume 90 mW/cm² (Figure 14.1) at a distance of 1.129 in. from the lens tip. If the assumption is made that 1×10^{-9} W/cm² is detectable and that the energy fall-off follows an inverse square law relationship, then Table 14.2 shows the transmission distance. This would correspond to a distance of approximately 1000 ft, with the location

TABLE 14.1
Distance versus Diverging Angle

Distance (mi)	Tan θ	θ (Diverging Angle)
1/16	0.0379	2° 10 ft
1/8	0.0189	1° 5 ft
1/4	0.00946	32 ft

TABLE 14.2
Distance versus Power/Unit Qrea

Distance	Power/Unit Area
1.129 in.	90 mW/cm^2
~4 in.	5.6 mW/cm^2
~16 in.	350 µW/cm^2
~64 in.	22 µW/cm^2
~256 In.	1.4 µW/cm^2
~1024 in.	87.5 nW/cm^2
~4096 in.	5.5 nW/cm^2
~16384 in = 1362 ft	0.34 nW/cm^2

TABLE 14.3
Comparison of Distance versus Units

Unit	Target Area 4X Lens Receiver	Target Area 1X Lens Receiver	Distance (ft)
50° T-1 3/4	214 ft diameter		400
16° T-1 3/4	175 ft diameter		1000
50° T-1 3/4		107 ft diameter	200
16° T-1 3/4		87.5 ft diameter	500

accuracy of the sensor a 175-ft-diameter circle. Converting to a flat window photocell without the 4X magnifying factor would reduce the range to approximately 500 ft, with a 86.5-ft-diameter target area. Table 14.3 summarizes the same type of calculation utilizing the broad emitting angle T-1 3/4 GaAlAs IRLED with and without the 4X magnifying lens on the photosensor.

14.2 THE PHOTOSENSOR

The photosensor usually consists of a photodiode and a high-gain amplifier. Figure 14.5 shows the block diagram of a commercial PIN photodiode and a high-gain amplifier.

If an input of 3 ΩW/cm^2 (from Section 14.1) is assumed with a PIN photodiode (0.108 in. × 0.108 in.) that has no magnifying lens, the minimum photocurrent should be 30 nA. The minimum input impedance of the amplifier is 40 K. This would create an input voltage of 1.0 mV. The amplifier is designed to function with 100 µV at the input. A ten-to-one safety factor is present.

There are other commercial amplifiers that are capable of performing the same function. The 40 kHz carrier is a carryover from the ultrasonic transmitter receiver used in early remote control systems for TVs. Figure 14.6 shows a pulse train from Chapter 13, Figure 13.1, where the 40 kHz carrier makes up each individual pulse.

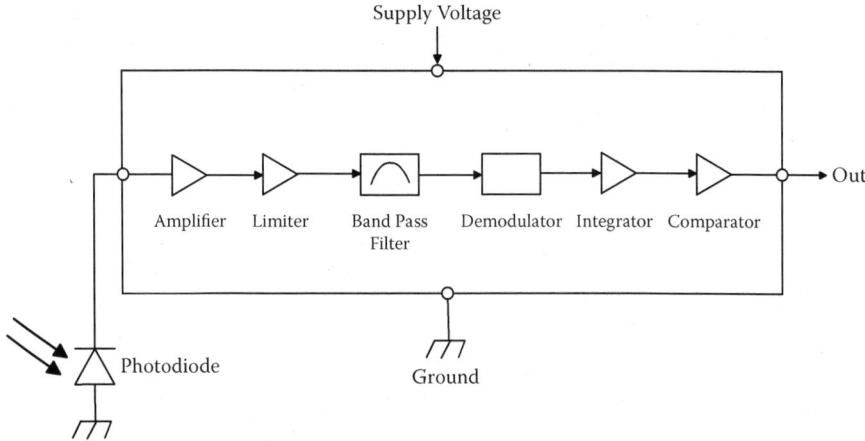

FIGURE 14.5 Block diagram of a pulse receiver. The relatively small current from the photodiode must first be amplified prior to further signal processing. This function, as well as signal processing circuitry, is available in various packages.

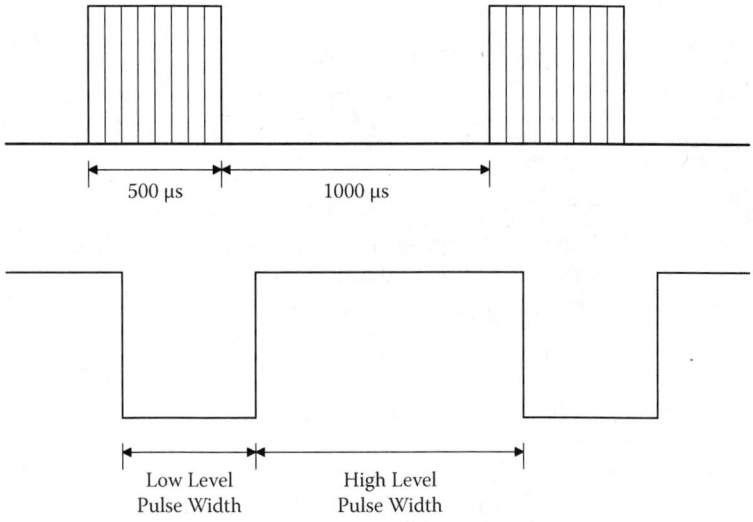

FIGURE 14.6 Pulse train utilizing a 40 kHz carrier. The choice of 40 kHz is a carryover from earlier ultrasonic remote control units (just above audio frequency, but low enough to minimize RFI).

Pulsed operation of IRLEDs and infrared receiver modules are commonly used for many useful applications. They can be used to control many functions of audio/video equipment such as the TV, VCR, CD, DVD, etc. They can also be used on other home appliances such as air conditioners and fans. Other uses include cable and satellite TV set top boxes and multimedia equipment, as well as sensor and light barrier systems.

15 Driving a Light-Emitting Device

15.1 INTRODUCTION

Light-emitting devices are generally separated into several groups. For this discussion we will talk about ultraviolet (UV, 100 to 400 nm), visible (400 to 700 nm), near infrared (NIR, 700 nm to 1100 nm), and a specialty group, vertical cavity surface-emitting laser (VCSEL, 850 nm). All of these devices are considered current sensitive, which means they work best with current as the main driving consideration and voltage as a secondary reference. All these light-emitting devices can be powered using either a pulse technique or with DC current. Figure 15.1 shows a typical schematic that is used to power these devices, utilizing a resistor as the current-limiting element.

Calculation of the resistor value for R_D is a simple application of Ohm's law. Table 15.1 provides a guide for the typical values for the forward current (I_F) and voltage drop (V_F) of the various light-emitting devices:

$$R = \frac{E}{I} = \frac{(V+)-(V_F)}{(I_F)} = \frac{(5.00)-(1.30)}{(0.02)} = \frac{(3.70)}{(0.02)} = 185\ \Omega\ (\text{for IRLED 890 nm})$$

Knowing the voltage across the resistor (V_R) and current passing through it (I_F) allows us to calculate a typical wattage (Power, P) for each type of device.

$$P = E \times I = V_R \times I_F = (3.70) \times (0.02) = 74\ \mu W\ \text{and a 1/8 W resistor should work well.}$$

The current through the device (I_F) will vary depending on several factors. These will be discussed in the next section.

15.2 VARIABLES

The V_F of a light-emitting device will vary from unit to unit. A typical IRLED with a center wavelength of 890 nm has a V_F of 1.30 V at an I_F of 20 mA and an ambient temperature of 25°C (77°F). Graph 15.1 shows the expected variation of I_F versus V_F versus temperature for a typical 890 nm IRLED. The graph configurations are typical for any of the standard emitting devices. As can be seen, the values of V_F and I_F vary greatly, and thus, the actual value of R_D is not critical. Also, we can see that the forward voltage decreases with temperature, thus increasing the forward current through the emitter and the relative optical power. Graph 15.2 shows the relationship between I_F, E_E (optical power radiated), and T_A (temperature). You can see the

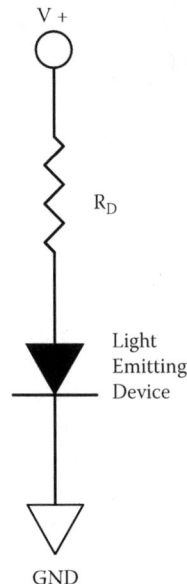

FIGURE 15.1 Drive circuitry.

TABLE 15.1
A Guide to the Typical Values for the Forward Current (I_F) and Voltage Drop (V_F) of the Various Light-Emitting Devices

Light-Emitting Device	Typical Forward Current (mA)	Forward Voltage (V)
IRLED 935 nm	20	1.25
IRLED 890 nm	20	1.30
IR VCSEL 850 nm	5	1.20
Visible LED red	20–350	1.9–2.7
Visible LED yellow	20–350	1.9–2.7
Visible LED green	20–350	3.0–3.7
Visible LED blue	20–350	3.0–3.7
Visible LED white	20–350	3.0–3.7
UV-A LED	10–350	3.0–4.4
UV-B LED	10–500	4.4–5.0
UV-C LED	20–350	6.0–8.0

variation as these parameters change. You should note that as temperature increases, the amount of E_E radiated from the device reduces. This is important to remember as we discuss the characteristics of a typical phototransistor.

Thermal considerations may be taken into account using Graphs 15.1 and 15.2. In order to calculate the junction temperature (T_J), we need to look at the power consumption of the device. With a forward current of 20 mA, the forward voltage

Driving a Light-Emitting Device

at 25°C is approximately 1.30 V, providing a power consumption of 26 mW (current times voltage, 0.020×1.30). A typical plastic-encapsulated emitting device will have a derating factor of approximately 1.33 mW/°C. This gives a junction temperature of 44.5°C with an ambient temperature of 25°C.

$$T_J = Re.Temp + \frac{Device\ power}{Derating\ factor} = 25°C + \frac{26.0\,mW}{1.33\,mW/°C}$$

$$= 44.5°C\ (for\ Plastic, IR-LED\ 890\,nm)$$

If the device is tested in pulse mode and used in constant-current mode, the designer should take into consideration the difference in junction temperature.

15.3 LINEAR OPERATION

Linear operation of a light-emitting device is dependent on many factors, including the thermal characteristics of the package, mechanical configuration of the LED chip (sometimes referred to as *die*) (actual device that is emitting the light), ambient

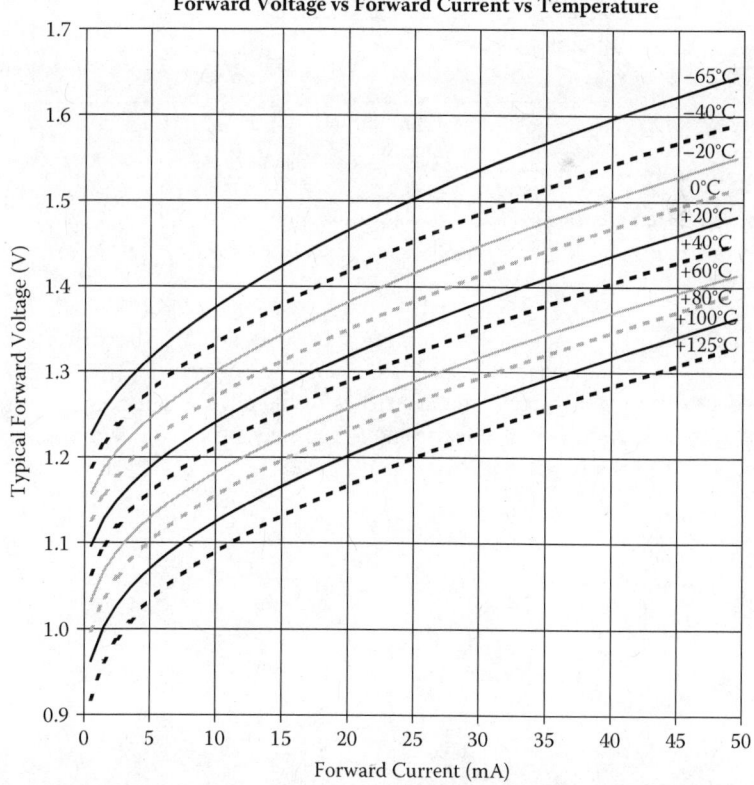

GRAPH 15.1 Graph for a typical IRLED 890 nm: I_F versus V_F versus temperature.

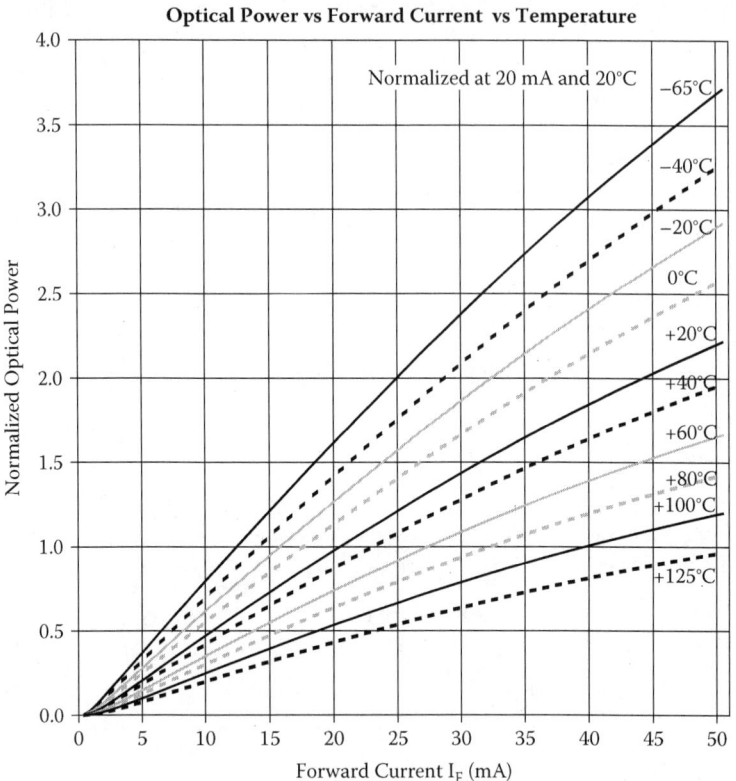

GRAPH 15.2 Graph for a typical IRLED 890 nm: I_F versus output power versus temperature.

temperature range, etc. As can be seen in Graphs 15.1 and 15.2, the forward voltage and optical power are linear in a specified area. The designer must reference each device data sheet for the actual characteristics of the device. The characteristics of some devices have very small linear regions and may be susceptile to temperature variations.

16 Interfacing to the Photosensor

16.1 THE PHOTODIODE

The P-N or PIN photodiode is usually utilized in applications where speed–response or linearity is required. The photodiode can switch in a few nanoseconds and respond linearly to input levels over nine orders of magnitude. The photodiode output is proportional to the level of illumination and is normally amplified. The input impedance of the amplifier used is normally low to minimize the RC (R—input impedance, C—diode capacitance) time constant of the system. The photodiode is typically operated in two basic modes: the photovoltaic mode and the photoconductive mode. In the photovoltaic mode, the photodiode is not biased, as shown in Figure 16.1. The output of a photodiode in a photovoltaic mode is typically in millivolts per optical power. In photoconductive mode, the photodiode is reverse-biased with a given voltage. The output of a photodiode, often called *reversed light current* (I_L) or just *light current*, is an output sink current that varies from the high picoampere to the low milliampere range. Figure 16.2 shows a photodiode in photoconductive mode.

Both circuits shown accept a current as the input and provide a voltage as the output. Therefore, the transfer function (output/input) of the preceding circuits has units of impedance (volts/ampere). As a result, both circuits are often called *transimpedance amplifier*.

The DC voltage output for the previous circuits is quite simply:

$$V_{OUT} = I_L \times R_F \qquad (16.1)$$

Hence, the DC amplification is only due to the feedback resistor value.

The voltage output increases as more light energy impinges on the photodiode. The value of R_F is very large (over 1 MΩ, typically), owing to the low photodiode light current level.

C_F is added for overshoot stability and compensation due to the photodiode capacitance and operational amplifier input capacitance. To calculate C_F requires more elaborate calculation. First, the operational amplifier and photodiode capacitance can be added for simplification:

$$C_{TOTAL} = C_{DIODE} + C_{AMPLIFIER} \qquad (16.2)$$

C_{TOTAL} or just C_T is added to the previous circuit as shown in Figure 16.3.

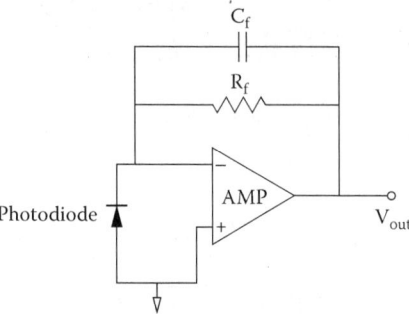

FIGURE 16.1 Photodiode in photovoltaic mode. The photodiode is unbiased; there is zero voltage across the photodiode. This mode is preferred for precise linear measurement of low light levels applications due to low noise level.

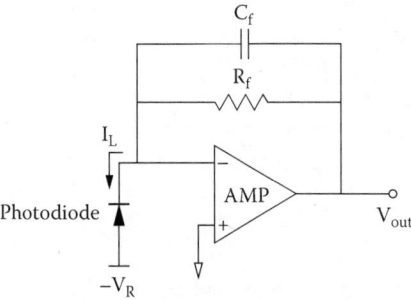

FIGURE 16.2 Photodiode in photoconductive mode. The photodiode is reverse-biased. This mode is preferred for high-speed applications. There is typically an increase in dark and noise current due to the reverse bias.

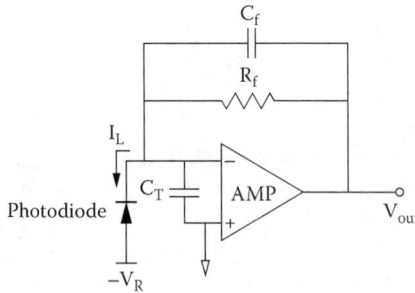

FIGURE 16.3 Transimpedance amplifier circuit. The circuit is useful for visualizing the photodiode and operations amplifier capacitance added to the circuit. The added capacitance (C_T) can make the circuit oscillate or overshoot.

Interfacing to the Photosensor

Usually, the diode's shunt resistance is very large; hence, it can be neglected in first-order analysis. There is also a diode resistance, which in most cases is very small (typically less than 100 Ω). The diode resistance may be neglected as long as it is lower than 500 Ω.

The transfer function of the circuit shown in Figure 16.3 is

$$H(j\omega) = R_F \times \left(\frac{1 + j\omega R_F (C_F + C_T)}{1 + j\omega R_F C_F} \right) \text{ with a pole and a zero at}$$

$$f_P = \left(\frac{1}{2\pi R_F C_F} \right) \text{ and } f_Z = \left(\frac{1}{2\pi R_F (C_F + C_T)} \right), \text{ where } \omega = 2\pi f$$

The preceding equations show how the gain of the circuit depends on the input frequency. Under sinusoidal steady-state conditions, the open-loop gain of the circuit falls at a constant rate, until it reaches the unity gain value or 0 dB. This behavior is inherent in all operational amplifiers. The equations for the pole and the zero frequencies (f_Z and f_P) are important to design a stable circuit.

Figure 16.4 shows the typical bode plot of an operational amplifier open-loop gain versus frequency along with the noise gain versus frequency plot of the circuit in Figure 16.3. The noise gain curve shows how at frequency f_z the noise gain starts to rise; if left without a feedback control, the noise will keep rising until it crosses the open-loop gain curve. For the system to be stable, the intersection of f_z and f_P needs to lie inside the open-loop gain curve. Outside the open-loop gain curve, the system is unstable. Inside, the system is unconditionally stable. From the preceding equations, f_p depends on the feedback network, R_F and C_F. Hence, a proper value

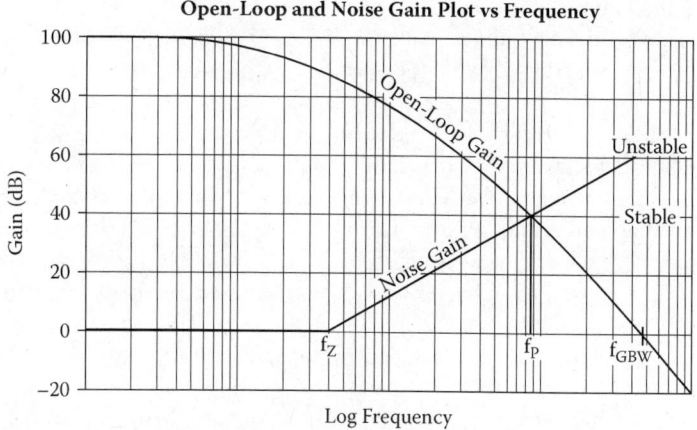

FIGURE 16.4 Open-loop and noise gain versus frequency. This plot is critical to determine the optimum value of the feedback network.

of C_F will make the system stable. If C_F is very large (system overcompensated), the circuit will be unconditionally stable, but it will limit the bandwidth. A limited bandwidth will work only for low-frequency signals. If C_F is very small, the system will be unstable. The optimum value of C_F provides the maximum bandwidth while maintaining a stable system. Such a value occurs when the intersection of f_Z and f_P crosses the open-loop gain curve, as shown in Figure 16.4.

The open-loop gain equation is the gain bandwidth product (f_{GBW}) with an asymptote at the frequency at which the open-loop gain crosses 0 dB. This curve can be found in almost all operational amplifiers' data sheets. To find the optimum value of C_F, f_{GBW} is divided by the pole frequency, f_P, and then setting the expression equal to the transfer function capacitance:

$$f_P \times 2\pi R_F C_F = \frac{C_F + C_T}{C_F} \tag{16.3}$$

When expanded, the preceding equations become a quadratic equation in C_F:

$$C_F^2 \left(F_{GBW} 2\pi R_F \right) - C_F - C_T = 0 \tag{16.4}$$

C_F can be solved utilizing the quadratic equation $x = \left(-b^2 \pm \sqrt{b^2 - 4ac} \right) / 2a$ with coefficients: $a = F_{GBW} 2\pi R_F$, $b = -1$, and $c = -C_T$.

$$C_F = \frac{1 + \sqrt{(-1)^2 - 4 \times (F_{GBW} 2\pi R_F)(-C_D)}}{2 \times (F_{GBW} 2\pi R_F)} \tag{16.5}$$

The new bandwidth is at the pole frequency f_P.

The benefits of the photodiode in photoconductive mode were explained in Chapter 4. From the circuit and application perspective, each mode offers benefits and challenges. In photovoltaic mode, there is no current flow through the feedback resistor, eliminating the dark current term. This eliminates error measurements (noise current), allowing low light signals to be detected. The photons impinging on the photodiode generate a voltage potential. This is the only excitation on the photodiode; therefore, the output–input relationship is linear, facilitating calculations. The photovoltaic mode is not suitable for high-frequency signals, because of the photodiode large capacitance in the depletion region (see Chapter 4 for further details). In photoconductive mode, the reverse voltage allows the depletion region to widen, decreasing the capacitance. As the capacitance decreases, the speed response of a given photodiode improves. This mode is best suitable for high-speed-sensing applications, such as fiber optics and detection of high-frequency-modulated light signals. A resistance–capacitance feedback network can be designed to filter the desired frequencies while maintaining a stable system. However, the reverse voltage helps to generate dark current, decreasing the linearity range. This particularly affects low light level detection.

16.2 THE PHOTOTRANSISTOR AND PHOTODARLINGTON

In the discussion in Part 2 on the transistor used as the photosensor, it was pointed out that the switching characteristics of the phototransistor are inferior to those of the conventional small signal transistor. This is due to the enlarged base area of the phototransistor, which increases the collector-to-base capacitance. Significant variables also affecting the switching time are the depths of the junctions, which are optimized for photon recombination, and the spacing or geometry of the transistor, which is usually larger on the phototransistor.

The phototransistor is normally used as a switch. When sinking current, it is considered "on," and when not sinking current, it is considered "off." Figure 16.5 shows a simple phototransistor as a switch schematic.

When energy is impinging on the base region, the transistor is in conducting mode. The voltage drop across the load is high, being $V_{CC} - V_{CE(SAT)}$. When the energy is blocked or removed, the only conduction current is the leakage current (I_{CEO}). The voltage drop across the load approaches zero. Figure 16.6 shows the voltage versus current for these two states.

The two conditions of the switch are thus specified. The voltage drop across the load must be large enough to actuate the switch in the "on" condition and must be small enough in the "off" condition to cause the switch to change state.

Typically, phototransistors are specified with a minimum on-state collector current output, or $I_{C(ON)min}$, and a maximum collector dark current, or I_{CEO}. These two parameters are often guaranteed based on specific test conditions. These two parameters are the key to calculating the correct value of the load resistor. However, they can change depending on the specific application and ambient light conditions. For example, high

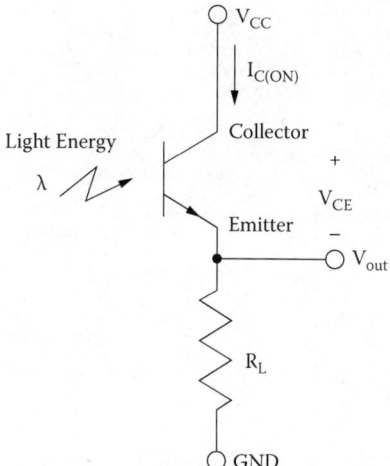

FIGURE 16.5 Simple schematic of the phototransistor operating as a switch. Perhaps the simplest circuit possible is to have a voltage source and load resistor (emitter to ground) in series with the phototransistor. Switching time is dependent on the value of R_L, and the speed of the phototransistor.

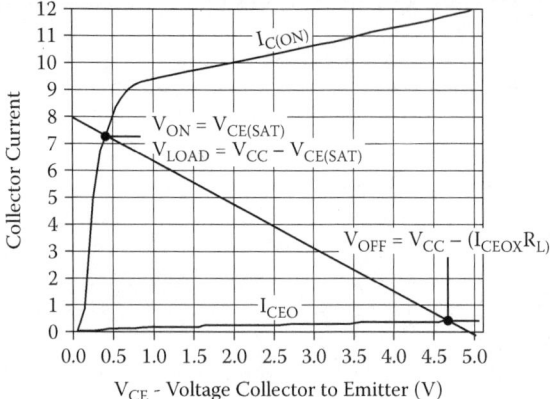

FIGURE 16.6 V_{ON} and V_{OFF} for a phototransistor operating as a switch. The transistor voltage drop is at its maximum when "off" (nonconducting dark state) and becomes minimal when turned to the "on" state.

light levels of energy would increase $I_{C(ON)min}$; this behavior was explained in Chapter 5. Furthermore, high light levels of ambient light condition would create a "noise" current, which, added to the dark current, would create a total leakage current in the application. These specific conditions need to be taken into account in order for a reliable switching operation. To have a successful switching characteristic, the minimum $I_{C(ON)}$ as well as maximum $I_{CEO} + I_C(noise)$ should be found per the specific application, taking into account the light energy controlling the photosensor and the ambient light conditions. Once the preceding two parameters are established, the value of the load resistor range is found with the following equations:

$$R_{Lmin} > \frac{V_{CC} - V_{CE(SAT)}}{I_{C(ON)min}} \qquad (16.6)$$

$$R_{Lmax} < \frac{V_{CC} - V_{CE(SAT)}}{I_{CXmax}} \qquad (16.7)$$

where I_{CXmax} = maximum $I_{CEO} + I_C(noise)$.

Based on the preceding equations, it is easy to see how a resistor that is very large would saturate the sensor at lower collector currents, making the sensor very sensitive to small light signals. A very large resistor would be a resistor approaching R_{Lmax} value. A larger R_{Lmax} resistor would prevent the switch from coming out of saturation: the sensor will remain in the "on" state at all times. On the other hand, a sensor with a load resistor lower than R_{Lmin} will prevent the sensor from entering into saturation; the sensor will remain the "off" state at all times.

Interfacing to the Photosensor

16.2.1 BASIC INTERFACE CIRCUITS FOR CMOS

The common-emitter circuit in Figure 16.7 will generate an output signal that transitions from high to low when the phototransistor detects light or infrared radiation. This is commonly referred to as an *inverting logic condition*. The common-collector circuit in Figure 16.8 will generate an output signal that transitions from low to high when the phototransistor detects light or infrared radiation. This is commonly referred to as a *noninverting logic condition*. These two circuit examples are compatible with all CMOS logic in the 3.3 V to 15 V versions. On-state collector current, $I_{C(ON)}$, is approximately 0.5 mA.

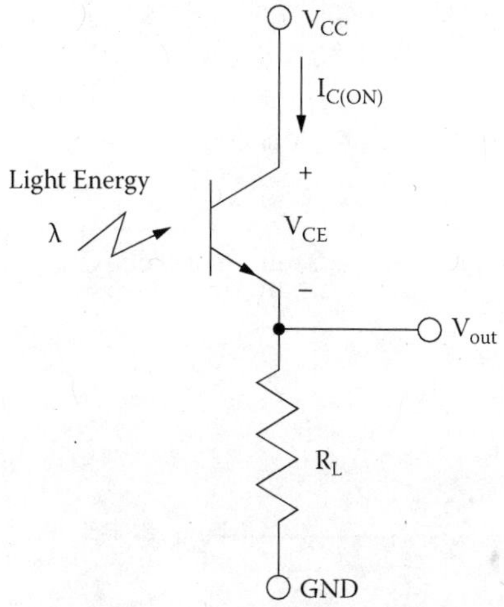

FIGURE 16.7 The common-emitter circuit. When the light energy is "off" or "blocked," the output is logic high.

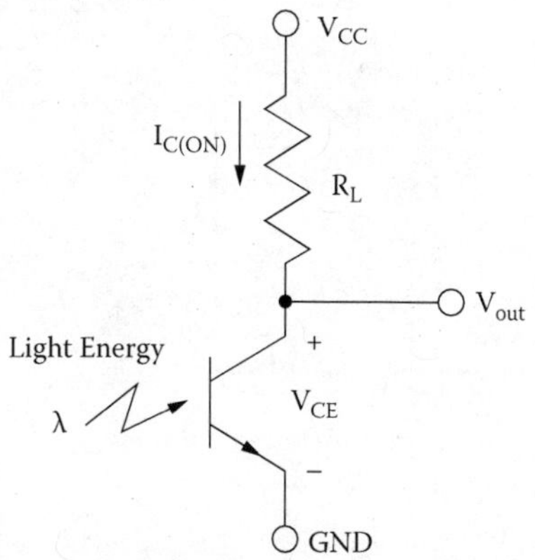

FIGURE 16.8 The common-collector circuit. When the light energy is "on" or "unblocked," the output is logic low.

16.2.2 BASIC INTERFACE CIRCUITS FOR TTL

Typical TTL interfacing requires a minimum of 1.6 mA sinking current (I_{GATE}) with a maximum voltage of 0.8 V at the input of the TTL gate for the "off state." For the "on state," the voltage must be greater than 2.2 V. This must hold true over the operating temperature range of the system and not exceed 0 to 70°C. Some phototransistors are not capable of sinking 1.6 mA of collector current ($I_{C(ON)}$); hence, additional circuitry is needed to meet the interface requirements. The transistor buffer, Figure 16.9, is a cost-effective choice to interface circuits with TTL devices. If designed properly, the circuit can provide optimum noise immunity and fast switching.

The selection of R_3 depends on the gain (β) of Q_1 and $I_{C(ON)min}$ from the phototransistor.

The addition of the 2N2222 lowers the output requirements of the phototransistor. R_2 is chosen as a current limit resistor for the phototransistor and must be large enough to provide adequate drive to turn the 2N2222 "on." R_3 is the pull-up resistor that provides a path to supply the I_{CEO} of the 2N2222 when V_{OUT} is high.

The equations for the preceding circuit are as follows:

$$I_{R3} = \frac{\left(V_{CC} - V_{CE(Q1)}\right)}{R_3} \tag{16.8}$$

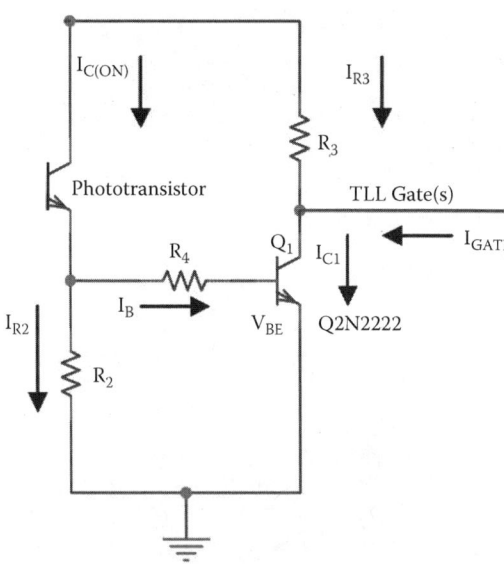

FIGURE 16.9 Improving the sink current with a transistor buffer circuit. An intermediate amplification stage may be as simple as the addition of a standard switching transistor circuit such as the 2N2222 and a few resistors.

Interfacing to the Photosensor

$$I_{C1} = (I_{R3} + I_{GATE}) \quad (16.9)$$

$$I_B = \frac{(I_{C1})}{\beta} \quad (16.10)$$

$$I_{R2} = (I_{C(ON)} - I_B) \quad (16.11)$$

$$R_2 = \frac{(V_{BE} + I_B R_4)}{I_{R2}} \quad (16.12)$$

Example using the preceding equations:

A phototransistor with an $I_{C(ON)min}$ = 500 µA

Q1 parameters:

$$2N2222\ (\beta = 60\ \text{min})\ V_{CC} = 5\ V$$

Typically, a resistor R_3 of 1 kΩ is used as a pull-up resistor for TLL interface. From (16.8):

$$I_{R3} = \frac{(5-0.4)}{1000} = 4.6\ \text{mA}$$

To support TLL interface, the circuit needs to be able to sink 1.6 mA = I_{GATE}. From (16.9):

$$I_{C1} = (4.6\ \text{mA} - (-1.6\ \text{mA})) = 6.2\ \text{mA}$$

From (16.10):

$$I_B = \frac{(6.2\ \text{mA})}{60} = 103\ \mu\text{A}$$

To choose the R_2 value:

$$I_{R2} = (500\ \mu\text{A} - 103\ \mu\text{A}) = 397\ \mu\text{A}$$

Typically, a resistance R_4 of 1 kΩ is used as a base-current-limiting resistor for TLL interface.
From (1.5):

$$R_2 = \frac{(0.7\,\text{V} + 103\,\mu\text{A} \times 1\,\text{k}\Omega)}{397\,\mu\text{A}} \cong 2.02\,\text{k}\Omega$$

Complete saturation of the phototransistor is not necessary; however, adjustment to R_2 may be required if sensor is too sensitive; for example, if the sensor is affected by ambient light, the R_2 value should be lowered.

The photodarlington is similar in output characteristics to the phototransistors with two exceptions. The $V_{CE(SAT)}$ characteristics of the photodarlington are higher than those of the phototransistor by one forward voltage drop of approximately 0.7 V. The cascaded structure of a transistor feeding a second transistor causes this problem. The switching times of a photodarlington are usually more than an order of magnitude higher than that of a phototransistor with the same effective load resistor. The added gain of the photodarlington, however, can be useful in certain application areas. This gain difference is normally a factor of slightly over 1 to greater than 250.

If the application requires the extreme upper end of sensitivity of a phototransistor (h_{FE} of 1000 to 1200), the cost may be prohibitive or the distribution may yield too few units to satisfy the volume. Photodarlingtons with h_{FE}'s in the 1000 to 3000 range are relatively easy to make and thus are far more suitable than high-gain phototransistors. The photodarlington is built with a common collector region. The simplified gain is nothing other than the square of the transistor gain. A photodarlington with a gain of 900 (30 × 30) is much easier to fabricate than a phototransistor with a gain of 900. Figure 16.10 shows the schematic for both a phototransistor and a photodarlington.

The second type of application involves the need for higher output current with a given energy level input. If the energy level is in the microwatts range, the high-gain photodarlington allows adequate gain for a usable output signal. This can be useful

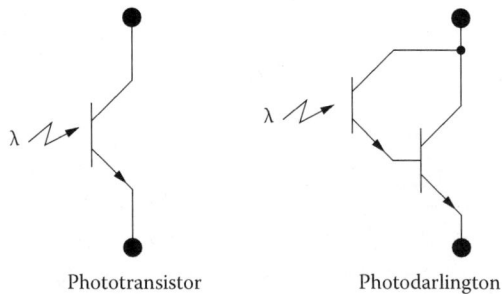

Phototransistor   Photodarlington

FIGURE 16.10 Schematic of a phototransistor and a photodarlington. The gain advantage of the photodarlington over the phototransistor may be by as much as a factor of 250. However, speed is usually slower by an order of magnitude. Phototransistors may be extremely sensitive to ambient light.

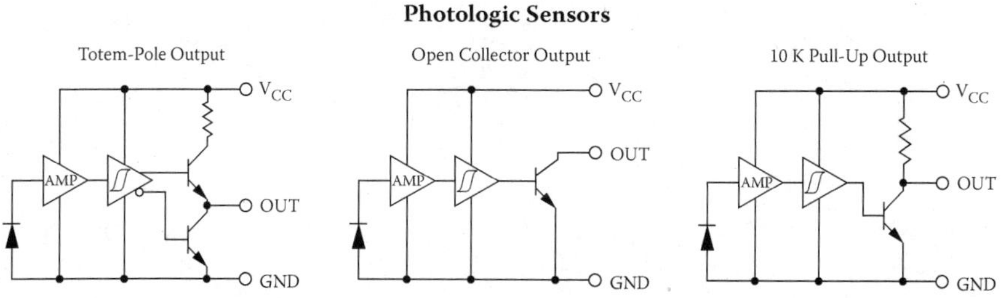

Interfacing to the Photosensor

Photologic Sensors

Totem-Pole Output   Open Collector Output   10 K Pull-Up Output

FIGURE 16.11 Schematics of common photointegrated circuits or Photologic sensors. Most of these sensors offer great speed and simple interface with logic circuits. Only the open collector, an additional component (a pull-up resistor) for proper operation. These sensors are available in buffer or inverted output configuration.

in applications such as a "touch screen," where the optical switches have separation between transmitter and receiver that exceeds the normal 0.050 in. to 0.500 in. If the input energy level is in the high-microwatt or low-milliwatt range, the high-gain photodarlington may be used to drive relays directly. There is clearly a place in the optosensor line for applications of the low-cost phototransistor and photodarlington.

16.3 THE PHOTOINTEGRATED CIRCUIT

The photo IC is the easiest unit to interface to because it is normally specified to be compatible with logic circuitry and is usually guaranteed to operate under specified conditions over a broad temperature range. Figure 16.11 shows three common schematics of Photologic® sensors, with buffer output configuration. The sensor output can be configured to go from low to high logic state (buffer output) when light energy strikes its sensing area. It may also be configured to go from high to low logic state (inverted output) when light energy strikes its sensing area. This type of sensor offers the advantages of the photodiode input and conditioning or logic circuitry output. The only major disadvantage is the premium price. The future trend will be toward availability of more special function options, decreasing price, and erosion of discrete photosensor applications as these units become more widely used.

17 Computer Peripheral and Business Equipment Applications

17.1 COPIERS AND PRINTERS

Copiers and printers utilize a varying number of interruptive or reflective optical sensors. A high-quality printer utilizes an optical interruptive- or reflective-type sensor to verify the presence of the paper travel in many locations to prevent damage to the system when the paper is not present. These locations may consist of paper out in each storage tray, paper being fed from the input tray, paper being ejected to the output trays, or internal paper travel from one element to another.

17.2 KEYBOARDS AND MICE

Keyboards and optical mice can be interfaced to a computer by using a near-infrared light source. This light source is typically a low-cost T-1 3/4 style package device (Figure 17.1).

It enables the designer to pulse the LED at a reasonable data rate with high forward currents up to 1.5 A without reducing the life expectancy of the assembly. The optical mouse may utilize a red visible LED to recognize movement of the mouse on a surface (see Figure 17.2). Besides using a red LED as the source of identification of movement, a trackball mouse may use a reflective optical sensor to identify movement of the trackball (Figure 17.3).

17.3 TOUCH SCREENS

The touch screen may use discrete-lensed IRLEDs and photosensors with convex magnifying lenses in front of the devices to focus the light beams. The ratio of current of several photosensors allows the ability to recognize a precise location of an object on the touch screen. By processing this information, specific actions can be taken as requested by the operator. The sequencing of the IRLEDs and photosensors allows the signals to more accurately identify with ambient light present, allowing the object breaking the light beam to be easily recognized in a variety of light environments.

17.4 DATA INTERFACE

Data may be communicated between computers using several methods. The use of routers and hubs connected through fiber-optic cables is a standard secure

FIGURE 17.1 Optical keyboard and mouse.

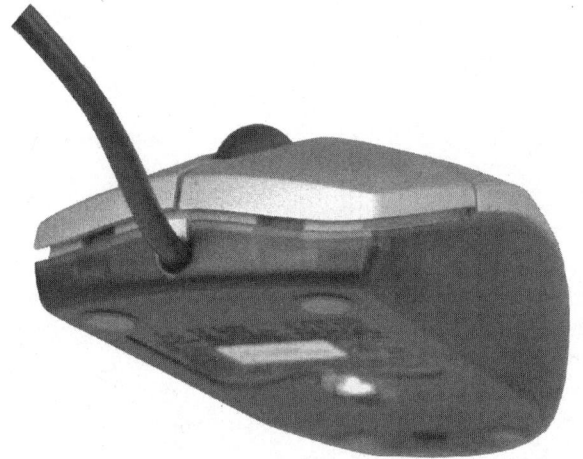

FIGURE 17.2 Optical mouse with red LED to light up the work surface to recognize movement.

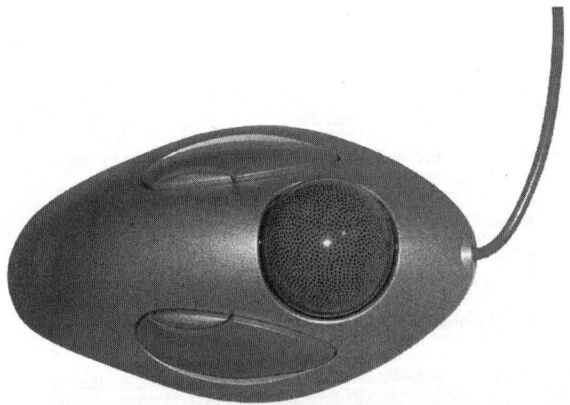

FIGURE 17.3 Mouse with trackball. Reflective assembly recognizes trackball movement.

Computer Peripheral and Business Equipment Applications

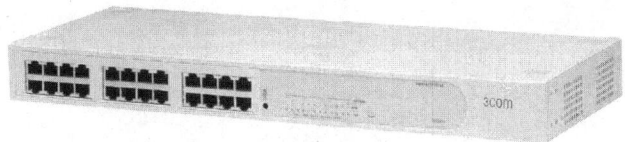

FIGURE 17.4 Computers use a variety of interface modules, including routers and hubs that utilize optical devices for data communications through fiber-optic cables.

FIGURE 17.5 Package availability for optical couplers/isolators. Isolation voltage typically ranges from 1,000 to 50,000 V.

configuration in many IT systems (Figure 17.4). Refer to Chapter 12 for more information.

17.5 CHECK/CARD READER

The leading or trailing edge of a check or credit card can be precisely located utilizing a slotted or reflective sensors; this enables the designer to begin the scan as a check or credit card is moved through the reader.

17.6 OPTICAL COUPLERS/ISOLATORS

Optical couplers (optically coupled isolators) are used to isolate one power system from another while transferring digital or analog information between them and to prevent ground loops. They are also used in switching power supplies to isolate the control circuits from the transformer primary. Isolation of current spikes or shorts can easily be accomplished by using optical couplers/isolators. Figure 17.5 shows some of the possible package variations for optical couplers/isolators. When electronic equipment of different manufacturers is tied together to create a network, optical couplers are used to isolate the output or input of one piece of equipment from another, preventing damage due to spurious signals or electrical shorts in the network. Isolation voltages typically range from 1,000 to 50,000 V for standard devices. The output may either be an analog signal or digital (TTL logic), depending on the application.

18 Industrial Applications

18.1 SAFETY-RELATED OPTICAL SENSORS

Discrete IR LEDs and photosensors can be utilized to form a safety shield to protect dangerous areas from accidental intrusion. Figure 18.1 shows the difference in light illumination between two basic types of emitters used for light curtains. The wider the light pattern, the easier it is to align the sensor array to the emitter array. The major drawback to using a wide light pattern is that the intensity of light on the photosensors is reduced considerably and the adjacent sensors will also be illuminated. Unless a VCSEL (vertical cavity surface-emitting laser) or additional lensing is used, more than one photosensor is illuminated with the highest intensity on the photosensor directly in line with the light beam of the emitter.

Two basic techniques are used on the emitter side of the light curtain. The easiest way to identify that an object is present is by interrupting the light path between the emitter and photosensor. By monitoring the total $I_{C(ON)}$ current from the photosensors and looking for a change in signal, small as well as large objects can be recognized. When a very small reduction of $I_{C(ON)}$ current is recognized, we can assume an object has interrupted some of the light beams. This technique is not immune to radiation of light from external sources. A more robust technique is to pulse sets of emitters and monitor individual or groups of photosensors (see Figure 18.2). By turning on specific emitters, the light patterns from each emitter can be easily distinguished by the photosensors and ambient lighting can be taken into consideration by monitoring the photosensor current when all of the emitters are off.

The emitter groups can be pulsed at very short intervals, thus reducing the radiation pattern on the photosensors from adjacent emitters. This allows us to monitor individual or groups of photosensors to recognize an object has interrupted a light beam. In this example, every group would be pulsed for 1 ms; therefore, it would take 7 ms (1 ms for each of the 6 groups plus 1 ms for ambient light monitoring) to complete a monitoring sequence.

Figure 18.3 shows the results when an object is placed between the emitter and sensor arrays. With at least some of the light beams reflecting off the target, the sensors will see a decrease in optical signal, resulting in recognition of the object.

If the average radiance over a specified period of time for either each photosensor or group of photosensors is referenced against the ambient light level for the area, ambient light interference may be reduced or ignored. This allows the light curtain to be used in almost any industrial application.

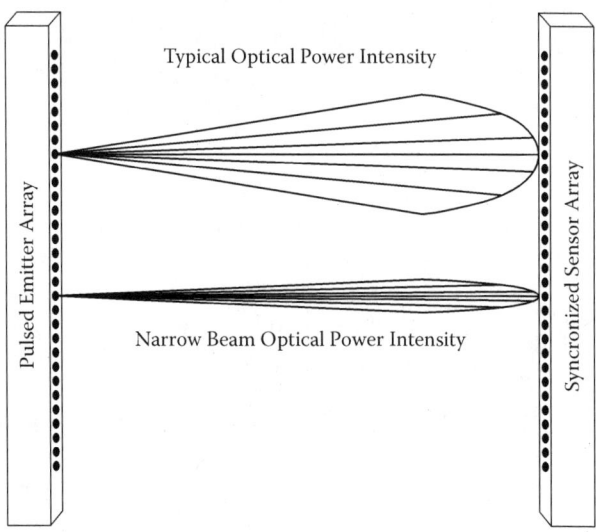

FIGURE 18.1 Emitter intensity light patterns.

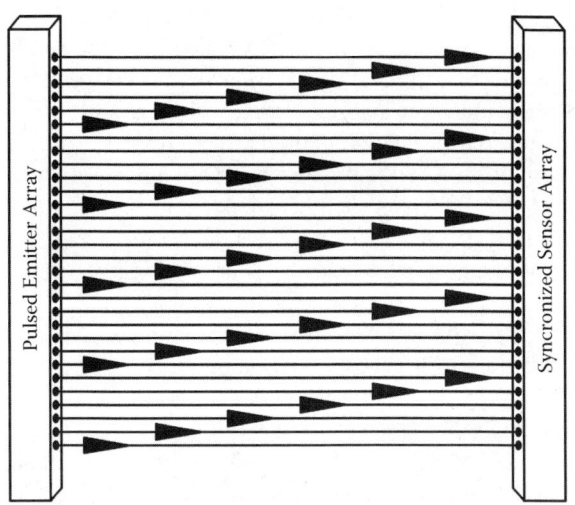

FIGURE 18.2 Emitter pulsing technique allows groups of emitters and photosensors to operate independently.

18.2   HAZARDOUS FLUID SENSING

Hazardous fluids such as gasoline require special handling and precise acknowledgement of the fluid level limits. An emitter and photosensor can be utilized to project a light beam down a quartz or plastic rod with a cone feature at the far end.

Industrial Applications

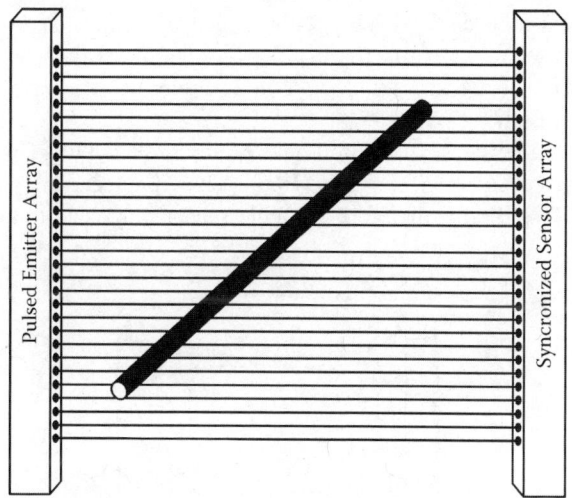

FIGURE 18.3 A rod breaking a beam of light between the elements of a light curtain.

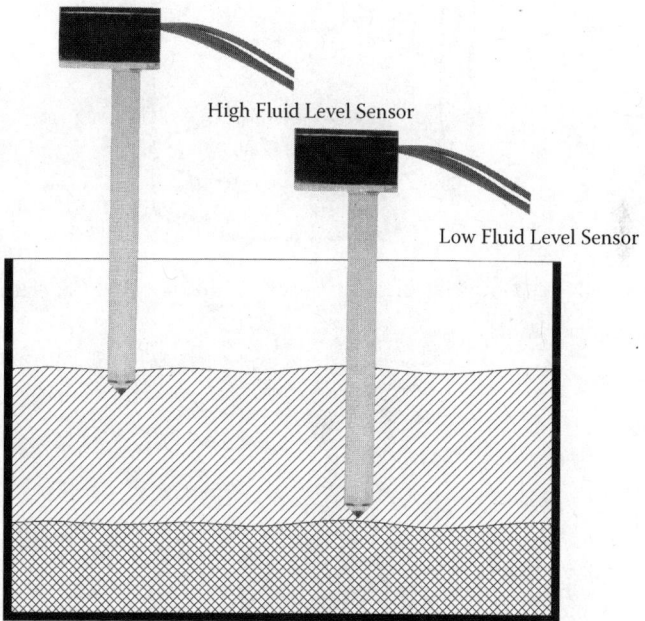

FIGURE 18.4 High and low fluid level identification utilizing a special reflective rod.

Without the tube inserted into a fluid, the light energy bounces from one side of the cone's 45° angle sides and reflects back to the photosensor. When the liquid level covers the base of the probe, the energy passes into the liquid, causing the switch to change states (see Figure 18.4).

234 Optoelectronics: Infrared-Visible-Ultraviolet Devices and Applications

The level of a fluid can also be acknowledged by using a sight tube and two sensors. One sensor will identify the low fluid level and the other will identify the high fluid level. A device similar to the OPB350 fluid sensor can be used for this requirement (see Figure 18.5).

Fuel or liquid flow can be measured by counting the revolutions per unit of time in a known volumetric area (see Figure 18.6). It may be necessary to ensure that a fluid is present in the tube, therefore a bubble sensor may need to be added. Both

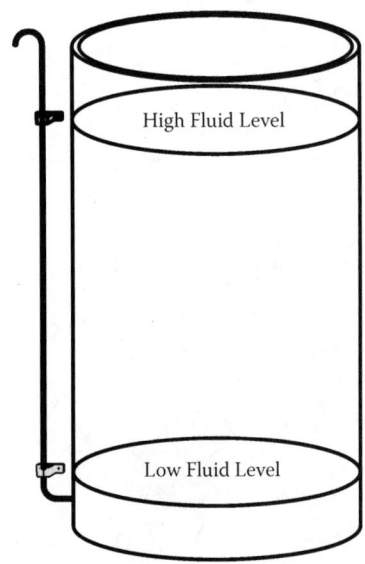

FIGURE 18.5 High and low fluid level identification utilizing an OPB350 fluid sensor.

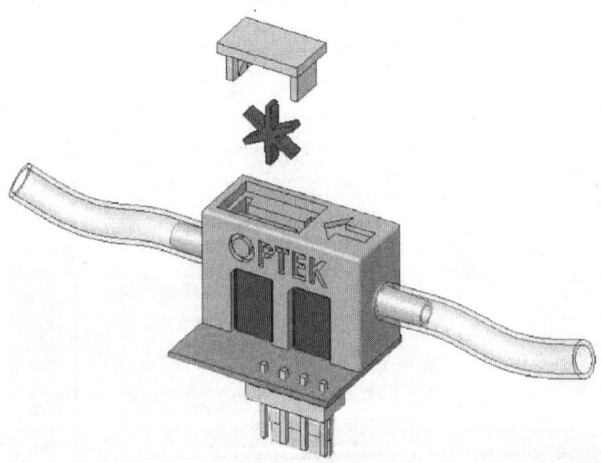

FIGURE 18.6 Flow sensor with bubble sensor.

Industrial Applications

of these functions can be accomplished with a slotted switch. The flow sensor may be either an analog or logical device, whereas the bubble sensor must be an analog device to be able to identify the difference between the no-fluid, fluid, and bubbles. In some cases, additional electronics will be added to optimize the calibration and operation of the system.

18.3　SECURITY AND SURVEILLANCE SYSTEMS

Figures 18.7 show a interruptive/transmissive system utilizing a single laser emitter and sensor. Both are in the same assembly. The light beam utilizes a reflective mirror to redirect the light signal around an area to be protected. The interruptive/transmissive systems may be used up to 500 linear feet while the reflective / reflective systems are limited to much less distance.

Figure 18.8 shows a reflective or proximity sensor system. Both configurations have all of the electronics located at a single point for each sensor location. Proximity sensors will identify any environmental change and may provide false signals due to animals, trash, or even quick temperature shifts in the sensitivity area. Reflective sensors send out a light signal and look for a reflection from any object it strikes. Reflective sensors typically look at a smaller area than a proximity sensor. Proximity sensors monitor the environment while looking for very small changes in the area being monitored while providing one output signal level for a static condition and another for a changing environment.

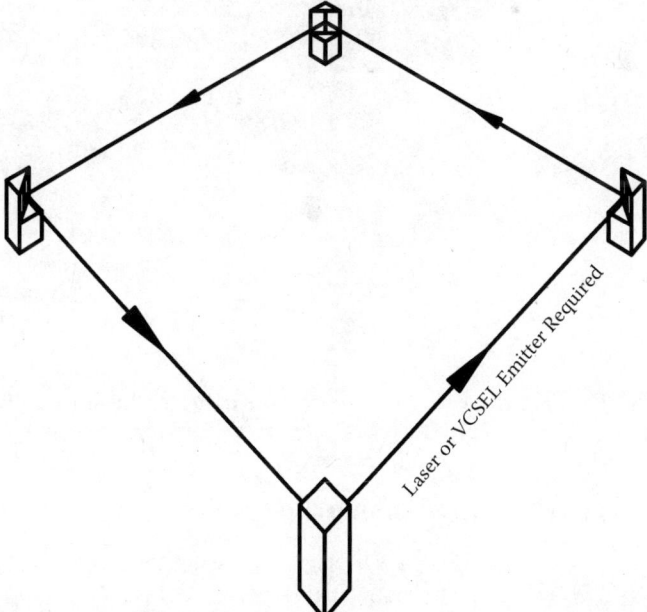

FIGURE 18.7　Security surveillance using a laser, mirrors, and sensor.

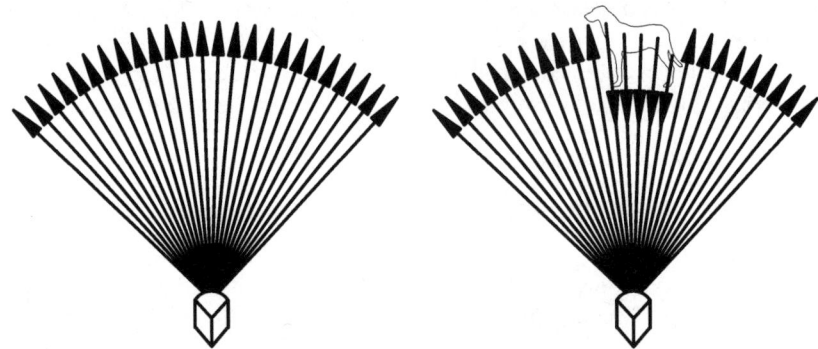

FIGURE 18.8 Security surveillance using a reflective or proximity sensor system.

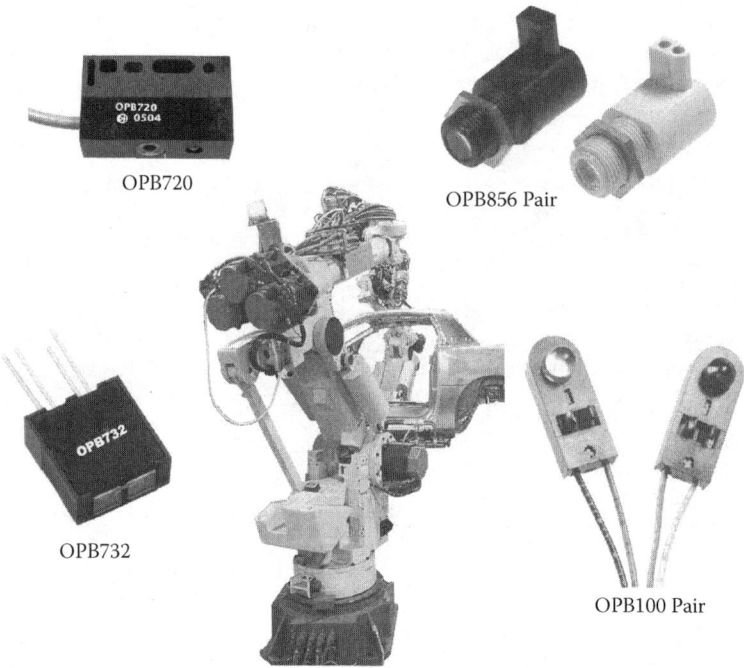

FIGURE 18.9 Robots may use reflective devices similar to the OPB720 or OPB732 for initial distance referencing or OPB100 and OPB856s pair for interruptive requirements.

18.4 MECHANICAL AIDS, ROBOTICS

Figure 18.9 shows a robot and a variety of optoelectronic components, including cameras, reflective switches both analog output (OPB732) as well as digital (OPB720), single- and dual-channel slotted switches (used as position or direction encoders), and individual components pairs (OPB100 or OPB856), which may be

Industrial Applications

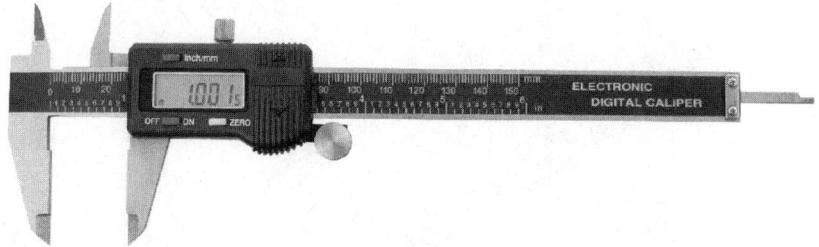

FIGURE 18.10 Digital caliper may use reflective switch for encoder application.

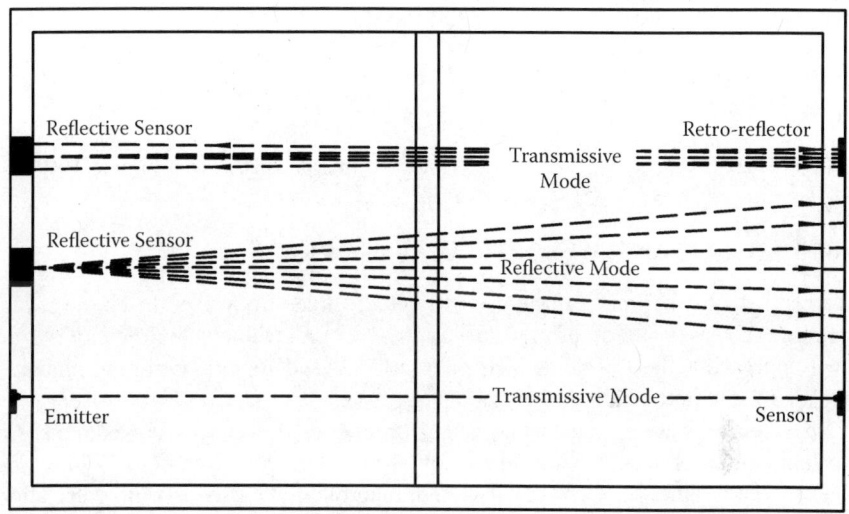

FIGURE 18.11 A window with three different types of optical recognition of an object interrupting the light signal, emitter, and sensor transmissive mode, reflective sensor reflective mode, and reflective sensor with retro-reflector transmissive mode.

used as beginning or end of travel and limit position indicators. Calipers may use a dual-channel reflective switch to determine both direction of travel and relative position. Normal design resolution is 0.0005 in. (0.0127 mm). Some encoder design applications have been able to improve this resolution (see Figure 18.10).

Figure 18.11 shows the use of an emitter and sensor pair located on either side of a entry, a reflective sensor and a reflective device reflecting a light signal off a retro-reflective material to identify the presence of something within the area being monitored. These solutions utilize an interruptive mode in which the light beam is broken, thus allowing the sensor to recognize an object or reflective where the light is reflected off an object. Both analog and logical sensors can be used for these applications. Examples of where these solutions can be used are entry to stores, conveyer belts, windows, doors, etc.

FIGURE 18.12 Typical optical couplers/isolators with isolation voltage from 1 to 50 kV.

18.5 MISCELLANEOUS OPTICAL SENSORS

Interruptive type sensors can be used to detect thread breakages in textile mills. A reflective sensor can be used to turn on a sewing machine when the cloth to be sewn approaches the needle. An encoder can be used for programming machine tools and for remote reading of gas pumps.

Photosensors have been used to speed up production in factories. For example: An infrared emitter is placed inside a beer can as an external array of sensors looks for holes or leaks. This is a much faster system than the previously used pressurization of the cans.

Industrial smoke detectors are similar to the ones used in the home. An IRED is pulsed and projects its energy into a chamber. When smoke particles are present, the energy is reflected to a photosensor, which trips an alarm when the level is exceeded. The industrial unit may actually measure the amount of smoke.

18.6 OPTICAL COUPLERS/ISOLATORS

Figure 18.12 shows some example of optical couplers/isolators. Isolation of power sources has been popular for many years. The optical coupler/isolator is the best way to get information from one power system to another. The optical coupler/isolator base components are a light-emitting device and a photosensor. The light emitting device sends an optical signal across an open space to the photosensor and transfers the information in either an analog or digital format. When the information is an analog signal, the ratio of input current referenced to the $I_{C(ON)}$ current of the photosensor is called current transfer ratio (CTR). This ratio is simply the

$I_{C(ON)}$ current of the photosensor divided by the forward current of the light-emitting device (I_F).

$$\text{CTR} = \frac{I_{C(ON)}}{I_F} = \frac{2}{10} = 20\%$$

With 10 mA driving the LED, we get 2 mA on the photosensor, providing a 20% CTR.

19 Automotive Applications

19.1 EXISTING APPLICATIONS

Automotive applications for automobiles continue to change as the market identifies new ways to use light-emitting and light-sensing devices (Figure 19.1). Optical devices have been used almost everywhere in the automobile, including in the engine compartment. The most promising areas for lighting in new automobiles are for all the exterior and interior lighting and light-sensing requirements. As the light intensity of devices increases, headlights will become more efficient and, by changing the white light wavelength, better identification of objects in the area of the vehicle can be made. Existing headlights utilize either a warm white (2,700 K to 4,000 K) or cool white (7,500 K to 11,000 K). As high brightness daylight (5,600 K) becomes available, objects should be more recognizable at night.

Figure 19.2 shows either a headlight or taillight assembly that can be illuminated using LEDs. The taillight would use an LED with a low-current mode for running and increase the current to enhance the light intensity for braking. When configured for a headlight, the emitter would be a group of LEDs on a heat-dissipating material to manage the thermal considerations.

Interior lighting is a prime location for LED lighting, given that most LEDs are directional lighting systems. LEDs are smaller, more efficient, and can be mounted much more easily than light bulbs. Heat-related problems for most high-power LEDs are solved by reducing the current drive while providing substantial area lighting.

Door closure is an important feature for any automobile and optical slotted switches provide a high-reliability solution. Figure 19.3 shows a door sensor design utilizing a slotted/transmissive switch.

With the addition of data transmission over plastic fiber, the need for large wire harnesses is eliminated. The only interface needed between systems in an automobile would be the plastic fiber and a positive-voltage wire. The voltage return would be the conductive main structure of the vehicle.

Some additional possible functions that could use optical devices are the following:

1. Door ajar (interruptive or reflective sensor)
2. Trunk lid ajar (interruptive or reflective sensor)
3. Hood ajar (interruptive or reflective sensor)
4. Oncoming headlight sensor (photo IC)
5. External lights operative (photosensor with possible fiber cable linkage)
6. Oil contamination (interruptive sensor with pulsed IRED and fiber optics to pick up point)

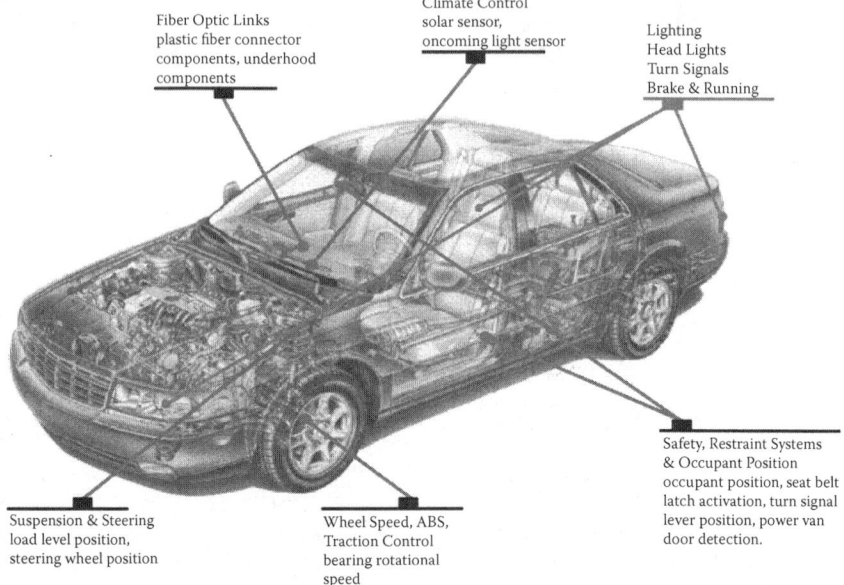

FIGURE 19.1 Automotive applications for optoelectronic devices.

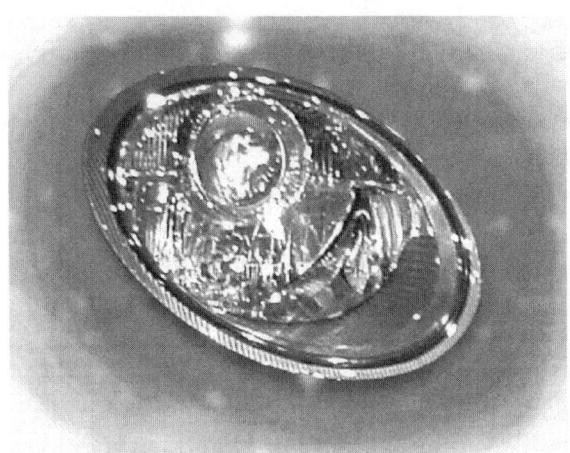

FIGURE 19.2 Headlamp or taillight assembly.

7. Short-range remote communication link:
   a. Releasing door locks by coded signal
   b. Communication within passenger compartment: door locks, window raise or lower, seat adjustment, mirror adjustment, other remote switch

FIGURE 19.3   Door closure optical assembly.

       control (rear window defoggers), hood opener, trunk opener, radio speaker sound transfer
- c. Data link within steering column for steering-wheel-mounted displays and controls
8. Long-range remote communication link:
   a. Car starting
   b. Air conditioning or heater on/off
   c. Windows up or down
   d. Lights on (locating of car or safety for intruders)
   e. IRED (radar viewing) used to locate vehicles in viewing dead zone
9. Sensor for determining driver drowsiness
10. Load leveler
11. Sensor to determine presence of glare ice on road
12. Sensor for determining buildup of ice in wheel well on interstate to prevent limited turn radius when exiting

The foregoing list is by no means all-inclusive. For each possible application, cost–benefit trade-offs will determine the actual feasibility.

20 Military Applications

20.1 MILITARY APPLICATIONS FOR OPTICAL SENSORS

Component requirements for military applications have changed (Figure 20.1), as commercial off-the-shelf (COTS) and commercial off-the-shelf with data and traceability (COTS Plus) are proliferating throughout the industry. The requirement for Qualified Products List (QPL) or Joint Army Navy (JAN) products is diminishing. COTS Plus products can be found in land, air, sea, and space applications today. Optical requirements have increased from near-infrared emitters and sensors to include visible and ultraviolet (UV) applications. Critical components may still require either JAN or COTS Plus products, whereas noncritical locations can use COTS devices.

A simple locking mechanism on a door of a spacecraft must be closed and secured to prevent loss of air. In this case, a reflective or interruptive switch can be used to ensure that the locking mechanism is in the proper position. Another example is the isolation of the power sources in the upper turret of a tank from the lower body while still retaining the ability to communicate between the systems. This can be accomplished with an optical air link using a series of near-infrared LEDs and photosensors system.

Actuation systems address the precise motion control requirements of spacecraft, aircraft, satellites, and missiles (see Figure 20.2). Functions requiring motion control include the following: optical filter selection, camera control, antenna positioning, fin positioning, solar panel orientation, and door open/close mechanisms. Actuation systems can use a combination of linear and rotary actuators to perform their motion-controlled functions. A slotted optical encoder assembly can be used to indicate the position of a rotary shaft, and a simple reflective or transmissive switch can acknowledge closure or position.

Shaped charges are placed around a hypersonic missile. Flight patterns can be changed by selective explosion of these charges. Optical couplers are used to isolate these firing circuits from the control electronics (see Figure 20.3).

Optical couplers (2N, 4N, and OPI series) are also used in switching power supplies in both ground control and flight hardware. They are used to isolate the control circuitry from different power circuits as well as minimize current spikes that may cause false data signals, causing the system to shut down or go through an initiation cycle.

The need for panel light brightness control in aircraft is far greater than in automobiles, both due to the number and type of instruments and a more rapidly changing environment. Using red, green, blue, and yellow (RGBY) LED arrays allows the cockpit lighting to be changed to optimize the ability of the pilot to read the instruments with a wide variety of ambient lighting conditions.

FIGURE 20.1 Military aircraft and tanks utilize a variety of optoelectronic devices for data transmission, power supply isolation, and cockpit visibility.

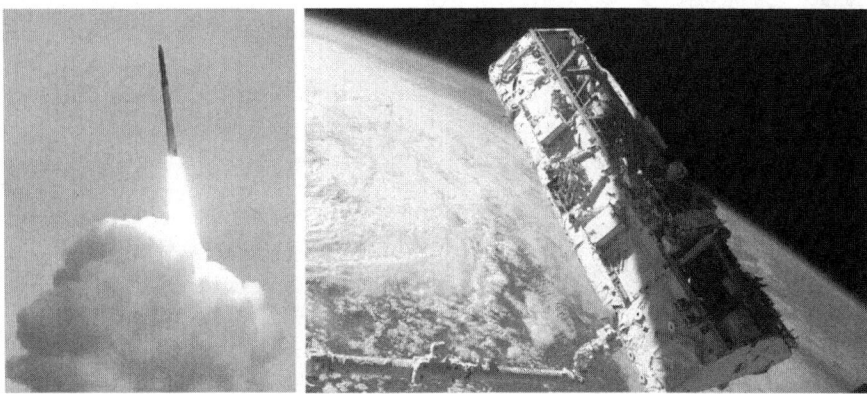

FIGURE 20.2 Spacecraft and satellites utilize a variety of optoelectronic devices for data transmission, power supply isolation, and cockpit visibility.

Other unique applications in the military environment include the following:

1. Near-infrared LED (NIR-LED) illumination for helicopter landing or running lights. A TO-46 NIR-LED operated at 100 mA and used with infrared conversion goggles will generate adequate energy to read a newspaper in a conference room with no other lighting. Such devices are also critical for blackout formation flying when used as running lights.
2. PIN quadrant diodes are used as homing sensors for guidance control. The photosensitive area is divided into four quadrants and, by use of a finely tuned infrared spot size, can perform automatic guidance. The control system forces the spot to the center or zero signal area.
3. Fighter plane pilots in high-gravity environments can utilize a reflective sensor mounted adjacent to the eye for supplemental aircraft control. By moving the eye, the pilot can generate coded signals, which can then be used to control certain aircraft functions.

Military Applications 247

FIGURE 20.3 Hypersonic missiles can use space charges for change of direction, with optoisolators providing the signal necessary to ignite the charge.

4. A visible LED array mounted on a vibrating member can generate signals similar to a miniature TV screen on a localized area of eyeglasses. This "head-up display" information could be used to keep the pilot informed of basic gauge position when he or she is not watching the instrument panel.

5. Digital altitude information can be obtained from an encoder mounted within the normal altimeter. This information could be read from a ground station without direct communication to the pilot.
6. The optical gyroscope is a unique design based on optoelectronic light signals to perform a gyroscope function.
7. Exterior light recognition. Military and civilian aircraft are required to provide a light signal visible for up to 10 miles in clear air space. These light arrays are made using high-power LEDs, both visible and near-infrared. The visible signals can be turned off for military zones while still providing a near-infrared signature to identify friendly or enemy aircraft.

21 Consumer Applications

21.1 TV AND GAME CONTROLS

Remote control of toys or TV games with optical sensors is quite common. The signals are typically transmitted using a near-infrared LED. The scheme used to transmit the information from a remote control to the receiving system such as a TV, DVR, stereo system, etc., is defined by the FCC. These schemes include the optical power levels as well as the code sequence for each key on the remote keypad or joystick. In general, a near-infrared emitter is pulsed with a high current transmitting a specified pattern of light pulses while the photosensor looks for very small changes in the ambient light level. When it recognizes a change, the following electronics checks the pattern and determines if it is a viable pattern. Different patterns are used by each manufacturer, allowing multiple remote or game controls to be used in the same room.

The basic operating principles of a joystick have changed; Figure 21.1 shows some principles that could be used for transmissive operation. Three or four transmissive- or reflective-type sensors are located at equally spaced intervals around the center point of the joystick. When the mechanism is moved in any direction, the amount of transmitted or reflected signal changes thus relates to the new position of the lever. If a small area detector is used, lensing may be required. This technique allows the system to identify the direction and amount of pressure applied. Another technique is to use interruptive-type sensors, locating them at the four compass points of the lever (see Figure 21.2). As the lever is moved, one or two of these infrared beams are broken. The direction of movement of the lever can be interrupted and the appropriate action can be taken. This scheme allows for direction but not force (speed).

The trackball (see Figure 21.3) operates with two interruptive sensors that look at rotating encoder wheels turned by the ball using the movement rods. In order to identify the direction of movement for the trackball, the optical sensor should have two photosensors and may have two emitters (dual-channel interrupter). The dual-channel interruptive sensors are located for "X" and "Y" movement and allow quadrature signaling to be generated, which signals forward or reverse movement of the ball. Counting the pulses per unit of time for both the "X" and "Y" channels provides the speed of rotation.

21.2 CD AND DVD DISCS

The laser beam is reflected from the grooves to recognize the signals written into the disc. The change in reflected signal allows the tracks to be identified as different signal conditions (high-"1" or low-"0"). A vary small variation for the reflected

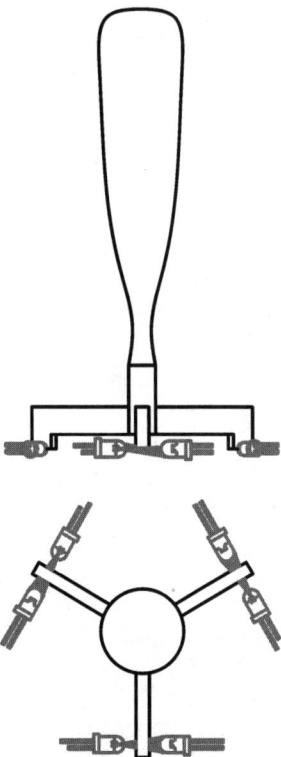

FIGURE 21.1 Joystick optical sensor. Both reflective or transmissive conditions can be used for this application. This configuration shows how a transmissive configuration is designed. The amount of current in each photosensor identifies the position of the joystick.

signal allows the system to identify if a "1" or "0" has been recorded. By narrowing the laser beam, more tracks and data can be recorded on a single medium.

21.3 DOLLAR BILL CHANGERS

Many moneychangers operate with an optical sensor principle. Calibrated near-infrared LEDs and UV LEDs with the appropriate photosensor are focused at key points on the bill. A bill is validated by the correct transmission of light from these key points as shown in Figure 21.4. Sophisticated currency validation equipment can identify the denomination and authenticity for different types of currency, U.S. $, British £, or European €, etc.

21.4 COIN CHANGERS

A coin changer offers a number of different applications for interruptive-type optical sensors or arrays. The coin changer must identify the difference between nickels, dimes, quarters, half dollars, and dollars for the United States and similar configurations for

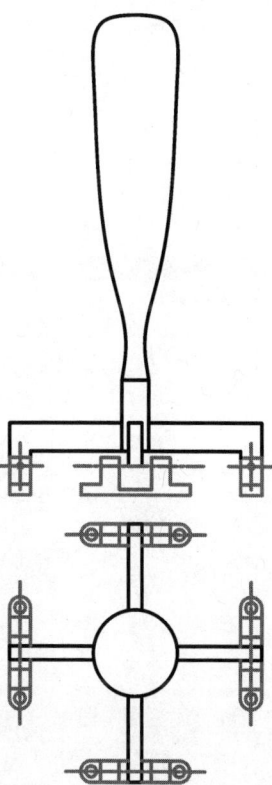

FIGURE 21.2 Joystick optical sensor. In this version, we use a slotted switch to identify if the joystick is moved in one of eight positions. The interruption of the light signal shows movement toward that switch.

other countries. In order to control the availability of coins for making change, coin "empty" and coin "full" sensors are required for each denomination. When the change slot is full, a coin diversion sensor is required to divert the coin to the reservoir box. A series of sensors can be used to read the elapsed time from leading edge to trailing edge and thus identify precisely the size of the coin. An additional sensor is used to establish that the coin has dropped and stayed in the machine. If the appropriate sequence of sensors is not correct, the machine will identify this as a problem and will open the coin return path and not perform the intended function. After a predefined time, the machine will reset and wait for identification of additional coins.

21.5 SMOKE DETECTORS

Smoke detectors are designed with either an ionization detection or optical detection scheme. We will discuss two principles that could be used in optical smoke detectors. Most optical smoke detectors use a near-infrared LED and photosensor 90° out of position with each other. This technique involves examining reflection of

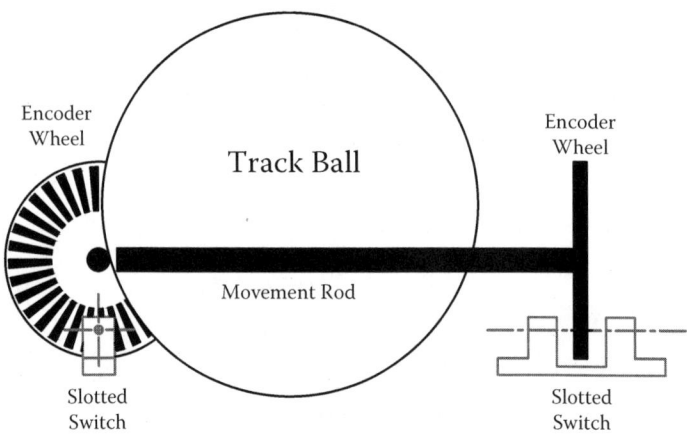

FIGURE 21.3 Trackball optical sensor. The trackball rotates the movement rod, identifying the direction of travel for the ball. As the rod is rotated, an encoder wheel breaks the beam between the emitter and photosensor in the slotted switch, identifying movement of the rod.

FIGURE 21.4 Bill with validation sensor.

the light from the LED off the smoke particles to the photosensor, and will identify large amounts of smoke (see Figure 21.5). The smoke detector efficiency can be greatly increased by changing the photosensor from a 90° reflective to a transmissive configuration (see Figure 21.6). This technique would allow the use of a photodiode

Consumer Applications

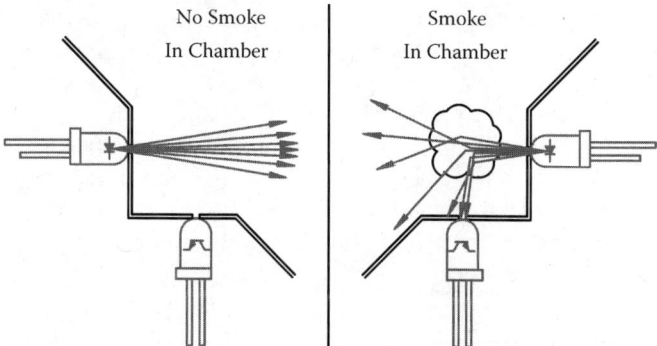

FIGURE 21.5 Optical smoke detectors' reflective mode. The principle of operation is one of the simplest optoelectronic applications. The light is emitted from an LED and as particles of smoke redirect the photons toward the photosensor, the smoke is detected.

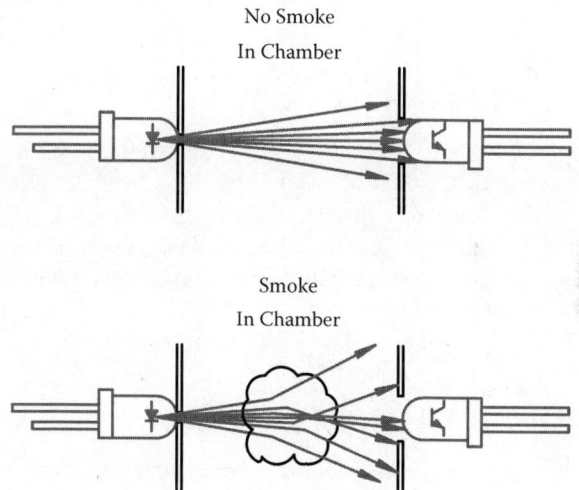

FIGURE 21.6 Optical smoke detectors' transmissive mode. The light is emitted from an LED directly to the photosensor. As particles of smoke redirect the photons, less light is seen by the photosensor and the smoke is detected.

with a transimpedance amplifier gain circuit and very low forward current for the LED; very small changes of the photodiode can be identified, providing a sensitive smoke detector.

21.6 SLOT MACHINES

Slot machines have coin sensors that work as noted in Section 21.4 as well as credit card sensors and encoders that provide the exact position of the rotating wheels. The credit card sensor utilizes either a reflective or transmissive device to recognize that the card has been properly inserted into the slot and starts the system looking for the magnetic signature. A magnetic sensor is used to read the magnetic strip as the card is pulled out, providing the unique code for each card. In addition to both of these applications, after either a button on the front panel or lever on the side of the machine is initiated, a microprocessor identifies a random solution and the wheels start rotating or display starts changing. If wheels are utilized, each wheel needs to be rotated several times until the proper solution is presented to the customer, showing either a winning or losing result. The exact position of each wheel is identified using an optical rotary encoder on the wheel.

21.7 CAMERA APPLICATIONS

Cameras are continually changing, and utilize more and more optoelectronic devices. A photosensor can be used to identify the amount of ambient light, thus determining if the flash needs to be used. The flash can be an LED with a center wavelength in the 5000 to 6000 K range. This will provide what is considered a "Daylight White" light, showing the best reflection for all the standard visible colors seen by the human eye.

21.8 OPTICAL GOLF GAMES

Interruptive- and reflective-type optical sensors can be used to measure the precise velocity and direction of a golf ball struck by a club. Pictures taken from various points on a golf course are shown on a cloth screen from the back. The golfer drives the ball, and optical sensors measure the velocity and direction. The ball then strikes the screen, while the computer calculates where the ball should come to rest. The appropriate picture then comes up. The ball is replaced on the tee, and the golfer swings again. The cycle repeats until all 18 holes are completed.

21.9 HOUSEHOLD APPLIANCE CONTROLS

Many household appliances are switching to LED lighting, glass or cup sensors, icemaker full sensors, and freezer frost sensors. The LED lighting replaces bulbs, reducing power consumption as well as heat. The LED replacement may not radiate as much area as the light bulb but with the reduction in power and heat, the LED is the preferred light source. Reflective switches can be used as a freezer frost control by looking at reflective light from a small nonfocused switch. As the frost builds up in the freezer, light will reflect from the ice on to the sensor of a reflective switch. When enough light is reflected back to the sensor, the defrost cycles is started. LED and photosensor pairs can be used to look across the icemaker and determine when the tray is full of ice by interrupting the light signal across the bin.

22 Medical Applications

22.1 INTRODUCTION

Optoelectronic devices have been used in medical applications for years, and advancements in optoelectronics have opened up many new application areas. Utilizing transmissive properties of body fluids, as well as transmissive properties of the human body allows the designer to recognize conditions today that could not be detected earlier.

22.2 PILL-COUNTING SYSTEMS

Counting of pills and other items is performed either by a person or an automated system. Figure 22.1 shows the light pattern for emitter and photosensor pairs spaced evenly around the funnel area in an interruptive/reflective mode for a funnel-type pill-counting system. When light shines from the emitter, it spreads out and bounces off the walls, illuminating more than one photosensor. Looking at very small changes in the total current of the photosensors allows the system to see pills anywhere in the funnel (see Figure 22.1). The major concern for any automated system counting pills is when two or more pills pass through the sensing area at the same time. If the system looks at each photosensor separately and as long as the pills fall through the opening such that one pill is offset from the other by at least half the pill size, the system could recognize the optical difference and count both pills.

22.3 ELECTRICAL ISOLATION SYSTEMS

Protection of the patient from electronic monitoring equipment with high voltages can be achieved with the use of optocouplers/optoisolators, discussed in Chapters 10 and 11. These devices allow separation of power supplies of different systems while allowing information to be transferred between the systems. This may be used for isolation of patient beds from other equipment, different pieces of equipment, or high-voltage defibrillators.

22.4 INFUSION PUMP APPLICATION

Infusion pumps are used for delivery of small quantities of drugs over long periods of time and for precise delivery of I.V. medication in critical medical care. They administer fluids in ways that would be prohibitively expensive or unreliable if administered manually by the nursing staff. These devices are also commonly called *syringe pumps*. As a worm gear, or shaft, slowly turns, it moves the plunger of a syringe, pushing the medication out (see Figure 22.2).

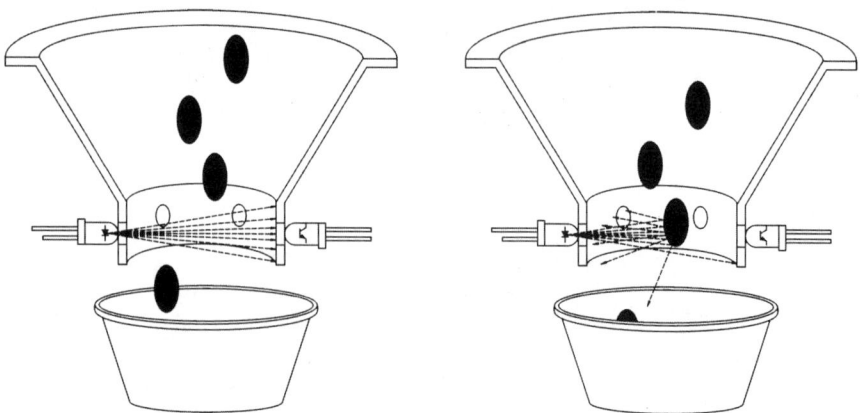

FIGURE 22.1 Pill-counting system using optoelectronic components. Pill-counting system using optoelectronic components with and without a pill in the system.

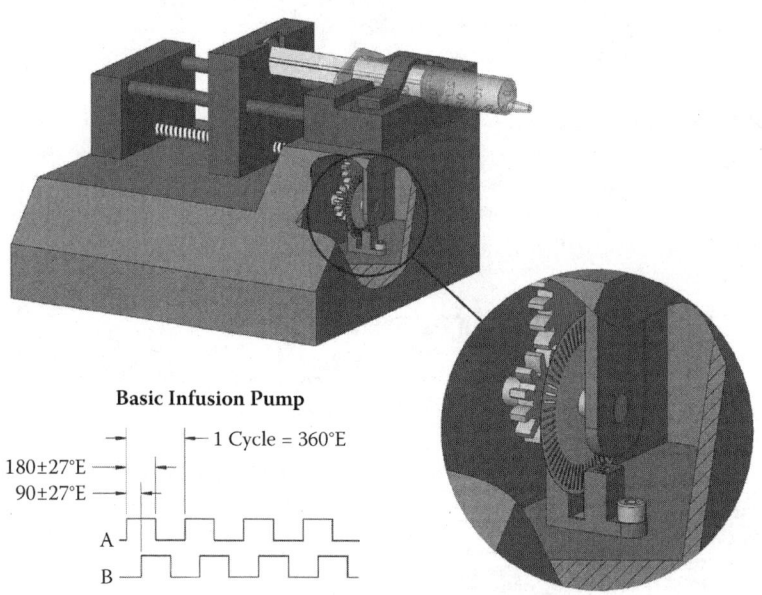

FIGURE 22.2 Typical infusion pump with optical switch counting revolutions and direction of movement.

An encoder wheel is driven by a set of gears from the rotating shaft. An optical sensor (encoder) precisely tracks the wheel movement with high precision using flags and holes. A dual-channel version allows the pump to be monitored for direction to identify insertion or extraction of fluids. Figure 22.2 shows how a dual-channel Photologic® or phototransistor output device, similar to the OPB822SD or OPB950, can be used to monitor the speed and direction of travel for the gears. A device similar to the OPB350 fluid sensor can be used to ensure a tube is in place with fluid in the tube.

22.5 HEMODIALYSIS EQUIPMENT APPLICATION

Hemodialysis is the most common method used to treat advanced and permanent kidney failure. In hemodialysis, the patient's blood is allowed to flow through a machine with a special filtration system that removes wastes and extra fluids. The clean blood is then returned to the patient's bloodstream. The hemodialysis instrument has three main jobs: pump blood and monitor blood flow for safety, clean wastes from blood, and monitor blood pressure and rate of fluid removal from the patient's body. An air and/or bubble sensor similar to the OPB350 checks the blood in the tubing to ensure that air does not get into the bloodstream. The same type of sensor can be use to detect the presence or absence of other fluids in the machine, such as purified water and other fluids. Figure 22.3 shows some of the locations where a fluid sensor can be used.

22.6 FLUID AND BUBBLE SENSING FOR MEDICAL APPLICATIONS

With optoelectronics, the designer can identify the following states: presence of a tube in the sensing assembly, whether a fluid is present in the tube, whether bubbles are present in the fluid and, in some cases, the type of fluid in the tube. In most situations, this is a very simple process, accomplished by evaluating the output level of an

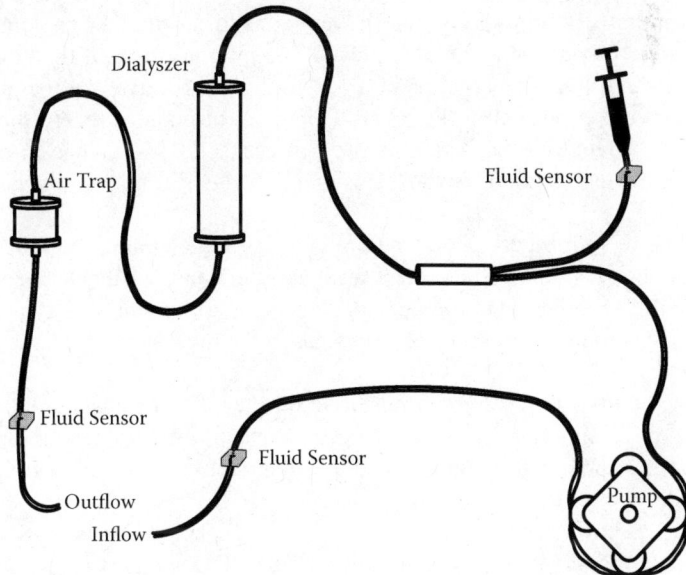

FIGURE 22.3 Typical hemodialysis flow with locations for fluid/bubble sensors.

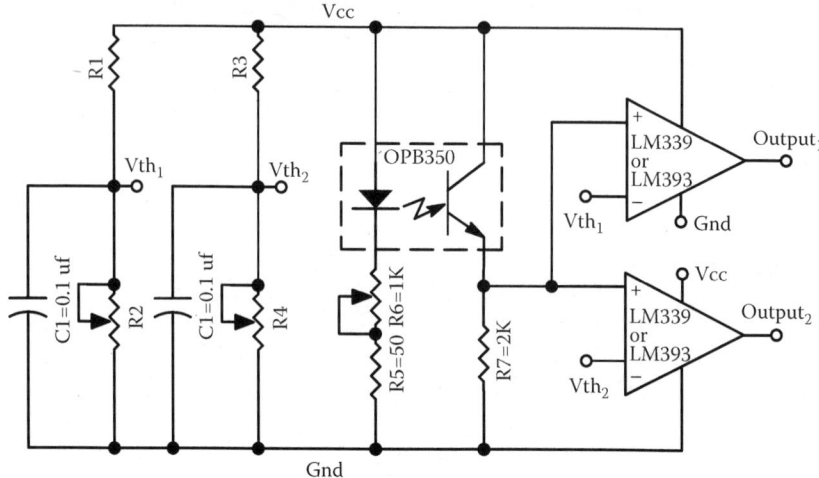

FIGURE 22.4 Typical schematic used for identification of a tube, bubbles, or fluid using discrete components.

optoelectronic sensor for each different expected condition by using a microprocessor or defined electrical circuitry (see Figure 22.4).

Optoelectronic devices consist of a light-emitting diode (ultraviolet, visible, or near-infrared led) and photosensor (photodiode, phototransistor, or Photologic device). The LED emits a light at a specified wavelength, and the photosensor receives the light and provides an output signal to be evaluated. As light is transmitted from the LED through a tube that contains air or another gas, the photosensor receives photons at a consistent level. The internal walls of the tube reflect some of the light, thus reducing the number of photons received by the photosensor. When a fluid is present in the tube, the number of photons received changes, resulting in a change in the output of the photosensor. This change in output is what allows the user to identify the states mentioned earlier as well as the type of fluid in the tube.

As can be seen in Figure 22.4, when monitoring the output current of the phototransistor across a resistor, a voltage level can be identified for different output levels by utilizing either a microprocessor or comparator circuit. The typical states that can be identified are presence of a tube, air in the tube, saline solution, or some hematocrit levels.

With the ability to vary the current drive on the LED, or change the sensitivity or gain of the photosensor, critical measurements of minor optical transmission changes of the fluid can be identified and different fluids such as hematocrit % can be detected.

Figure 22.5 shows some expected output levels of a phototransistor with different percentages of blood in a fluid (hematocrit %). The left line on the graph is with the lowest LED current, and the right line shows the highest LED current level. The user will have to evaluate different tube types to provide a suitable signal level.

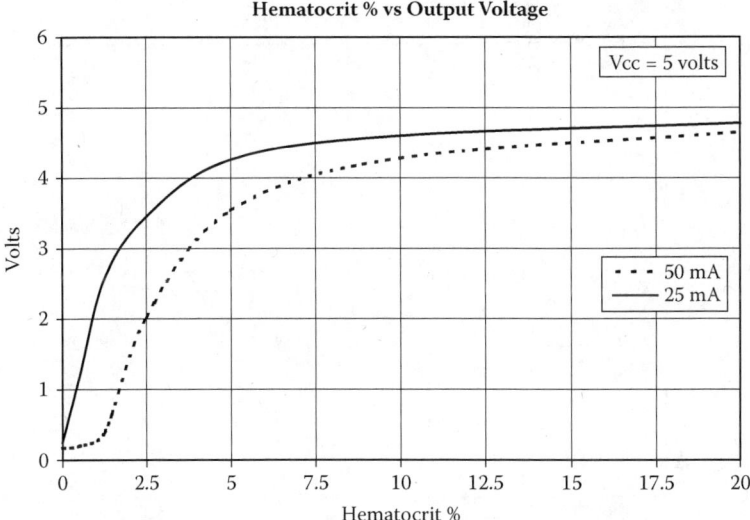

FIGURE 22.5 Typical hematocrit % versus output voltage using a fluid/bubble sensor and 1/16 O.D. tube.

This all sounds great. Now, what are the areas of concern? Different types of tubes provide different amount of light through the tube (optical transmission of the tube), and not all devices are manufactured identically. The power of the LED changes as well as the gain of the photosensor. Temperature also changes the expected output levels. The LED's age, and so light emission decreases over time. (You can expect 5% to 20% degradation over 100,000 hr.)

The following are solutions to these concerns:

Identify the type of material for the tube to be used that will keep the optical transmission properties consistent.

With the ability to use microprocessor or other sensing device across the phototransistor load resistor, the designer can program the current drive of the LED to provide the expected output level for a known start-up state. This is typically without a tube in the unit; thus, setting the base point and all other output levels are identified as a percentage of this level (see Figure 22.6 for some typical output ratios).
Example 1: Dry Tube = 2.9 and No Tube = 1.9, giving a ratio of 2.9/1.9 = 1.53
Example 2: Saline = 1 and No Tube = 1.9, giving a ratio of 1/1.9 = 0.53
Example 3: Blood = 4.9 and No Tube = 1.9, giving a ratio of 2.58.

By adding a start-up test of a clear tube, you can compare the percentage change in output, thus checking the type of tube being used.

The temperature coefficient of the LED goes in the opposite direction than the phototransistor, thus reducing the effects of temperature change. Also, the expected use of this device for medical applications reduces the probability

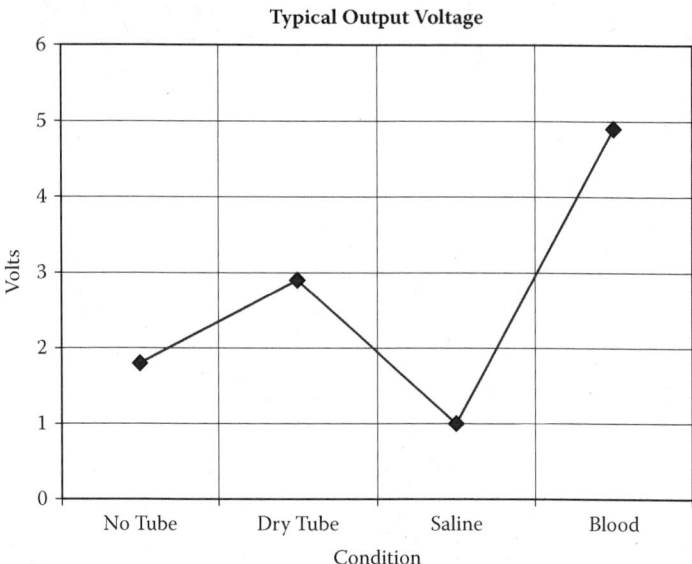

FIGURE 22.6 Typical output voltages for no tube, tube, saline solution, and blood using a fluid/bubble sensor.

of large temperature extremes, and calibrating the device just before use almost totally eliminates this effect.

The typical LED life expectancy of an LED driven at 20 mA over 100,000 hr of continuous operation is about 20%. This is over 10 years of continual operation, and by calibrating the device just prior to use, this problem is also eliminated. Lowering the LED drive current reduces the degradation of the system.

22.7 INTRAVENOUS DROP MONITOR

Intravenous injection rate can be easily controlled by a transmissive or interruptive sensor. The curved leading edge of the drop bends the IR energy away from the sensor. By monitoring the output for a given time interval, rate of fluid drops can be monitored because the amount of light striking the photosensor changes as the drop passes by, thus making it possible to very accurately monitor the size and duration of drops of the fluid, as shown in Figure 22.7.

22.8 PULSE RATE DETECTION

Two configurations are used today to monitor the pulse rate of a patient (Figure 22.8). By locating a reflective sensor adjacent to a blood vessel, a change in the blood vessel, when the heart pumps causes a change in signal level to the reflective sensor.

Medical Applications

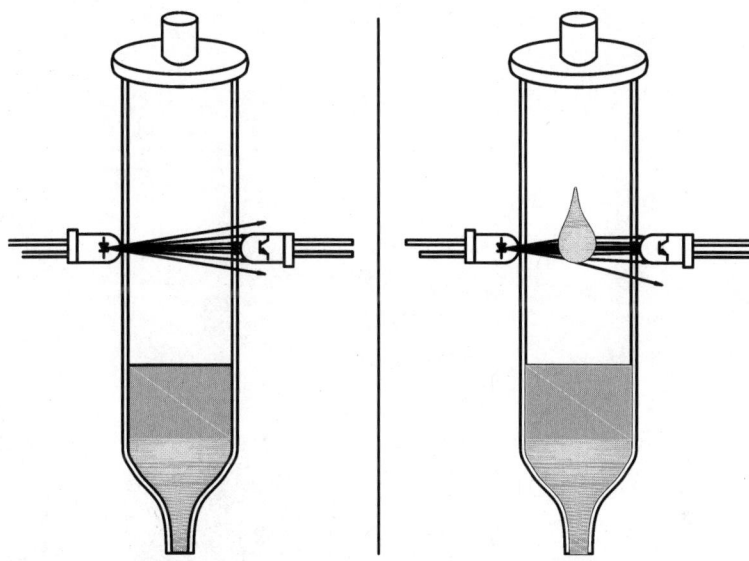

FIGURE 22.7 Intravenous drop monitor.

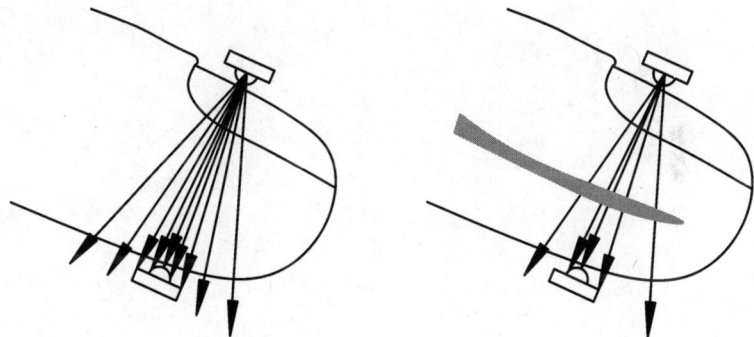

FIGURE 22.8 Pulse rate detection.

Because near-infrared light penetrates several centimeters into the skin, this system can be used anywhere that a blood vessel is near the skin of the patient.

Fingertip pulse rate monitors utilize an interruptive monitoring technique in which an LED is placed on one side of the finger and the other side is monitored with a photosensor. As the photosensor output is modified owing to the change in transmissivity of the blood vessels, the pulse rate can be easily calculated. Because of the unique requirement of having to place the devices on both sides of the fingertip, a flex circuit is used. Both techniques can utilize visible as well as near-infrared-emitting devices.

23 Telecommunications

23.1 TELECOMMUNICATIONS OVERVIEW

Telecommunications is a very broad topic, and we will cover some basic areas in which optoelectronic devices are used. We will cover some of the basic areas in which optoelectronic devices are used. To begin understanding this area, we need to know the seven layers of Open Systems Interconnection (OSI), which is the basic model for describing communications and networking protocols for telecommunications (see Table 23.1).

With the inclusion of phone and video signals using the same medium, we need to include the broadband network of the television industry. All the data are now being transmitted using fiber-optic as well as wire and wireless systems. The broadband and telecom industries are divided into several segments. Table 23.2 shows most of the areas where optoelectronic devices are used.

These industries are further broken down into distance segments, with the local area network (LAN) being the largest industry and wide area network (WAN) the second in utilization of equipment. Long haul is the third segment; covering the entire world is limited to only a few providers.

A local area network can range from a single building to a large campus.

A wide area network typically encompasses a small city to a large metropolitan area.

Two basic configurations of fiber-optic cables are available in the industry, plastic and glass.

Plastic optical fiber (POF) is made with a core (low loss area) and jacket (protective covering), and is designed to have the lowest attenuation (optical loss) at 650 nm wavelength. It is typically used in controlled environments with relative short distances (100 m or less) and data rates no higher than 100 Mbps.

Glass fiber-optic cables are broken into two groups, multimode and single mode. Table 23.3 shows some of the basic properties needed to identify which cable should be used in which application. A glass fiber-optic cable is made with a core (low loss member), cladding (high loss member), and jacket (protective covering) and is either single mode (designed for use with lasers) or multimode (designed for use with LEDs). As an example, the multimode fiber is referred as 50/125 consisting of a 50 micron core and 125 micron cladding, whereas a single-mode fiber of 9/125 references a typical core of 9 micron and cladding of 125 micron diameter. The jacket material is not normally referenced in this part of the specification. The typical temperature range for a glass fiber-optic cable is −40°C to +85°C.

In order to properly interface with an optical emitter and sensor, the cable must be inserted into a connector. Connector types continue to change. Some basic connector configurations are ST, SC, FC, LC, and MTRJ. The industry original connector types of SFR and SMA are rarely used today.

TABLE 23.1
The Seven OSI Layers

Layer	Name	Type of Optical Equipment
7	Application	None
6	Presentation	None
5	Session	None
4	Transport	Long-distance communication—Lasers and high-speed diodes with TIAs, single-mode fiber-optic cable
3	Network	Local area networks—Switches and routers that use VCSEL, LED, and diodes with TIAs
2	Data Link	Bridges and switches that use VCSEL, LED, and diodes with TIA
1	Physical	Fiber-optic cable, hubs, repeaters, etc., that use VCSEL, LED, and diodes with TIAs

TABLE 23.2
Application Areas for Optoelectronics in Telecommunications and Broadband

Technology	Speed Range	Optical Devices
Various	9.6–56 kbps	LEDs,
DS0	64 kbps	Photodiodes,
DS1/T-1	1.544 Mbps	Phototransistors,
E-Carrier	2.048 Mbps	Photologic®
T-1C (DSIC)	3.152 Mbps	
Token Ring/802.5	4 and 16 Mbps	
DS2/T-2	6.312 Mbps	VCSELs, LEDs,
E-2	8.448 Mbps	Photodiodes, Phototransistors
Ethernet, 10Base-F	10 Mbps	VCSELs, Photodiodes with TIA
E-3	34.368 Mbps	Laser Diodes,
DS3/T-3	44.736 Mbps	VCSELs,
OC-1/STS-1	51.84 Mbps	Photodiodes with TIA
Ethernet, 100Base-F	100 Mbps	Laser Diodes,
FDDI	100 Mbps	VCSELs,
T-3D	135 Mbps	Photodiodes with TIA
E-4	139.264 Mbps	
OC-3/SDH/STM-1/STS-3	155.52 Mbps	
E-5	565.148 Mbps	
OC-9/STM-3/STS-9	466.56 Mbps	
OC-12/STM-4/STS-12	622.08 Mbps	
OC-18/STM-6/STS-18	933.12 Mbps	

(Continued)

TABLE 23.2 (Continued)

Technology	Speed Range	Optical Devices
Gigabit Ethernet	1 Gbps	Laser Diodes,
OC-24/STM-8/STS-24	1.24416 Gbps	VCSELs,
SciNet	2.325 Gbps	Photodiodes with TIA
OC-36/STM-12/STS-36	1.86624 Gbps	
OC-48/STM-16/STS-48	2.48832 Gbps	Laser Diodes,
OC-192/STM-64/STS-192	9.95328 Gbps	VCSELs,
OC-256	13.271 Gbps	Photodiodes with TIA

TABLE 23.3
When to Use Single-Mode and Multimode Fiber-Optic Cables

Fiber size	Type	Light Wavelength	Minimum Bandwidth	Distance	Attenuation
62.5/125	Multimode	850 and 1310	200 MHz/km	@100 Mbps—2 km	~3 db/km
			500 MHz/km	@1 Gbps—220 m	~1 db/km
50/125	Multimode	850 (VCSEL) and 1310	2000 MHz/km	@100 Mbps—2 km	~3 db/km
			500 MHz/km	@1 Gbps—220 m	~1 db/km
9/125	Single mode	1310 1550	~ 100 THz (terahertz)	Up to 100 km	~0.40 db/km
					~0.25 db/km

In order to use a fiber-optic cable, it must be bent. Either the plastic or glass fiber-optic cable will withstand a relatively sharp bend radius. Fiber cable should have a bend radius no less than 15 times the outside diameter of the cable. It is preferable to have the bend radius as large as possible to minimize loss of signal.

Part 7

Visible and Ultraviolet Technologies

24 Visible-Light-Emitting Diodes (VLEDs)

24.1 INTRODUCTION

Visible light, that portion of the electromagnetic spectrum (400 to 780 nm) lying between ultraviolet (UV) and infrared (IR), allows us to see what we are doing, can excite us by imparting brilliant color to objects, and even change our moods (Figure 24.1). If you have ever experienced the aurora borealis or a fantastic sunset, you will understand. Fire (thousands of years ago), gas discharge (1907), incandescent bulb (1879), and even fluorescent tubes (1938) only allowed us to see in the dark and if these were filtered, color could be realized. Now, with visible LEDs (VLEDs), not only efficient white light, but even more brilliant colors can be experienced. When mixed together, millions of different colors (wavelengths) can be generated.

In 1907, a "curious phenomenon" was noted by H. J. Round when investigating current passage through SiC (carborundum). This highly impure non–P-N junction combination emitted a yellowish green light, and thus began the LED story. Like many great discoveries, several decades of research and development were required to commercialize this "curious" phenomenon. Here we are, 100 years later, with a disruptive and explosive technology that is about to replace incandescent and fluorescent lamps. Solid state lighting using VLEDs is also better for the environment.

Traditional lighting consumes about 30% of the electricity generated and 20% of the fossil fuel. So, replacement of conventional light bulbs and fluorescent tubes by energy-efficient and mercury-free solid-state visible LEDs will save multi-terawatt-hours of energy worldwide each year.

24.2 WHAT IS AN LED?

An LED, whether it is UV, visible, or IR, consists of a layered semiconductor chip that is packaged to facilitate maximum light output, allow for circuit connections, and provide basic thermal management of the semiconductor. The overall efficiency of a white LED is about 18% compared to 10% for incandescent and 22% for fluorescent tubes. The balance (82%) is lost as wasted energy in the form of heat (Figure 24.2).

24.2.1 CHIP (DIE)

AlInGaP semiconductor materials are used for red, red-orange, orange, and amber both in low- and high-power LEDs. The InGaN material is better suited for the UV, blue, cyan, greenish yellow, green, and blue source for white phosphor LEDs (Figure 24.3).

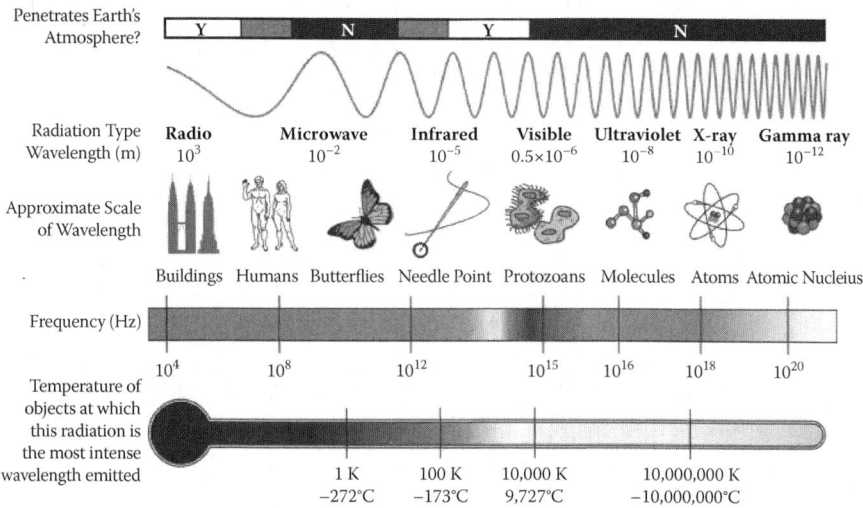

FIGURE 24.1 The electromagnetic spectrum.

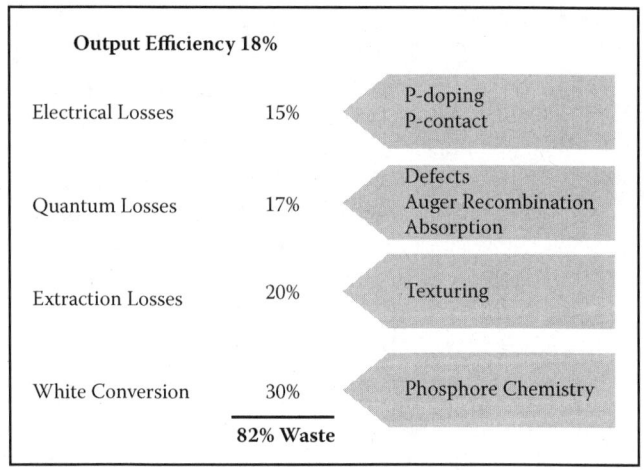

FIGURE 24.2 Efficiency losses in white VLED.

A typical InGaN chip can consist of as many as 8 layers, all aimed at maximizing performance. Chip layers are typically a sapphire (Al_2O_3) or SiC substrate, a metal reflecting layer, a thin proprietary gallium nitride (GaN) buffer, GaN epitaxy, and an InGaN active junction surrounded by layers of AlGaN cladding (Figure 24.4). Lastly, the semiconductor chip is capped with another GaN layer that has undergone specialized surface treatment to improve photon release from the structure. As the crystal structure in the InGaN active layer is grown, magnesium (Mg) atoms are doped into the mix as one part in 10,000, aiding in the generation of light.

Visible-Light-Emitting Diodes (VLEDs)

Material	Color	Wavelength (nm)
InGaN	Blue	470
	Cyan	505
	green	525
	Yellow-Green	570
	Yellow	589
AlInGaP	Amber	595
	Orange	605
	Red-Orange	615
	Red	625

FIGURE 24.3 Typical visible wavelengths.

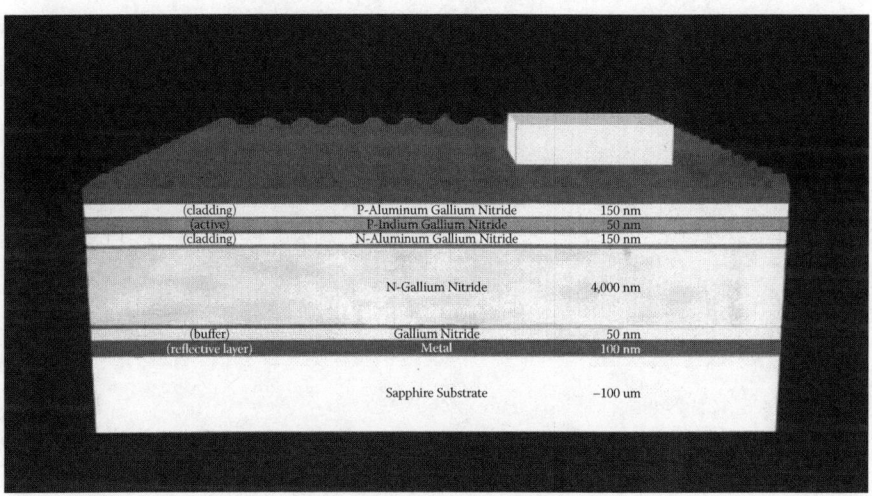

FIGURE 24.4 Layers of typical InGaN chip.

Die size determines the current capability and can be increased by how well the packaged VLED and overall system is thermally managed (Figure 24.5).

24.2.2 Packaging

The important packaging features are heat dissipation (thermal resistance), maximizing light extraction, and creating the radiation pattern (Figure 24.6). Because of this, new LED packaging and processes are developing all the time—about as fast as the dies are changing to attain the ultimate target of 200 lm/W. However, if the die is too bright, then the current can be reduced, dropping the power dissipation and, subsequently, the heat generated.

Size Type	Dimensions Square		Current (mA)
	eng (mils)	metric (um)	
small	10 to 14	250 to 350	20 to 50
medium	20 to 30	500 to 800	100 to 250
large	40	1000	350 to 1500
X-large	80	2000	1500

FIGURE 24.5 Die size versus typical maximum current applied.

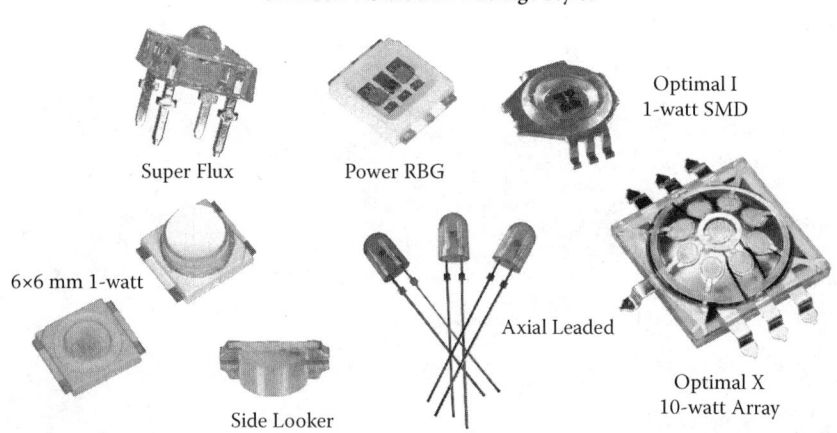

FIGURE 24.6 Common VLED package styles.

24.2.3 Phosphor Coatings

Phosphor has become the industry standard for white light conversion from a blue or UV LED. The efficiency of phosphors is limited by many issues and parameters, especially thickness of the coating for excellent color temperature (CCT) and white light efficiency (Figure 24.7). The efficiency is fundamentally limited by what is called the "Manley–Rowe" condition, because each high-energy photon is converted to a low-energy photon. It is this even coating of phosphor over the die that gives a white light without a yellow outer ring, and several patented approaches are available:

- Over top of die only
- Top of die and sides only
- Cavity fill
- Remote phosphor
- Phosphor in glass over wafer

Visible-Light-Emitting Diodes (VLEDs)

Phosphor Application Methods

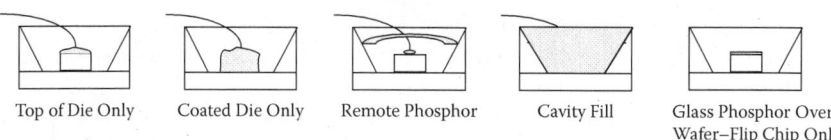

Top of Die Only   Coated Die Only   Remote Phosphor   Cavity Fill   Glass Phosphor Over Wafer–Flip Chip Only

FIGURE 24.7 Phosphor application methods.

24.2.4 Secondary Optics

Secondary optics are any optical lens or diffuser placed over the LED but not a part of the original manufactured device. Typical optical efficiencies of these secondary optic systems, depending on design and supplier, are around 90%.

24.3 GENERATING LIGHT

The generation of light is a complex quantum physics process dependent upon the various layers within the atomic makeup of the chip. As current flows through the layers of the chip, the free electrons that constitute that current ultimately combine with atoms in the active junction layer and emit light. (Refer to Chapter 1, which describes the radiative recombination phenomena that generate photons of light in all directions). As mentioned before, LEDs are currently about 18% efficient, and about nine distinct issues are responsible for the 82% inefficiencies. Doubling the efficiency of the LED is the equivalent of reducing the thermal management challenge by half. Many factors affect the performance of LEDs; the more efficient the LED, the lower the cost for thermal management. We will discuss the following factors:

1. I^2R losses—As the electrons move throughout the semiconductor material, they encounter ohmic resistance that in turn produces heat. This heat manifests itself at the LED junction.
2. Nonradiative recombination—When the exchange of an electron fails due to imperfections in the crystal growth structure, a photon is not emitted. Instead, a phonon is emitted, and this phonon emission causes structure vibration that is converted into heat, which lowers the efficacy of the LED and contributes to a larger amount of heat.
3. Electron leakage P-layer—The chip construction includes an AlGaN barrier that the electron must overcome in order to participate in the radiative recombination. This barrier also serves the dual purpose of preventing wandering electrons from backtracking away from the active junction of the chip and thereby wasting its energy without creating a photon. When the electron reaches the P-N junction but leaks back into lattice of substrate without generating a photon, energy is again wasted.
4. Auger dissipation—A three-particle process whereby an electron and hole recombine, but transfers the energy to a third carrier (electron or hole). The third particle relaxes, emitting the energy as wasted heat. Auger

recombination, being a three-particle process, increases with the cube of the carrier concentration.
5. Polarization—The polar nature of nitride semiconductors can give rise to internal electric fields that separate electrons from holes in the light-emitting InGaN region, reducing the radiative efficiency. These electromagnetic fields are created by stresses in the crystalline structure.
6. Threading dislocations—Imperfections in the crystal create pathways for electrons and holes to release their energy as heat rather than light. These dislocations are particularly problematic in the nitride-based LEDs because of the difficult interface between the aluminum oxide crystalline structure and the adjoining GaN surface.
7. Contact resistance—The resistance between metal–semiconductor interfaces adds further heating through added ohmic heating. Nitride LEDs are particularly susceptible to this issue because of the large band gap and difficulties with P-type doping.
8. Output coupling efficiency—Light extraction is difficult because of the higher refractive index of semiconductors than the surrounding air. Photons incident beyond the critical angle suffer total internal reflection owing to dramatically different refraction characteristics of the chip-to-air interface and become trapped in the chip. These photons will be reabsorbed and lost as wasted heat.
9. Phosphor efficiency—Phosphoring has become the industry-standard method for white light conversion of blue/UV die. The efficiency of these phosphors is limited by many issues and parameters, especially quality, thickness uniformity, placement, mixture, and homogeneity. Efficiency is fundamentally limited by what is called the "Manley–Rowe" condition because each high-energy photon is converted to a lower-energy photon.

24.4 OPTICAL, THERMAL, AND ELECTRICAL MEASUREMENTS

24.4.1 MEASUREMENT OF PARAMETERS

Photometry is the measurement of visible light based on the response of the average human observer. Colors are the wavelength, and the brightness of that color is the intensity of light. The most accurate method for measuring color is by using a spectroradiometer performing a complete spectral power distribution of the source being measured from which all photometric, radiometric, and colorimetric parameters can be mathematically calculated (Figure 24.8).

The optical radiation (W/nm) emitted by a source is read by the combination of an integrating sphere and a computerized spectroradiometer system. Spectral density in W/nm interval defines the spectrum of this light source, and flux (W) can be thought of as the fundamental quantity of radiometry. Currently, meaningful measurements of these parameters are broken down into geometric and spectral aspects of a light source. The four most common geometric aspects of a light source are flux, intensity, luminance, and illuminance. The spectral aspects are color, peak

Visible-Light-Emitting Diodes (VLEDs)

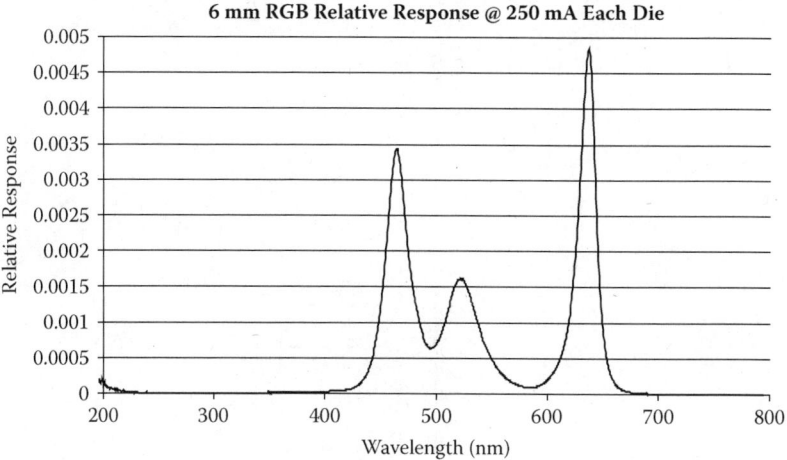

FIGURE 24.8 Relative spectral response of an RGB LED.

or dominant wavelength, in chromaticity coordinates as well as color temperature (K), and a color-rendering index. Other data collected at the same time are V_F and current applied, beam angle at half-power points, efficacy, and efficiency numbers. Figure 24.9 is a typical data collection printout from the spectroradiometer.

Flux is the total amount of light emitted from a source in all directions. Luminous flux (lumens) is measured by placing a spectroradiometer at the exit port of an integrating sphere and measuring the visible light energy per second emitted by the source. The luminous flux is then calculated by mathematically integrating the power values with the CIE-defined photopic value for each wavelength. One watt of radiant flux at the peak photopic wavelength of 555 nm (peak eye response) is equivalent to a luminous flux of 683 lumens.

Intensity (cd) is measured into a Condition B tube and is actually the irradiance of the source over 10 cm into a 1 cm^2 aperture. This standard CIE-127 method gives a good mechanism for comparing data sheets and is an easy test method for manufacturing.

Luminance is the visual perception of how bright a source is at some distance and angle. It is usually measured with a spot meter and directly reads out in luminance units (cd/m^2 = nit) or ($lms/m^2 sr$ = nit).

Illuminance (lumens/m^2 = 1 lux = 1 ft-candle) is the flux per unit area falling on a particular surface. When designing light fixtures for an office area, this characteristic is needed to determine if sufficient light is falling onto a desk surface according to a defined specification.

Dominant wavelength (nm) of a color stimulus is defined as "the wavelength of the monochromatic stimulus that, when additively mixed in suitable proportions with the specified achromatic stimulus, matches the color stimulus considered." In more general terms, this is the color the eye actually sees regardless of the peak wavelength emitted by the LED. Finding the dominant wavelength of an LED is a simple geometrical problem.

Optical Measurement Lab

Date/Time: 8/8/08 12:49 PM 5 Cool White Stars

#	2° 1931 x	2° 1931 y	Color Temp K	Dom λ nm	Half Band nm	Peak Spec Val W	Peak λ nm	Total Rad Flux W	Purity	Center λ nm	General CRI	Efficacy lm/W	Total Lum Flux lm	Forward Voltage V	Forward Current mA
1	0.3054	0.3122	7126	481.2	25.0	2.25E-03	453.1	1.84E-01	0.113	455.6	70.682	50.1	57.7	3.289	350.0
2	0.3005	0.3071	7551	480.5	25.0	2.38E-03	453.1	1.88E-01	0.134	455.7	71.300	50.4	58.4	3.310	349.9
3	0.3075	0.3170	6926	483.1	22.9	2.48E-03	450.2	1.98E-01	0.101	452.4	68.511	53.5	63.3	3.380	350.0
4	0.3022	0.3105	7366	481.5	25.3	2.28E-03	453.1	1.86E-01	0.125	455.6	70.908	50.6	58.1	3.285	349.9
5	0.3032	0.3101	7308	481.0	24.8	2.31E-03	453.1	1.86E-01	0.122	455.6	70.991	50.2	58.0	3.304	349.9
	0.3038	0.3114						Average					59.1	3.314	
	0.0027	0.0036						Stdev					2.4	0.039	

FIGURE 24.9 Optical measurement printout.

Visible-Light-Emitting Diodes (VLEDs)

Using the CIE 1931 chromatic chart (Figure 24.10) and the chromatic coordinates supplied by our colorimetric system, a line is plotted through these coordinates and the achromatic stimulus CIE standard illuminant E with $x = y = 0.3333$ to the spectrum locus (edge of curve). At this intersection with the curve, the dominant wavelength is read. For example, for a green LED of peak coordinates of $x = 0.2835$ and $y = 0.6870$ and the stimulus at $x = y = 0.3333$, the projection to the curve intersection is ($x = 0.2801$ and $y = 0.7117$), or 547 nm.

Correlated color temperature (K). In the late 1800s, British physicist, William Kelvin, heated a block of carbon. The block started out at a dull red to yellow and ended up as a bright bluish-white. This is what one sees when the stove heating elements glow red and light bulbs glow. However, colors are matched to a standard blackbody, and it is not the actual temperature that a filament burns. In measuring LEDs, any set of chromaticity coordinates that are the most similar to the blackbody standard is described as having that color temperature (Figure 24.11).

Color rendering index (CRI) is a subjective method of determining how well a light source renders color to the average observer (scale is 0 to 100). It is based on the average response of a group of human subjects as to how accurately the colors appear when compared to the same colors under either tungsten or daylight sources. This is the capability of the light source to render the true colors of an object. A CRI around 70 is normal, and 85 is suitable for even the most demanding applications such as paintings.

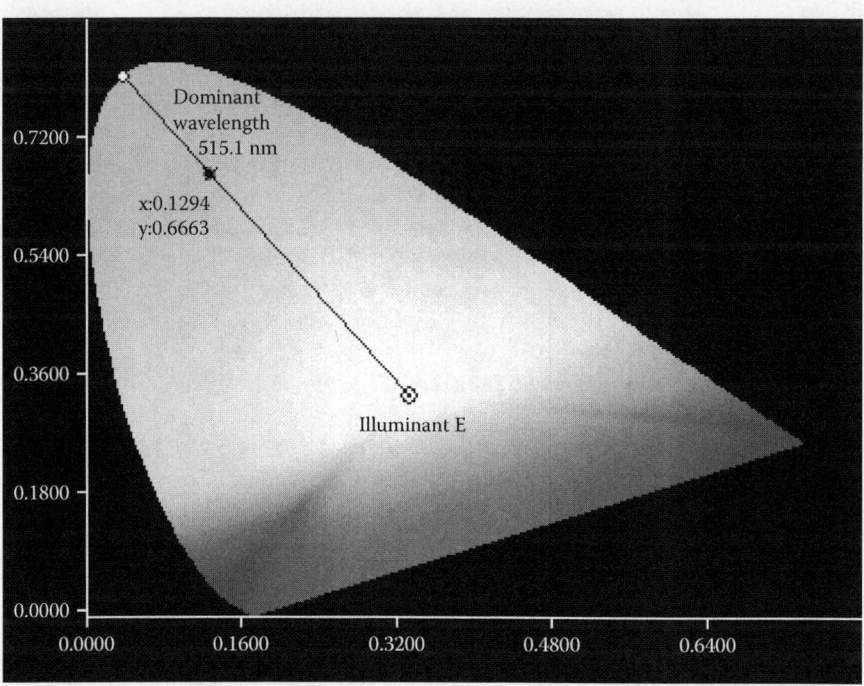

FIGURE 24.10 Chromaticity chart 1931.

Color Temperature of Familiar Light Sources	
Degrees Kelvin	Light Source
1700 - 1800	match flame
1850 - 1930	candle flame
2000 - 3000	sun @ sunrise or sunset
3000	tungsten lamp 500W - 1khrs
3200 - 3500	quartz lights
3200 - 7500	fluorescent tubes
5000 - 5400	sun - direct @ noon
5500 - 6500	daylight (sun plus sky)
6000 - 7500	sky overcast
8000 - 10000	sky partly cloudy

FIGURE 24.11 Color temperature of familiar light sources.

V_F (V) and I_F (mA) are recorded forward voltage drop across LEDs at specified constant current (I_F) drive. V_F matching can be important when LEDs are configured in a parallel circuit. The lowest V_F device will draw the most current and, subsequently, will be brighter.

Beam angle (degrees) is the measured angular goniometric results plotted to represent how much intensity is at a specific angle. This creates a radial plot to visually represent the beam angle of the tested LED. Half-power points are determined, and this defines the visual beam angle (Figure 24.12).

Efficacy (lm/W) = Luminous flux (lm) of LED/input electrical power (W) is a benchmark target of 200 lm/W and a measurement to determine attainment of that goal. It is also a good measure of different suppliers' VLED devices.

Radiant efficiency = Radiant power (W) of LED/input electrical power (W) is the ratio of total power input to the usable power output of the VLED as a percentage. This efficiency measurement is useful for comparison of different VLED sources and the technology itself.

Luminous efficiency (lm/W) = Luminous flux (lm) of LED/radiant power (W) of LED is another measure of visible light compared to the energy of that light.

VLED data sheets convey typical parameter results based on 25°C testing using a very short pulse time. This test method does not take into account any heating affect (junction temperature increase) on the light output, dominant wavelength, or V_F shift. Several reasons exist for this test approach: (1) it makes the LED supplier's part look brighter, (2) correlates to the production line tester, and (3) junction temperature of the tested part is easily verifiable. The lack of standards for measurement techniques and definitions has led to confusion among the users of LEDs. Parametric values at higher T_j need to be determined from graphs supplied on the data sheets. This interpretation by the end user must correlate with the desired application junction temperature measurement. Optical measurement equipment can measure these parameters over time and temperature and

Visible-Light-Emitting Diodes (VLEDs)

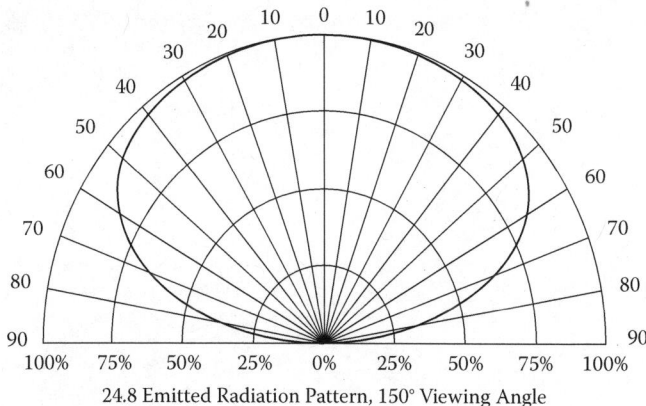

24.8 Emitted Radiation Pattern, 150° Viewing Angle

FIGURE 24.12 Visual viewing angle for 150° device.

graphically present junction temperature results more like the application. Graphs can be produced for V_F, total luminous flux (TLF), and dominant wavelength (λ_D) versus I_F, T_j, and degradation versus time. Measurement of these characteristics is more demanding when arrays are used and the matching of the LEDs is critical.

Figures 24.13 through 24.17 show typical VLED responses that have been discussed, including shifts due to changes in T_j and I_F.

24.5 LABORATORY TESTING

Building a comprehensive solid-state lighting laboratory is no small undertaking; it requires the acquisition of the latest state-of-the-art equipment and is a capital-intensive investment. For LED technology to enjoy the huge growth predicted in the solid-state lighting market, these test laboratories must provide quantifiable and repeatable test results that establish long-term, stable reference points to answer the unknown questions in the adoption of this technology. The LED maker must be a keen engineering company capable of understanding the customer's end objective. There is little doubt that LED technology has the potential to replace nearly all traditional methods of generating light.

An optical measurement laboratory is an essential tool for a supplier of LED components and assemblies. Visible light and its detection by the human eye and how the eye interprets this small band of electromagnetic radiation will be studied for quite some time photometrically. In this visible world, many unknowns exist for the unsuspecting designers and buyers of such a component. Although knowing that the light is green, and that light output is bright and throws a large beam, these observations are hard to convey to another human in such a manner as to ensure it can be replicated again and again. Attempting to replicate identical original lighting characteristics by human input without quantifiable data is extremely difficult. Standards must be set to ascertain that expectations can be met repeatedly.

Electro-optical specifications and requirements change with the availability of materials that are constantly evolving, and are not evolving simultaneously or all at

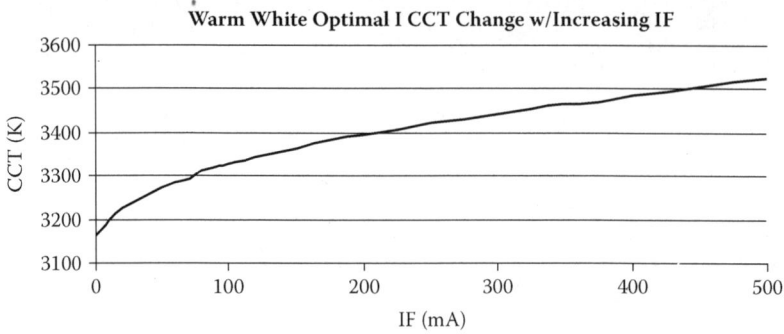

FIGURE 24.13 CCT versus I_F for white LED.

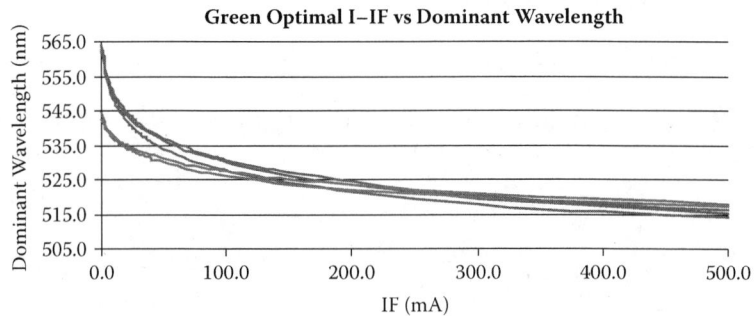

FIGURE 24.14 Dominant wavelength versus I_F.

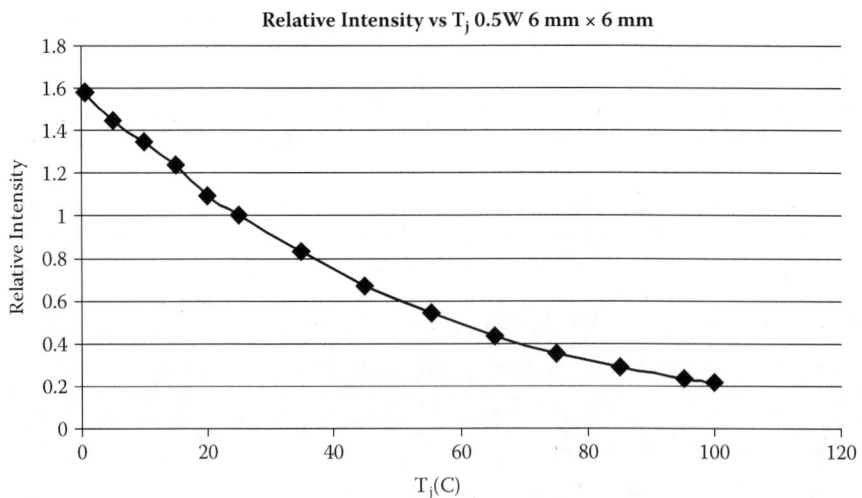

FIGURE 24.15 Intensity versus T_j.

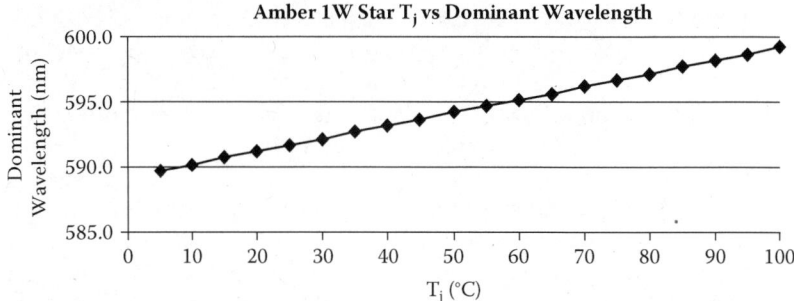

FIGURE 24.16 Dominant wavelength versus T_j.

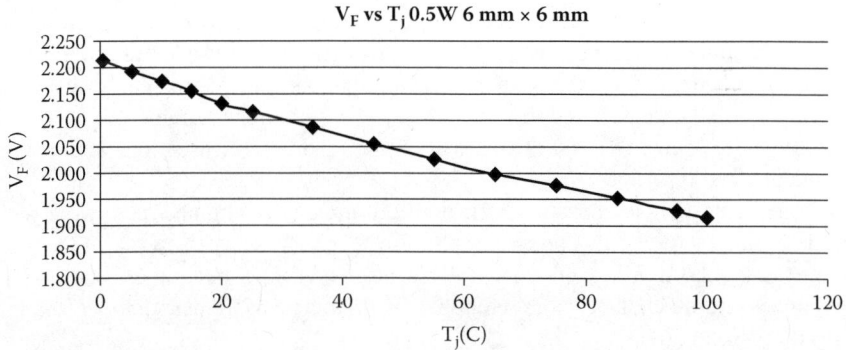

FIGURE 24.17 V_F versus T_j.

the same rate. These material changes extend beyond the chip or die made from the compound semiconductor that is the heart of the light source. The modern power LED is "upgraded" with new lens materials and with approaches in packaging at the component level that will not just result in a maximum amount of light but also in the certainty that a sufficient level of light output will be available after an extended time period. It is the certainty of lumen maintenance that will justify the acceleration of LED technology and solid-state lighting for both new LED-specific designs and for "retrofit" or replacement of sometimes century-old lighting technologies. The key to widespread adoption of LED lighting is validation of illumination equivalent to established light sources and long-term reliability.

Lastly, LEDs and LED light engines can fail, and it is necessary to have the proper failure analysis equipment capable of analyzing failures down to the die level. Determining just the failure mode without finding the cause does not contribute to finding an irreversible fix. LEDs have the same failure mechanisms as other semiconductor devices such as thermal and electrical overstress, insufficient die attachment and wire bonding, packaging, and lensing construction. The best fix to these is, as always, good quality control during the manufacturing process. Analysis of components and assemblies down to levels requiring the use of a scanning electron microscope (SEM) and its attached EDS (energy dispersive system) for material analysis is an essential part of complete validation. At the macro level, cross-sectioning and/or chemical etching capability, ovens and curve tracers, a variety of optical microscopes and cameras, material expansion equipment, and X-ray fluorescence tools make up the laboratory of the optoelectronics companies that are established as an engineering extension of the customers they serve.

An engineering-oriented optoelectronics company recognizes and responds to the multiple material and process issues encountered by the system designer.

By taking any incandescent or fluorescent light of interest and using a large 1-m integrating sphere, one can determine the color temperature (CCT) and brightness (TLF), which makes it possible to suggest a VLED component or array to replicate this output (Figure 24.18). Once the new LED assembly is prototyped, it can be checked using the same optical measurement system to verify its performance to the original characteristics.

For those applications where a complete specification does not yet exist, comparisons are made with multiple engineering designs. For example, a customer has decided that he or she likes this color, this emission pattern creating a uniform glow behind a frosted material, and this brightness. The optical measurement system will translate these characteristics into numbers, which then can be used for specification parameters and replicated.

Thus, the best choice for an electrical or lighting design engineer is to work with an LED manufacturer who approaches this emerging market with a view of the final system assembly in mind and who can provide that customer/engineer access to an optical measurement laboratory equipped with the measurement and evaluation tools discussed herein.

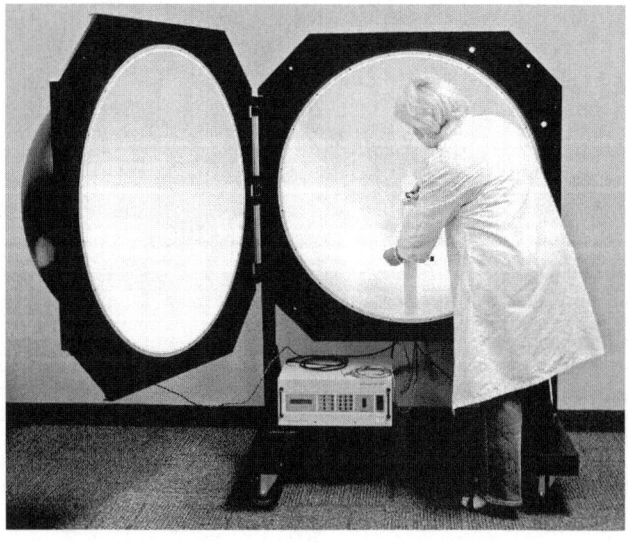

FIGURE 24.18 Integrating sphere.

Visible-Light-Emitting Diodes (VLEDs)

A spectroradiometer system is used to measure radiometric and photometric characteristics. It comprises the following:

- Integrating spheres
- Goniometer
- Condition A and B tubes
- Thermal electric cooler (TEC)
- Constant-current and pulsing power supplies

An infrared thermal imaging camera is used to optimize thermal management designs.

Scanning electron microscopes and X-ray analyzers are used to analyze failures generated during manufacturing, from the applications, or during reliability and environmental testing.

Accurate data accumulated from each piece of equipment during the testing regime can make the difference and enable a designer or manufacturer to justify bringing a new lighting system to the marketplace.

24.6 RELIABILITY

24.6.1 LIFETIME AND OPERATION OF LEDs

Reliability of LEDs is dependent on several factors, including proper thermal management, packaging design, and process control during manufacturing. When considering LED reliability, it is important to understand lumen maintenance and mortality rates when used in terms comparable to traditional lighting conventions.

The lighting industry has defined the useful life of a light source by denoting specific points at which percentages of the population can be expected to fail. These points essentially describe the mortality rate of a light source or what is known as a B-lifetime chart.

Point B10 (Figure 24.19) for a light source specifies the time when 10% of the population is expected to fail; similarly, B50 is the point at which 50% of the population is expected to fail. Typically, these charts were used to describe the behavior of incandescent lights, which display little change in output until a catastrophic failure occurs such as an explosion and the result is a broken bulb.

In LEDs, light output decreases over time, and LEDs rarely suffer from catastrophic failures. A catastrophic failure in LEDs could be a parameter change in electrical or optical specifications that prevents the LED from lighting. As a result, the term *lumen maintenance*, describing the light output in lumens over time, is used as a metric to describe LED lifetime. According to the Alliance for Solid-State Illumination Systems (ASSIST), two coordinates can be used to describe the lifetime of an LED: L70 and L50. L70 is the time taken to get to the point at which the human eye can detect a noticeable reduction in light levels, which is at 70% lumen maintenance, or a 30% reduction in total lumen output of the LED. Likewise, L50 is the time taken to get to 50% lumen maintenance.

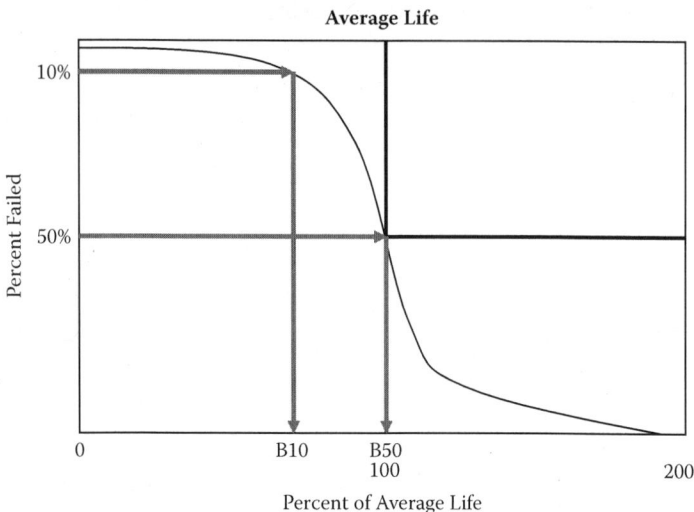

FIGURE 24.19 Average life curve.

24.6.2 FACTORS AFFECTING LED LIFETIME

From a design standpoint, the amount of current driven through the device is just one of the factors that affect how much time it takes to reach these lifetime points. Driving the LED at its full maximum rated current will result in a lifetime that is considerably shorter than when the LED is driven at a lower current value (Figure 24.20).

Similarly, operation at lower junction temperatures (Figure 24.21) will increase the overall lifetime, whereas operation at higher junction temperatures results in a decreased life span.

Packaging of the LED should also be considered when discussing LED reliability. Choosing the proper materials for the package casing, bonding wire, encapsulate, epoxy, metal interconnects, and lens all affect the overall reliability of the device as well as the reliability of the LED chip itself.

When in operation, the LED chip generates excessive heat that must be dissipated in order to reduce the possibility of damage to the LED chip as well as the overall package. As a result, thermal resistance of the device plays a key role in determining LED lifetime because dissipating the heat quickly will result in a longer operating lifetime.

Is 50,000 to 100,000 hour operating lifetime for LEDs real? Yes, theoretically, as long as the design is sound, stringent quality control is in place in all the manufacturing processes, and LEDs are operated within the maximum rated guidelines. Extensive study has shown that there can be considerable variation among VLED suppliers over the long term in light performance and color shifts. The related two graphs are derived (Figures 24.22 and 24.23) from 6000 hr performance data, across several LED manufacturers, and indicate one supplier with excessive color temperature shift over time. Light degradation appears excessive with a different supplier even though initially it was the higher performer for output. The graphs also

Visible-Light-Emitting Diodes (VLEDs)

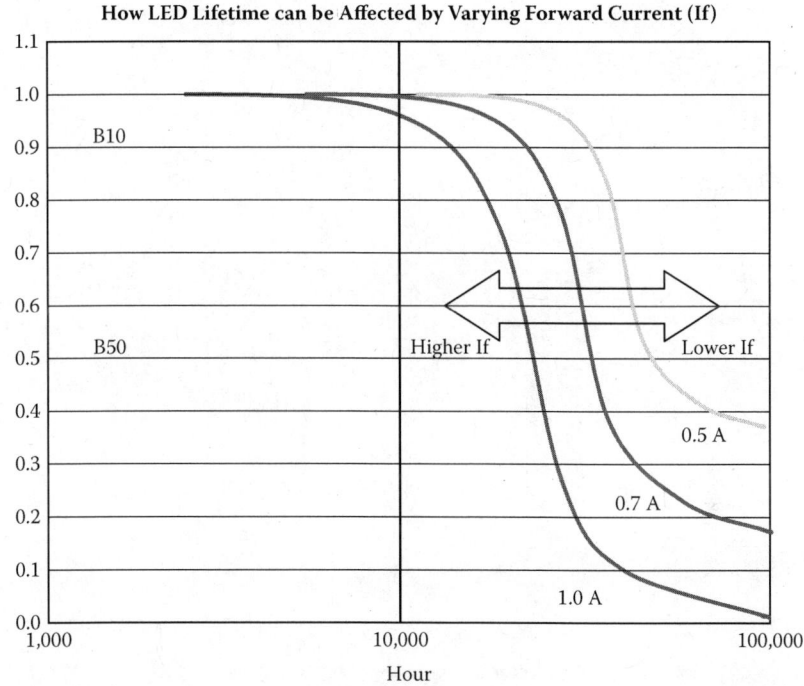

FIGURE 24.20 Lifetime versus forward current.

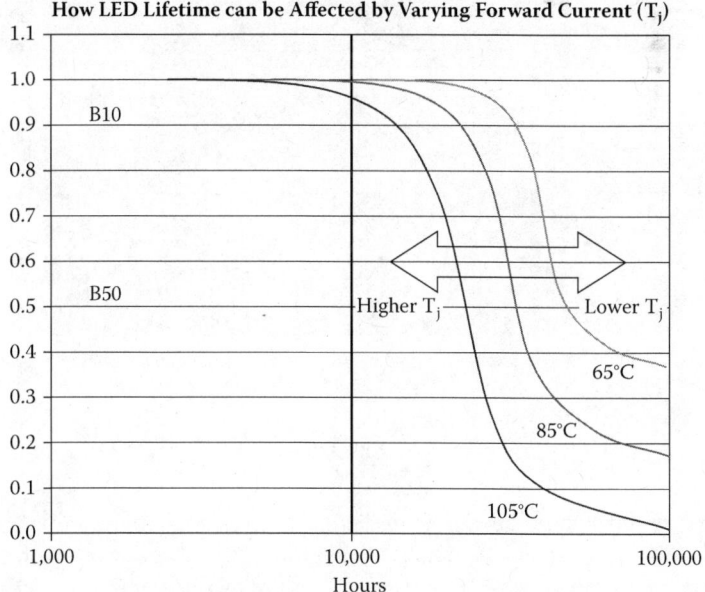

FIGURE 24.21 Lifetime versus junction temperature.

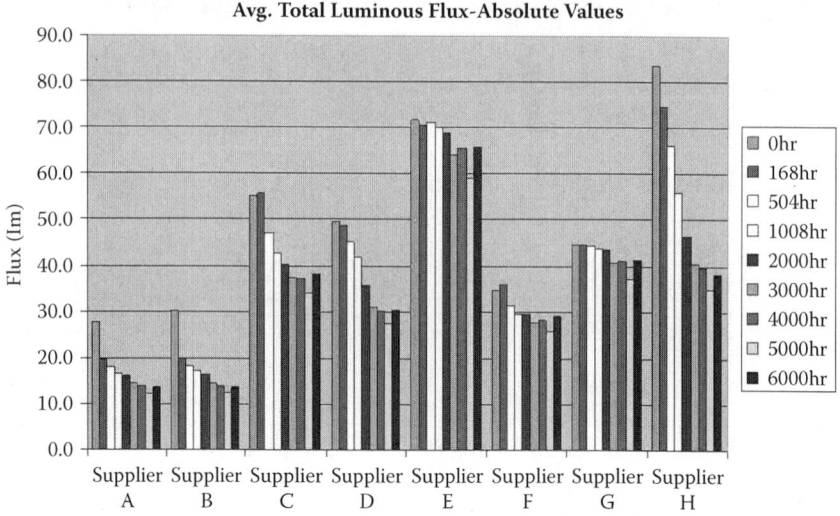

FIGURE 24.22 Benchmarking—total luminous flux results 6000 hr.

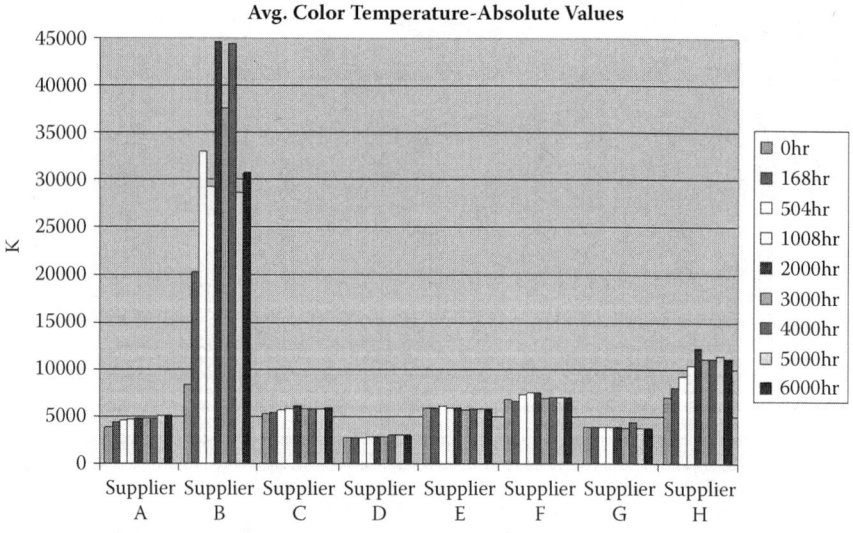

FIGURE 24.23 Benchmarking—color temperature (K) over 6000 hr.

demonstrate good stability over time for both characteristics from a few suppliers. It is the selection of these suppliers' parts that Optek will use and work with the supplier to improve their process with data and failure analysis equipment to identify root cause and irreversible fixes.

24.7 APPLICATIONS

Architectural. Flexibility in styling and aesthetics has allowed designers to initiate projects not possible with other technologies. Advances in color performance and color rendering index of white LEDs redefine lighting a building or landscape. This creativity is supplemented by tremendous energy cost savings and the fact that corporate and brand logo recognition is greatly enhanced by the "color on demand" attributes of LEDs.

Channel letter/contour lighting. Piping light around a building or retail/corporate signage enhances name and brand identity. LEDs in diffuser tubes as linear "daisy chains" create highly versatile neon looks without the enormous comparative installation and maintenance cost. The replacement of neon is not only economical but environmentally sound, making it one of the first niche markets.

Consumer portables. The number of LEDs in cell phones, GPS, personal music systems, and flashlights is reduced as brightness per component increases, and other materials such as light guides, light pipes, and diffusers are incorporated to spread the light to all keypad buttons instead of lighting them individually.

Landscaping. Only those with a vested interest in the replacement bulb business could hold back residential landscaping and special task lighting with LEDs. With consumer pressure and response to the variety, longevity, and flexibility of installation, this specialized lighting has grown quickly.

Retail display. Requirements are well suited to the variability and availability of LED color temperature, for example, in a jewelry store a cool color temperature is best for diamonds, whereas a warm white is best to display gold.

Automotive applications. VLED usage has accelerated at differing levels depending on the country doing the design and the car or truck model. High-brightness LEDs first interested carmakers as replacements for miniature incandescent bulbs used as instrument clusters, and light piping often enabled their use as dashboard indicators. Interior solutions for map lights, dome lights, glove box lights, and eventually all interior lighting, continue to grow, with many carmakers designating their own unique brand colors. Exterior LED use began with linear arrays that created center-high mount stop lights (third brake lights) and then became a styling tool for creating rear combination lights (stop, back up, and brake lights). Fog lights and daylight running lights were next, with the biggest challenges optically and thermally in the creation of forward lighting (head lights) replacement. The first production cars with LED forward lighting appeared in 2008. Additionally, special lighting in rearview mirrors or as turn signal repeaters embedded in mirrors have become possible and widespread.

LCD and TV. VLEDs have grown into a backlighting source. In early liquid crystal displays (LCDs), the LCD application actually replaced LEDs in some instances (first watches and calculators). The LCD later became powerful enough to used in a television and would be replaced by LEDs.

Signage and electronic billboards. Video advertising often compensates by use of a sensor to regulate the drive currents and offset degradation of an LED used to form pixel groups. Additionally, the use of LEDs outdoors has prompted the makers of LED encapsulate materials to create LED lenses that negate the effect of UV sunlight radiation. One material evolution is tied to the next to create the best-case benefit. The stability improvements of LEDs have prompted even greater government support, whereby installation of variable message signs (VMS) control traffic and provide safety and other messages to the motorist.

Fountains, pools, and spas. Here VLEDs not only add safety and security but also create emotional stimulation.

Entertainment and stage lighting. An early niche application of HB LEDs continues to grow as more power and color manipulation become available.

Aircraft, train, boat, and automotive mood lighting. These markets developed with modules replicating the replaced technology, but quickly began to take new shapes and allow wider use owing to the flexibility of form factors that make possible new styling and appearance not available with bulbs.

Traffic and pedestrian signals. This VLED technology was adapted early in large part due to government support of initiatives reducing energy consumption, and the desire to reduce greenhouse gases.

The foregoing uses show the difference in usable or needed light and emotional or pleasure-centered lighting. The direction of the lighting applications, and the tuning and manipulation of colors, is only limited by the human imagination.

The first factors that drive a market to change are not always the one-to-one cost ratio or even near parity of cost options. As evidence, SSL breakthroughs in outdoor applications were first in delivery of solid and tangible benefits of energy savings, color and brightness uniformity, safety, and ecological advantages, making the total cost of ownership broadly accepted and appreciated. Driving the usage of LED designs is the opportunity to innovate and deploy to the consumer new styling for cars, buildings, signage, and general lighting. Undoubtedly, cost and market price points matter greatly, and as the component and assembly costs continue to drop as LED efficacy increases, continued expansion reaches new levels for the consumer. Government incentives promise to reward manufacturers for the attainment of specific benchmarks that are required for solid-state lighting to truly become the standard for general illumination in an acceleration to reach all the inherent benefits. The emergence of standards established by the Department of Energy's Caliper and Energy Star program helps to advance its credibility, leading to accelerated growth of solid-state lighting.

25 Ultraviolet Electromagnetic Radiation

25.1 OVERVIEW

Ultraviolet electromagnetic radiation, commonly known as UV, is currently employed in many industries and applications. The market for UV equipment is conservatively estimated at over $5 billion. UV electromagnetic radiation is used in a variety of applications, such as germicidal air and water purification, surface disinfection, currency validation, medical, military, industrial curing, instrumentation, effect lighting, and forensic analysis.

The predominant method used to produce UV electromagnetic radiation today is based on tube technology developed nearly 100 years ago. The emerging UV LED technology has an opportunity in the coming years to provide a competitive technology in a manner similar to ongoing events in solid-state lighting using visible LEDs. The currency validation market, which will be discussed in more detail later in this chapter, is now in the early stages of transition to UV-A LEDs. UV LEDs will be an enabling technology in the future to drive new innovative applications.

This chapter begins with a basic review of fluorescent and UV tube technology. Chapter 25 will discuss the germicidal effects of UV electromagnetic radiation on DNA and the UV dosage level required to inactivate many lethal varieties of microorganisms such as anthrax, *E. coli,* and influenza. UV lamps have been used for germicidal purposes for almost 100 years. The construction and operation of UV LEDs will be reviewed along with advanced UV LED packaging concepts. The chapter will then conclude with a discussion of the various market applications for UV technology.

25.2 THE UV SPECTRUM

The UV spectrum lies between the visible light range the human eye can detect and x-rays, as shown in Table 25.1. The term *ultraviolet* refers to all electromagnetic radiation with wavelengths in the range of 10 to 400 nm. In addition, there are several classifications within the UV range: UV-A, UV-B, UV-C, and Vacuum UV.

UV-A. The UV-A range includes wavelengths from 315 to 400 nm, which have the least amount of energy. Wavelengths in the UV-A range are used for currency validation, industrial curing, phototherapy, and for forensic/analytical instruments. UV-A wavelengths from 315 to 345 nm are used for sun tanning and are a suspected cause of premature aging of human skin. Most of the UV-A range cannot be seen by the human eye. UV that is at approximately 385 to 390 nm and below cannot be

TABLE 25.1
Electromagnetic Wavelength Spectrum Including Ultraviolet

Electromagnetic Radiation Spectrum

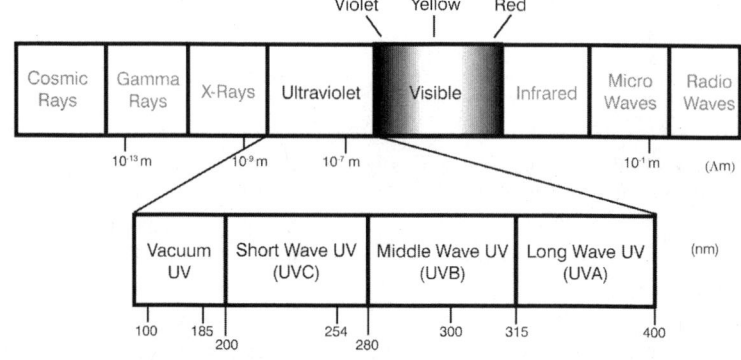

detected by the human eye. Therefore, it is essential to take precautions to protect your eyes and skin when working with UV light sources.

UV-B. The UV-B range refers to wavelengths from 280 to 315 nm. These wavelengths are more hazardous than UV-A wavelengths, and are largely responsible for sunburn. The UV-B range is used in forensic and analytical instruments and for the more recent narrow-band UV-B phototherapy skin treatments for psoriasis (308 to 311 nm). UV-B does not penetrate as deeply in the skin as UV-A; however, the deadliest types of skin cancer (malignant melanomas) start in the epidermis, an upper layer of the skin. UV-B is largely blamed for these cancers, although shorter UV-A wavelengths are considered possibly cancer-causing as well.

UV-C. The UV-C range refers to shorter UV wavelengths, usually 200 to 280 nm, and is sometimes referred to as the *deep UV range*. Wavelengths in the UV-C range, especially from the low 200s to about 275 nm, are especially damaging to microorganism's DNA. UV-C is often used for germicidal applications for water, air, and surface decontamination. The Earth's atmosphere absorbs most of the UV-C radiated by the sun.

Vacuum UV. This range, from 10 to 200 nm, has the shortest wavelengths and highest energy level and is absorbed by the atmosphere. The strong absorption of vacuum UV in the Earth's atmosphere is due to the presence of oxygen. State-of-the-art semiconductor photolithography processes seek to use shorter UV wavelengths to manufacture the next generation of smaller IC chips.

Until recently, lamp technology was the only way to produce UV-A, UV-B, and UV-C electromagnetic radiation. With this in mind, we will begin our technical discussion with a brief discussion of fluorescent lamps and proceed to a review of UV germicidal lamps that dominates today's billion dollar market. The UV lamp discussion will ensure that the reader gets a basic understanding of the existing UV lamp technology that has been in place for many years and the characteristics of these UV lamps. The chapter concludes with a technology review of UV LEDs, including UV applications, advanced packaging concepts, and market applications.

Remember to use extreme caution in working with any UV light source! As noted, earlier UV wavelengths are not seen by the human eye. Never look directly into a UV light source or UV light from a reflected surface or through an optical system. Always wear UV-rated eye-protective safety glasses and proper skin covering when working with any UV light source.

25.3 FLUORESCENT LAMP

The fluorescent lamp is predominantly a UV-C light source that uses an inorganic triphosphor coating on the inside of the lamp to fluoresce and convert the UV-C electromagnetic radiation into white light as seen by the human eye. A cutaway illustration shows the construction for a portion of a fluorescent lamp as shown in Figure 25.1. The emission of light from a fluorescent lamp operates along some of the same principles of photon emission discussed for LEDs in Chapter 1. The main components of a fluorescent lamp are the glass tube that is coated on the inside with triphosphors, a small amount of mercury, and the cathode filament, which when electrically heated emits electrons and inert gas, which is usually argon contained inside the tube at a low pressure. When AC power is applied to the fluorescent lamp cathode, there is an electric discharge through the argon gas

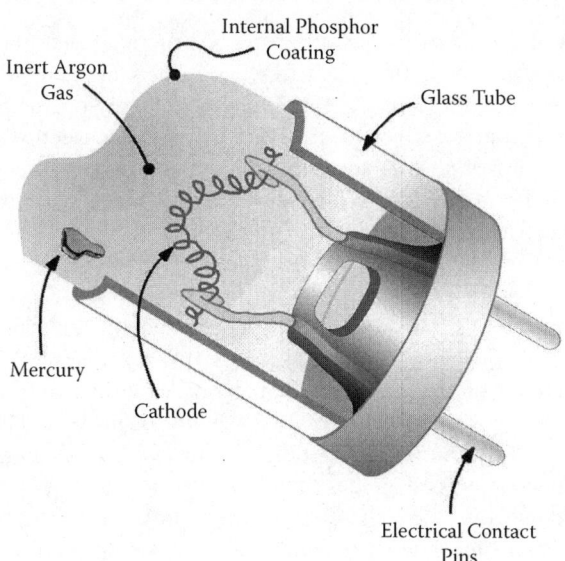

FIGURE 25.1 Fluorescent lamp.

and mercury vapor; thus, the fluorescent lamp is also known as a *gas discharge device*. The cathode is usually made of coiled-coil or a single coil of tungsten wire that is coated with an emission mix. The emission mix is made from barium, strontium, and calcium oxides, and these materials are specifically selected for high electron emission capability at both low energy levels and low temperatures. However, the emission mix oxide coating is sputtered away (evaporated) with UV lamp use. Eventually, the emission mix on the cathode is consumed over the life expectancy of the lamp measured in hours. When the cathodes are electrically heated, the emitted electrons collide with the atoms of argon gas, causing the inert gas contained in the tube to ionize. The resulting plasma formed by impact ionization results in current flow from one end of the tube to the other. As the temperature increases, the liquid mercury vaporizes and begins colliding with the high-energy plasma. The electrons of the mercury atom are then excited into higher-energy levels. When these unstable higher-energy electrons return to their original energy level, a photon of light in the UV range is released. The process of photon emission resulting from an electron returning to its normal energy state for the UV lamp is similar to what was described for semiconductor LEDs in Chapter 1. The wavelength for light emission is based on the specific type of atom being excited into a higher energy state and amount of energy released. In a low-pressure fluorescent lamp filled with mercury, the main spectral line of light emission is at 253.7 nm (germicidal) with a secondary spectral emission at 185 nm (ozone generation) and several additional bands of visible light. The significance of these two UV emission bands will be discussed more fully later in this chapter.

25.4 TYPES OF UV LAMPS

Low-pressure UV lamps are very similar in construction to the fluorescent lamp shown in Figure 25.1. The noticeable difference is there are no phosphors applied to the inner tube of UV lamps. Low-pressure UV lamps operate at an internal pressure ranging from 2×10^{-5} to 2×10^{-4} psi with an electrical input power of approximately 0.5 W/cm. A second category of low-pressure lamps are the low-pressure high-output (HO) types that offer higher UV output as a result of increased electrical input power ranging from 1.5 to 10 W/cm. In practice, UV lamps need an initial burn-in period to stabilize, because of impurities contained in the lamp.

Electrical power levels for low-pressure UV germicidal lamps with a peak emission wavelength of 253.7 nm can be as little as 1 W up to several hundred watts (HO types) depending on the specific manufacturer. Accordingly, the UV power output can range from less than 1 W up to 65 W depending on the specific UV lamp type. Generally, the rated operating life, specified in hours for a UV lamp, is the point where the UV power output degrades to 80% of the initial UV power output. This provides a safe point at which to change out the low-pressure lamps to ensure that the system operates as intended over its operating lifetime. However, this can vary from manufacturer to manufacturer, and some do specify a 50% degradation point. The operating life for a low-pressure UV lamp is typically in the 8000 hr range at 80% degradation from the initial power UV output rating. Low-pressure UV lamps with a longer effective life (12,000 hr) are common and result from improvements made in the design and material selection.

The material selection for the UV lamp is also critical because the transmission and reflectivity characteristics can change drastically for materials used in the UV-C range. Materials of choice for the UV lamp tube are fused quartz, or for better performance, a high-purity synthetic fused silica that minimizes the optical transmission losses through the tube. Absorption loss for the type of glass used in a UV-C lamp can range from 5% to 15%. There is also the need to address whether the end user requires the ozone-generating 185 nm band. If the UV lamp is to be used in a germicidal application where ozone generation is not desired, the lamp tube may be doped to block the 185 nm ozone-producing wavelength from escaping from the UV lamp or by using a soft glass that blocks the 185 nm wavelength. However, if the UV lamp is being used in decomposing and treating groundwater contaminated with volatile organic compounds (VOCs) or wastewater, the tube is generally not doped, because the ozone created by the 185 nm emission band is effective in this oxidation application. Low-pressure UV lamps are generally considered monochromatic, with a 253.7 nm wavelength, and they have a peak efficiency around 80°F to 100°F depending on the manufacturer. The spectral response of a low-pressure high-output UV lamp is shown in Figure 25.2.

Medium-pressure UV lamps also use mercury vapor; however, these lamps operate at very high internal pressures ranging from 2 to 200 psi. In addition, the lamp operates at a much higher power density, with electrical power inputs ranging up to 250 W/cm. Thus, with the increased vapor pressure and input power, medium-pressure lamps operate at much higher temperatures than low-pressure lamps, ranging from 600°C to 900°C. Medium-pressure lamps have increased power and temperature, resulting in more electrons being excited, producing polychromatic emissions. The wavelength emissions from a medium-pressure UV lamp typically are from the UV-C to the visible range (200 to 600 nm), as shown in Figure 25.3. However,

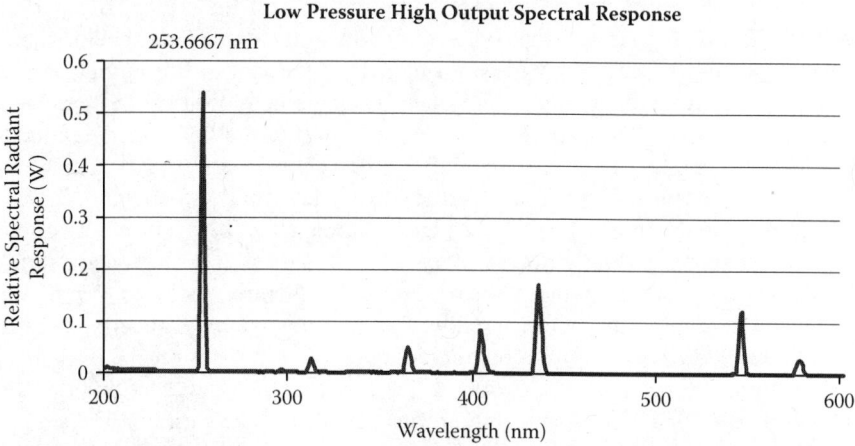

FIGURE 25.2 Spectral response for a low-pressure high-output UV lamp.

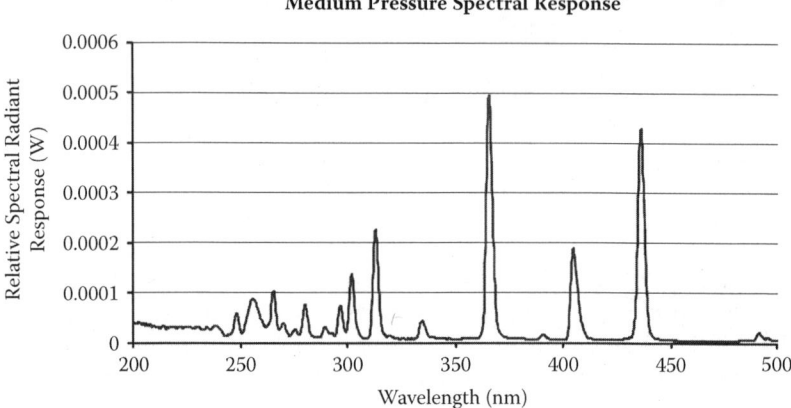

FIGURE 25.3 Spectral response for a medium-pressure lamp.

the efficiency of medium-pressure lamps is much lower than a low-pressure lamp for this much higher UV output. In certain applications, it may be desirable to have fewer UV lamps with a higher output in exchange for lower efficiency. As seen in Figures 25.2 and 25.3, both low- and medium-pressure UV lamps have some wavelength emission in the visible range. The band of spectral emission can be increased into the IR range using other materials such as gallium or iron in conjunction with mercury at a higher vapor pressure in the UV lamp. Medium-pressure lamps need from 5 to 10 min to reach the proper operating temperature. In the event of a loss of electrical power, these types of lamps require a similar amount of time to cool down before restarting the lamp.

25.5 UV LAMP AGING

As UV lamps age, the decrease in UV lamp output over the typical lifespan of 8,000 to 12,000 hr can vary somewhat among manufacturers. The specific manufacturer should be consulted for information on the end-of-life output of UV lamps. The decrease in UV output is generally accounted for in the design phase to avoid having the output reduced to the point where the UV water treatment system, for example, would become ineffective. The most conservative approach is to size the system well before the end of life of the lamp UV output and change out the UV lamps on a regularly scheduled maintenance basis. In practice, the UV power output is generally closely monitored with sensors to ensure that the minimum UV power output is maintained. Each manufacturer specifies UV power output degradation somewhat differently. Certain manufacturers specify it as the point at which 80% of the initial output (degradation of 20%) is reached, whereas other manufacturers specify it as 50% of the initial output (degradation of 50%). UV lamps need to be clean and free of dust at all times because the UV output will be diminished and converted into heat, thereby lowering the efficiency of the UV lamp. Light-emitting diodes also experience output degradation over time; this was discussed in Chapter 3.

25.6 ROHS AND WEEE CONCERNS FOR UV LAMPS

Environmentally toxic materials such as mercury, lead, and cadmium are some of the materials to which restrictions apply for commercial use. End product producers and customers through the supply chain are now requiring that suppliers provide material declarations that inform them of the presence and amount of these RoHS (Restriction of the use of Certain Hazardous Substances) materials. In Europe, there are additional requirements under EC Directive 2002/96 on Waste Electrical and Electronic Equipment (WEEE). This directive addresses the prevention of waste electrical and electronic equipment, and the reuse, recycling, and other forms of recovery of such wastes so as to reduce the disposal of waste in Europe. Lamp manufacturers are financially responsible for proper labeling of their product and, ultimately, the collection of lamps for recycling from private households deposited at various collection facilities for recycling.

In the United States, the National Electrical Manufacturers Association (NEMA) has begun a labeling initiative for CFLs (compact fluorescent lamps) that went into effect on April 15, 2007. The voluntary commitment will limit the mercury content and also guide users of lighting products by Internet or telephone to appropriate contacts for advice regarding disposal of spent mercury-containing lamps (fluorescent, compact fluorescent, and most HID). At present, there are 40 different fluorescent-lamp-recycling companies registered and operating in the United States and Canada according to NEMA (National Electrical Manufactures Association), which has issued recycling guidelines to lamp manufacturers, wholesalers, and commercial facilities.

25.7 FAILURE MODES IN UV GAS DISCHARGE LAMPS

UV lamps are the predominant technology used in industry today; however, these lamps do have several drawbacks. Although UV lamps are able to generate considerably higher power output levels than today's existing UV LEDs, there are several drawbacks of UV lamps, such as the following:

- Mechanically, lamps are fragile and susceptible to breakage.
- Mercury-based lamps are environmentally unfriendly, incorporating RoHS material.
- UV lamps have working life spans defined in hundreds to thousands of hours, due to the depletion of the emission mix and/or mercury.
- Medium-pressure lamps operate at very high temperatures; in applications such as photochemical curing, the high temperature can pose a problem for polymer and the substrate material being cured.
- UV lamps are prone to gas leaking from the tube owing to thermal stress cracking the glass, the metal seals in the tube, or the glass itself. Lamp explosion is possible in medium- and high-pressure lamps.
- UV lamps are susceptible to temperature variation. Depending on the manufacturer, low-pressure lamps have an optimum output with an ambient temperature of 25 to 30°C. Temperatures above or below this optimum temperature range will reduce the UV output; amalgam-type UV lamps can be used to somewhat reduce the temperature effect.

25.8 OVERVIEW OF UV APPLICATIONS

Some typical effects caused by UV electromagnetic radiation are the following:

- Many organic and inorganic materials fluoresce under UV light.
- Certain materials experience chemical reactions when exposed to UV light. Certain types of adhesives and coatings undergo polymerization or "curing" with UV light.
- DNA can be inactivated by UV radiation, and this is utilized in germicidal applications. The germicidal range is roughly between 225 to 300 nm, with 260 to 265 nm being the most effective to inactivate DNA.
- UV-A and UV-B electromagnetic radiation is used for phototherapy to treat certain types of skin conditions.

Table 25.2 provides an overview of some UV applications and provides the main UV classifications most widely used for these applications. We will discuss each of these effects and UV applications in more detail in this chapter. As stated earlier,

TABLE 25.2
UV Applications and Market Segments

UV Application	Market Segments	UV-C (200–280 nm)	UV-B (280–315 nm)	UV-A (315–400 nm)
Air disinfection	Hospital–commercial–home–car–plane	X	O	
H_2O disinfection	Municipal–well–residential–pool–spas	X	O	
Disinfection—other	Food–beverage–document–wafer fabrication	X	O	
Validation/Security	Passport–stock certificate–currency			X
Industrial curing	Adhesives–coatings–paints–flooring–inkjet and digital printing	O	X	X
Medical	Skin dermatology–instrumentation		X	X
EPROM erasure	Reprogramming EPROM ICs	X		
Forensic/Leak detection	Crime scene–engine oil–A/C–brake fluid detection–scientific analysis		O	X

Note: X defines primary UV classification used for these applications; O defines secondary UV classification used for these applications.

UV lamps are the primary UV light source used for many of the applications shown in Table 25.2. However, as UV LEDs have become commercially available in the last several years, UV-LEDs have started to replace UV lamps for some of these applications.

25.9 GERMICIDAL EFFECTS OF UV LIGHT

UV germicidal technology has been established in Europe for nearly 100 years. The first use of UV light to disinfect drinking water occurred in 1910 in France, using mercury-based lamps as the UV-C light source. Around the same time, UV-C light from mercury-based lamps was being used to disinfect the air of pathogens such as tuberculosis. These applications were based on the key discovery in 1877 by Dr. Arthur Downes and Thomas P. Blunt of the germicidal properties of direct sunlight. They correctly identified the increasing germicidal effectiveness (ability to inactivate pathogens) with shorter electromagnetic wavelengths (from visible blue, to violet, and then to ultraviolet electromagnetic wavelengths).

Downes' and Blunt's work revealed that ultraviolet wavelengths had the most germicidal effect on bacteria and other organisms. In the late 1800s, Danish scientist Niels Finsens utilized light of various wavelengths in phototherapy medical applications, for which he won the Nobel prize in 1903. More recently, the U.S. Environmental Protection Agency (EPA) has recognized the use of ultraviolet electromagnetic radiation as a proven technology to inactivate pathogenic microorganisms without forming regulated chlorinated disinfection by-products in public water supplies. UV can also be used to disinfect surfaces and is used in the food, beverage, medical, and semiconductor industries, to name a few, to maintain a sterile environment. The following text will explain these germicidal effects, starting with a very brief review of biology.

All living organisms contain nucleic acids, the two most commonly known being deoxyribonucleic acid (DNA) and ribonucleic acid (RNA). As you might already know, DNA provides the genetic code information for all living organisms to develop and function. RNA, in turn, facilitates translation of the genetic information of DNA into proteins. Generally, DNA is a double-stranded helix structure, as shown in the "before section" of Figure 25.4, and is measured in angstroms. The individual rungs of the DNA ladder shown in Figure 25.4 are made up of nucleotides. The nucleotides of DNA have nitrogenous bases of adenine, cytosine, guanine, and thymine, and the nucleotides of RNA have bases that consist of adenine, cytosine, guanine, and uracil.

Nucleic acids (DNA and RNA) readily absorb UV electromagnetic radiation, especially in the range of 240 to 290 nm. The UV absorption in DNA peaks at around 260 nm (Tsuboi, 1950), which is very close to the primary emission line of a low-pressure mercury lamp at 253.7 nm. Further examination showed that nucleic acids in DNA absorb 10 to 20 times the amount of UV electromagnetic radiation as equal weights of the protein component of DNA, whereas the sugar and phosphate components of DNA do not absorb UV above 210 nm (Jagger, 1967). It should also be noted that both the rate of absorption and peak absorption occur at different levels for each of the nucleotides of DNA (adenine, cytosine, guanine, and thymine),

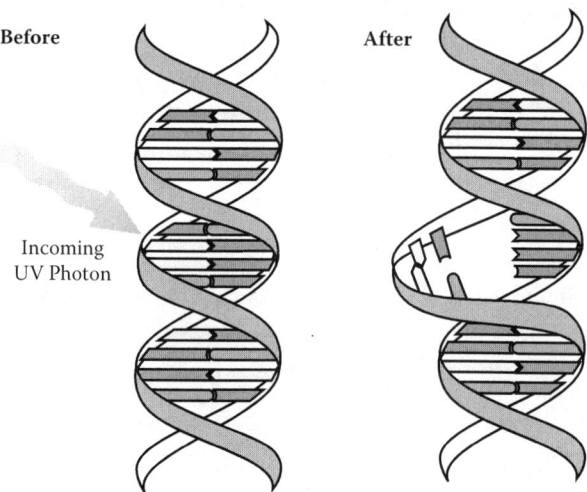

FIGURE 25.4 DNA damage event from UV-B. (Courtesy of David Herring and NASA.)

and the pyrimidines (thymine and cystosine) have been shown to be much more sensitive to UV electromagnetic radiation (Davidson, 1965). Three possible pyrimidine dimers that can be formed in DNA are thymine-thymine, cystosine-cystosine, and thymine-cystosine. The absorption of UV light by nucleic acid (three types of pyrimidine dimers) is what leads to alterations in the genetic material, the smallest of which can ultimately lead to the death of a living organism. A microorganism that cannot replicate is not capable of infecting a host.

The Earth has been exposed to UV radiation for millions of years. In some cases, UV has played a helpful role in forming Vitamin D, which is essential, and likewise a harmful role in causing sunburn, skin cancer, and cataracts. UV harms DNA in different ways.

An illustration describing one method of how UV can alter DNA is shown in the "after UV-B" exposure portion of Figure 25.4. In this common damage event, adjacent bases bond with each other instead of across the nucleotide ladder. This creates a bulge, and the distorted DNA molecule does not function properly. If the distorted DNA molecule cannot produce the correct proteins, the cell could die. Over millions of years, living cells have adapted to an environment exposed to UV-B electromagnetic radiation and have evolved by employing an enzyme in an attempt to repair the damaged DNA. These enzyme-driven microbial repairs can be derived from light energy (photorepair) or chemical energy (dark repair). However, as the length of time for UV exposure increases, so too is the risk of an incorrect DNA repair.

Exposure of DNA to a higher-energy-level UV-C light source coupled with the fact that this is where the DNA peaks in absorbing UV energy (240 to 290 nm) will result in even greater levels of molecular damage. DNA with increased levels of disruption to cellular processes due to incorrect repairs are more likely to be inactivated and will possibly die. High-energy UV-C radiation from a typical low-pressure

Ultraviolet Electromagnetic Radiation

mercury lamp emitting at 253.7 nm is very effective at inactivating viruses, bacteria, mold, and protozoa that can be harmful to humans. Some very lethal pathogens, such as those for anthrax, typhoid fever, diphtheria, cholera, dysentery, salmonella, and tuberculosis, can be inactivated at energy levels measured in millijoules per square centimeter.

A scanning electron microscope image of the waterborne *Giardia* protozoan parasite is shown in Figure 25.5. This dangerous parasite affects human's small intestines and is transmitted in animal fecal matter in the form of a cyst that is ingested in contaminated water supplies. *Giardia* infections can be fatal to individuals who have compromised immune systems. The infectious *Giardia* cyst can survive for weeks to months in the wilderness and is sometimes referred to as *Beaver Fever* because the source of the infection originated from contaminated beaver ponds. This particular waterborne parasite is highly resistant to both water chlorination and water that has been treated with ozone. However, the *Giardia* parasite is highly susceptible to UV radiation. Many public water systems are implementing UV-based treatments to prevent or address *Giardia-* and *Cryptosporidium*-contaminated water.

Many health officials worldwide are concerned about the potentially pandemic situations posed by the Avian influenza virus (H5N1), which causes bird flu. Health officials are taking steps to develop a vaccine before any major outbreak of bird flu occurs. The effects of the SARS virus from a few years ago on the worldwide economy and resulting loss of life are only part of the reason for these preemptive steps.

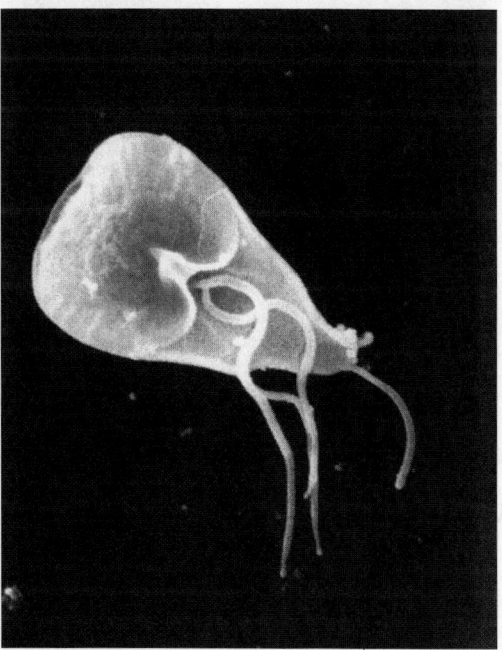

FIGURE 25.5 SEM image of *Giardia* protozoan parasite. (Courtesy of the CDC.)

More importantly, the action is based on the worldwide Spanish flu influenza (H1N1) pandemic that occurred between 1918 and 1920 that is shown in Figure 25.6. The Spanish flu mortality estimates ranged upward of 5% of the human population (50 to 100 million), and 400 million people worldwide were infected at the time. A greater proportion of the Spanish flu deaths occurred in healthy young adults than is normally associated with influenza, in as little as 1 to 2 days. Avian "bird flu" is a more virulent influenza strain with high fatality rates. If one considers the greater travel speeds and higher amount of international travel of today when compared to 1918, the pandemic concerns appear to be warranted. UV radiation can inactivate and kill the Avian flu virus, and measures can be taken to install UV systems in hospitals, office buildings, planes, and homes to minimize the spread of a pandemic influenza.

As the relative size of the target organism increases, generally so will the amount of UV electromagnetic radiation required to cause disruption to cellular processes. Figure 25.7 provides an overview of the relative sizes of DNA, a virus, bacterium, and a multicellular protozoan.

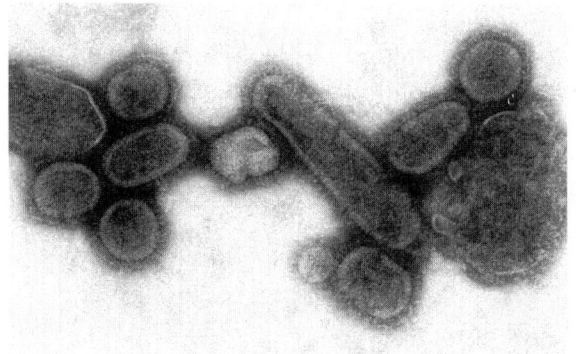

FIGURE 25.6 Recreated 1918 Spanish flu influenza virus. Courtesy of the CDC.

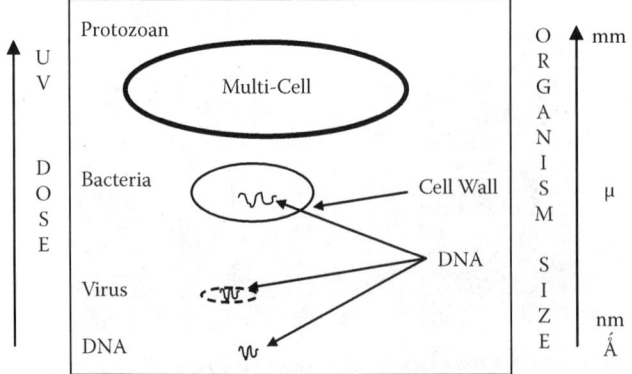

FIGURE 25.7 UV dose versus relative size.

The amount of UV required to inactivate a specific target organism involves many different factors in addition to the relative size of the target. The specific DNA chemical composition and, accordingly, the amount of UV absorption will vary between the DNA of a virus, bacteria, mold, or protozoan. The different rate of UV absorption in DNA is based on the nucleotides of DNA (adenine, cytosine, guanine, and thymine), and the pyrimidines (thymine and cystosine) have different rates of UV absorption. The particular shape of the microorganism will help determine the varying amounts of UV required to cause cellular damage. Possible shapes include, but are not limited to, spherical, spiral, rodlike, or filamentous and should also include other construction factors (cyst). In particular, keep in mind that UV must be able to strike the microorganism in order to inactivate the target, which is challenging in a very large UV air or water treatment system. Scattering can also be a factor; when the size of the target microorganism is much less than that of the UV wavelength, then Rayleigh scattering is present. When the target microorganism is larger than the wavelength, then empirical adjustments are generally made to account for this, including the shape of the target microorganism (rodlike versus spherical). Harmful microorganisms can withstand considerably more UV radiation in water than in dry air. Consequently, higher dosage levels are required to kill the same type of pathogen in water than in air.

The amount of cellular destruction is a function of both the time and the intensity of the UV electromagnetic radiation. A longer exposure time at a lower UV intensity level can be as effective as a short exposure time at a higher UV intensity. The UV dose is the product of the UV intensity (I) expressed as energy per surface area (microwatt-second/cm²) and dosage residence time (T) in seconds.

$$DOSE = I \times T = \mu W\text{-}s/cm^2 \ (mJ/cm^2)$$

In an actual UV germicidal application, there are many factors that must be considered. In UV water treatment systems, some of these factors include the following:

- Chemicals that are present in the water, such as iron, sulfites, and nitrides, which absorb UV
- Water turbidity (suspended solids), which diminishes UV dose levels
- Flow rate (seasonal effects) and distance of the microorganisms to the UV light source
- Specific types and amounts of target microorganism in the water

All of these factors need to be taken into consideration when establishing the UV dose. UV water treatment facilities are becoming more common in the United States, though the use of UV water treatment facilities has been more widespread in Europe. UV treatment systems do not create toxic by-products known as DBPs (disinfection by products). Moreover, UV systems do not use toxic and corrosive chemicals such as chlorine and ozone. The U.S. Environmental Protection Agency (EPA) has established the UV dose requirements for water treatment facilities. The EPA (2006) has developed detailed procedures for the validation testing of UV water treatment systems under the LT2ESWTR guidelines, which were established in November 2006 to ensure the system is working properly. After the system validation is completed, there are also guidelines in place for the safe operation and maintenance of these facilities. The UV dose requirements shown in Table 25.3 are for public water systems (PWS) for *Cryptosporidium, Giardia,* and viruses. The UV dose levels are expressed in millijoules for each target microorganism.

The EPA has established maximum concentration levels permitted for these harmful microorganisms and also specified the UV dosage needed to achieve an order-of-magnitude reduction in the concentration of the target microorganism. The concentration levels shown for the various log inactivation levels in Table 25.3 for the *Giardia* cyst are presented in Table 25.4. As you will note, the doubling of the UV dose will increase the destruction of the target pathogen by a factor of 10. Therefore, a 4 log reduction results in the destruction of 99.99% of the target pathogens, as shown in Table 25.4. The EPA guidelines are based on a UV light source operating at 253.7 nm.

As we discussed earlier, UV light will inactivate DNA-based microorganisms given a sufficient UV dosage, because the UV light breaks down DNA on a cumulative basis. As an example, as air circulates through the ductwork of an HVAC system containing UV light, the UV inactivates target microorganisms such as mold, bacteria, and viruses to improve the air quality for humans. If a target microorganism is not effectively eradicated on the first pass through the HVAC ductwork, the UV light will continue to break down the target microorganism's DNA on the second

TABLE 25.3
UV Dose Requirements

Target Pathogens	Log Inactivation (UV Dose in mJ/cm^2)			
	1.0	2.0	3.0	4.0
Cryptosporidium	2.5	5.8	12	22
Giardia	2.1	5.2	11	22
Virus	58	100	143	186

Source: Courtesy of the EPA.

TABLE 25.4
UV Dose for *Giardia*

UV Dose (mJ/cm^2)	Reduction in Number of Live Organisms (%)
2.5	90.0
5.8	99.0
12	99.9
22	99.99

pass and subsequent passes through the system. UV electromagnetic radiation can also be used in HVAC (heating-ventilation-air conditioning) systems to control the growth of mold colonies formed on the HVAC cooling coils. The mold forms because the cooling coils are generally in a dark, wet, and warm environment that is ideal for growing mold. The air treated with UV light helps to reduce incidences of inhaled pathogens for persons who reside or work in indoor environments. The Pennsylvania State University's Department of Architectural Engineering has an extensive airborne pathogen database and has done research on UV HVAC systems. This research has shown feasible payback periods for those who have installed UV systems to increase overall HVAC system performance based on higher efficiency, lower usage of electricity, and lowering employee absenteeism due to illness from greatly improved air quality having less microorganism (virus, bacteria, mold).

The opportunity for semiconductor UV LEDs in the coming years is that the technology can be optimized to produce emissions at a specific wavelength that is optimal for the particular application. In a germicidal application, the 265 nm wavelength is the most efficiently absorbed by the three possible pyrimidine dimers of DNA, thus increasing the amount of microorganism inactivation when compared to the conventional 253.7 nm wavelength emitted by low-pressure UV lamps.

25.10 ULTRAVIOLET LED

The UV LED consists of a P-N junction formed by two dissimilarly doped semiconductors with an abrupt change. The UV LED is formed by introducing dopants to the semiconductor base material to add either free electrons in the negative or N-type regions or to create holes to attract electrons, enabling current flow in the positive or P-type regions. When no external voltage is applied, the electrons near the junction zone flow until equilibrium is reached. This gives rise to a built-in potential barrier; however, there is no current flow. When an external electric voltage bias is applied across the junction that counteracts the potential built-in barrier, current can be made to flow, and when the holes from the P-type and electrons from the N-type meet at the junction and combine in a radiative process, a photon of light is released for direct band-gap materials, as shown in Figure 25.8. Indirect band-gap semiconductor materials are not efficient for light emission, because the holes and electrons combine indirectly. Similar to visible and infrared LEDs, there are also nonradiative recombinations in UV LEDs that take the form of heat known as *phonons*.

The wavelength of the light emitted is a function of the band-gap energy of the semiconductor materials used in constructing the UV LED. The band-gap energy refers to the energy threshold that electrons must pass to flow across the junction. The band-gap energy is also inversely proportional to the wavelength of light emitted from a semiconductor. Therefore, in order to emit light in the UV region, semiconductor materials with higher band-gap energies should be selected. In the near-UV to UV-A range down to 360 nm, a suitable semiconductor material would be gallium nitride (GaN) with a band-gap energy of 3.4 eV and a typical forward voltage (V_f) of 3.8 V. The substrate material selected for UV-A LEDs is generally sapphire, based on the lower cost when compared to silicon carbide (SiC) or aluminum nitride (AlN). By adjusting the materials used and the doping (i.e., the depositing of N-type

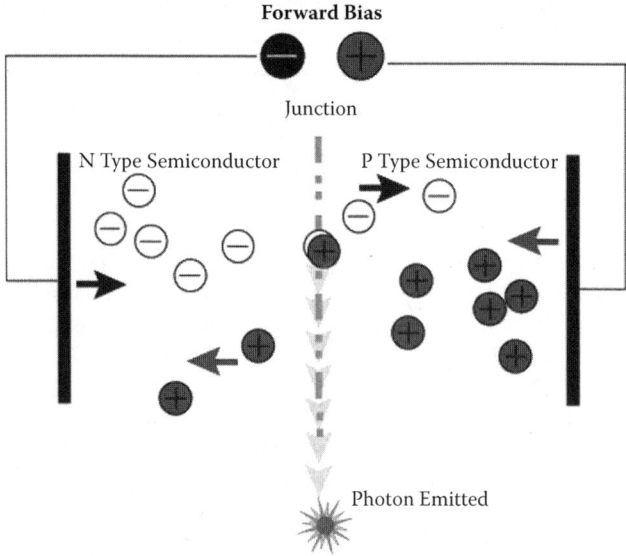

FIGURE 25.8 UV LED P-N junction.

or P-type impurities into the GaN), a wide range of wavelengths is possible, such as the wavelengths recently developed in the UV portion of the electromagnetic spectrum.

Deep UV (DUV) LEDs would require semiconductor material with a higher band-gap energy, such as AlN with a band-gap energy of 6.2 eV and a typical forward voltage (Vf) of 8.5 to 10.0 V. Increasing the amount of aluminum added to the semiconductor material, the shorter the output wavelength. Typically, magnesium (Mg) is used as the P-type doping, and silicon (Si) is used as the N-type. However, with increased aluminum concentration in the semiconductor material, it is more difficult to grow the structure. This is a result of the crystalline defects and impurities that can more easily occur in AlN. The defects and impurities lead to premature performance degradation, shortened operating lifetimes, and lower overall wallplug efficiencies of the LED. The defect and impurity issues are the challenges in developing DUV LEDs. However, UV LED material, process, and thermal performance are improving for this relatively new technology and are expected to improve dramatically in the coming years.

A state-of-the-art DUV LED structure by Nippon Telegraph and Telephone is shown in Figure 25.9. The aluminum nitride (AlN) UV LED device operates at 210 nm and is the world's shortest wavelength DUV LED device successfully demonstrated at this time. The structure is built upon a silicon carbide (SiC) substrate material. The light emission characteristics for the NTT DUV LED are shown in Figure 25.10.

UV LED structures can be fabricated using MBE, MOCVD, or HVPE processes discussed in Chapter 2. The HVPE manufacturing process is less complex than MOCVD or MBE and offers lower operating costs because it uses pure metals

Ultraviolet Electromagnetic Radiation

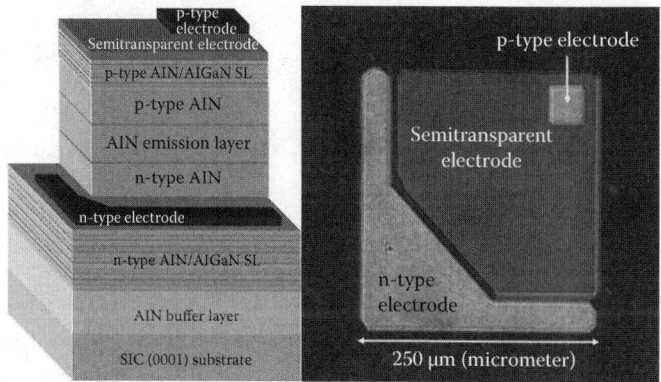

FIGURE 25.9 AlN DUV LED structure. (Courtesy of NTT Basic Research Laboratories.)

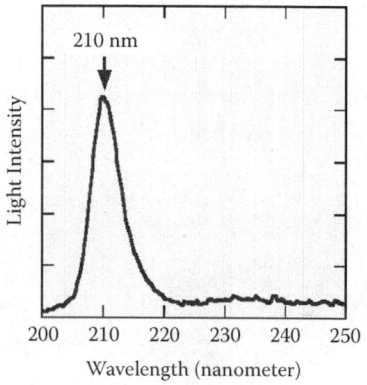

FIGURE 25.10 Emission characteristics of 210 nm AlN DUV LED. (Courtesy of NTT Basic Research Laboratories.)

in place of the more expensive metal-organic precursor gases. The HVPE process uses HCl (hydrogen chloride) gas flowing over heated Group III metals to form metal chlorides. The metal chlorides react with Group V metal hydrides to form III–V compounds. In the case of gallium nitride (GaN), the metal hydride is NH_3 (ammonia). An advantage of this process is the capability of growing thick GaN layers, which reduces the material defect density when compared to the MOCVD or MBE processes. However, one of the process drawbacks is the challenge of growing high-brightness devices. Hybrid growth processes that merge the best attributes of the MOCVD and HVPE reactor are starting to surface.

25.11 UV PACKAGING TECHNOLOGY

Material properties can drastically change in the UV spectrum; materials that were good reflectors in the visible and IR range can have high absorption rates in the UV range, making the material unsuitable in UV applications. Examples of metal packages include Kovar™ (a type of steel), which has a much higher absorption in the UV range and is generally replaced with a nickel or aluminum plating or material. The same is true of lens materials; many types of clear plastic lens material used for IR LEDs turn brown over time owing to UV exposure. Glass lens material used for IR LEDs tends to have unacceptably high absorption in the UV range. Some materials of choice in UV applications will be aluminum for reflectors and fused synthetic silica or fused quartz as the lens material to minimize UV absorption losses.

Several manufacturers offer commercial UV LEDs that are housed in metal can packages. The TO-18 and TO-46 are two very common package styles that were first developed about 40 years ago and are hermetic. Figure 25.11 shows a UV-A LED in a metal can TO-46 package. Figure 25.12 shows the metal header of a TO-46 with the domed lens cap removed, exposing the wire-bonded UV-A LED chip and a protective Zener diode chip. As you will note, the anode and cathode wire bond pads are located on the four corners of this large chip (1 mm × 1 mm) to ensure current spreading. It is very important to ensure there are no voids in the conductive die attach epoxy underneath the UV-A chip, which can cause stress fractures in the chip or catastrophic failure due to overheating as shown in Figure 25.13. Overheating can cause the gold wire to melt and even vaporize the metallization on the anode or cathode pads.

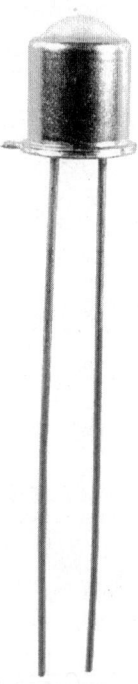

FIGURE 25.11   UV LED in domed lens TO-46 metal can package.

FIGURE 25.12 UV LED and Zener diode chips on header.

FIGURE 25.13 UV LED die attach and wire bond.

Plastic packages for UV LEDs should substitute silicones in place of the water clear epoxies generally used for IRLEDs and visible LEDs, because the UV emission will degrade the epoxy package, reducing the power output. A eutectic die attach is preferred for UV die attach for chips emitting below 350 nm because the UV emission can break down the conductive die attach epoxy.

Companies participating in the industrial UVA curing market using LEDs in place of medium-pressure UV lamps have developed higher-density UV-A/visible blue LED arrays to increase the UV power output in curing applications. Traditionally, adhesives and coatings have been processed using primarily heat to produce a reaction to cause the materials to harden. Often, this process is slow and energy intensive and in many cases involves the emission of volatile organic compounds (VOCs). Processing using UV light is fast, and most VOC emissions can be eliminated with minimal heating in the material being cured. Peak irradiance levels have now reached the 1.2 W/cm² and 100 W level for the largest arrays (multiple LEDs).

These power output levels for UV LED curing systems have developed using innovative techniques such as microreflectors plated directly to the ceramic substrate that also act as heat spreaders coupled with active cooling methods using water and Peltier technology to achieve high-power density LED arrays. These techniques allow for up to 1000 UV chips to be mounted in a packaged array.

One of the next generation of high-density packaging technology now includes silicon wafer-scale packaging. This innovative packaging technique processes silicon wafers into single-cavity package for UV LED, as shown in Figure 25.14. In this method, the silicon cavity is etched into multiple sites on the wafer much as an IC is, and then the UV LEDs or other LED chip is then placed into the silicon package. The left side of Figure 25.14 shows the wafer with individual package "chip" sites, and the right side shows an enlarged picture of a single silicon package. The silicon package can be used with the newer Flip-Chip or the existing wire bond technologies for chips up to 2.5 mm². The technology was originally developed for MEMS technology and then adapted for use with LEDs. More importantly, the LED chips will have an excellent thermal conduction path owing to the extremely thin wall of the silicon cavity. The cavity can also be used as a reservoir to hold a silicone coating applied on the LED chip. This process is used to ensure a long operating life of the LED chip, by providing a buffer layer to minimize external stress.

The largest LED chip on the market today measures slightly larger than 1 mm², so the silicon cavity package will be able to accommodate future LED chips as they increase in both size and power output. More importantly, the LED chips will have an excellent thermal conduction path owing to the extremely thin wall of the silicon cavity, with a thermal resistance of 4 to 5°C/W. The silicon-cavity-packaging technique

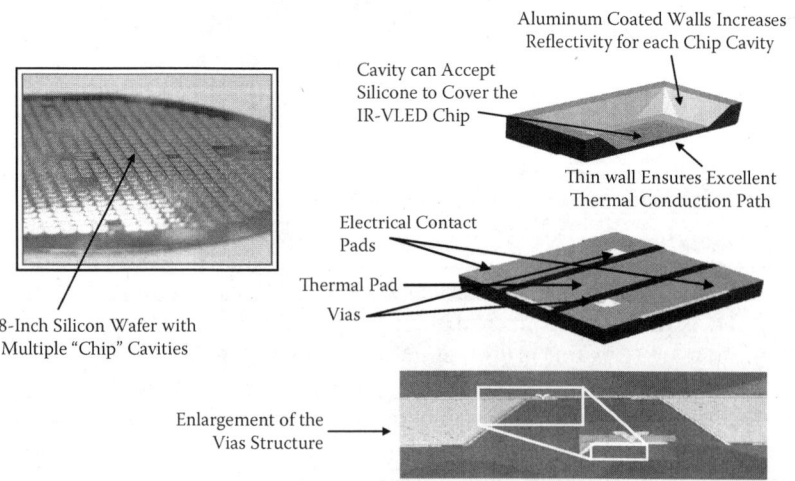

FIGURE 25.14 Silicon cavity packaging. (Courtesy of Hymite A/S.)

is applicable for IR, UV, and VLED chips at present. The aluminum reflector is ideal for UV LEDs with up to 92% reflectance, which is sputtered onto the silicon cavity in a vacuum. Anodic bonding techniques for attaching a fused silica glass cover are being developed to create a hermetic seal for this package, which would allow for UV-C LEDs on a wafer-scale level in applications such as water purification that would also be able to take advantage of the flowing water as a heat sink. Figure 25.15 shows the assembly process for a silicon cavity-type package.

25.12 CURRENCY AND DOCUMENT VALIDATION APPLICATIONS

Protecting the integrity of paper currency and other important financial documents such as stock and bond certificates against counterfeiting is fundamental to a sound monetary system. The U.S. Treasury Department and, specifically, the Secret Service Bureau, was established in 1865 by Congress for the purpose of controlling counterfeiting. The mission was to prevent and prosecute counterfeiting activity and thus maintain the public's confidence in the nation's currency. Over the years many different features were used to deter counterfeit U.S. currency. In 1861, the first circulation of paper money issued by the federal government occurred to finance the Civil War. These non-interest-bearing demand bills were green in color, and the popular nickname "greenbacks" has been in use since that time. Many additional anticounterfeiting measures have been taken since the first currency bills were issued such as the paper texture, paper weight, embedded fibers, intricate images, and serial numbers. Stock and bond certificates also adopted these features.

The U.S. Treasury Department has recently completed the security upgrade of U.S. currency that was initiated with the twenty-dollar bill in 2003 and completed with the release of the $5 bill in 2008. The new anticounterfeiting measures implemented include watermarks, new colors, microprinting, and security thread that emit a different color under ultraviolet radiation based on the specific denomination. The color-coded stripe can be seen by holding the bill in front of a strong source of white

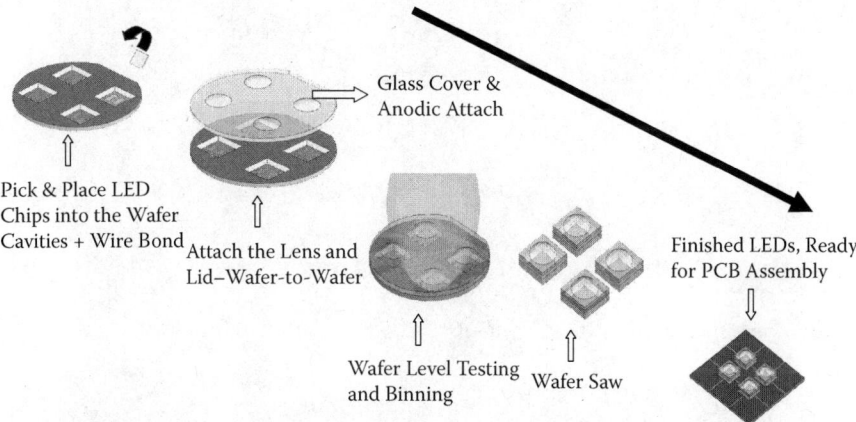

FIGURE 25.15 Silicon cavity process flow. (Courtesy of Hymite A/S.)

light. However, when illuminated with UV-A light, the security thread glows a bright color: blue for $5, orange for $10, green for $20, yellow for $50, and red for a $100 bill. Figure 25.16 shows U.S. and British currency illuminated with fluorescent lighting and also with 365 nm UV-A light emitted from UV LEDs in a dark room. UV-A LEDs are now being investigated as replacements for mercury-based UV tubes. U.S. passports and many credit cards have implemented UV threads and materials in their anticounterfeiting efforts. A practical application is to include a UV-A LED emitter into a cell phone, allowing consumers to conveniently validate the integrity of their currency. These measures will greatly increase both the technical challenge and financial costs to forge currency and financial instruments, thus maintaining the integrity of the worldwide monetary system.

25.13 INDUSTRIAL UV CURING APPLICATIONS

UV curing uses electromagnetic radiation in place of a solvent that is traditionally evaporated with heat or left to dry, leaving behind the solid ink or coating material. The environmental standards in place today greatly restrict the use of solvents or volatile organic compounds (VOCs), which evaporate into the atmosphere and are harmful to humans. UV curing uses UV electromagnetic energy (photoenergy) to start a chemical reaction known as *photopolymerization*. UV curing uses a mixture of fillers, wetting agents, monomers, oligomers, and photoinitiators that creates polymer chains almost instantaneously when UV light is introduced to the mixture. The photoinitiators comprise a very small portion of the mix but readily absorb the UV energy, creating free radicals, which begins the polymerization process. Interestingly, the monomers, oligomers, and photoinitiators all coexist without reacting with each other, until the photoinitiators are exposed to light of the correct wavelength and

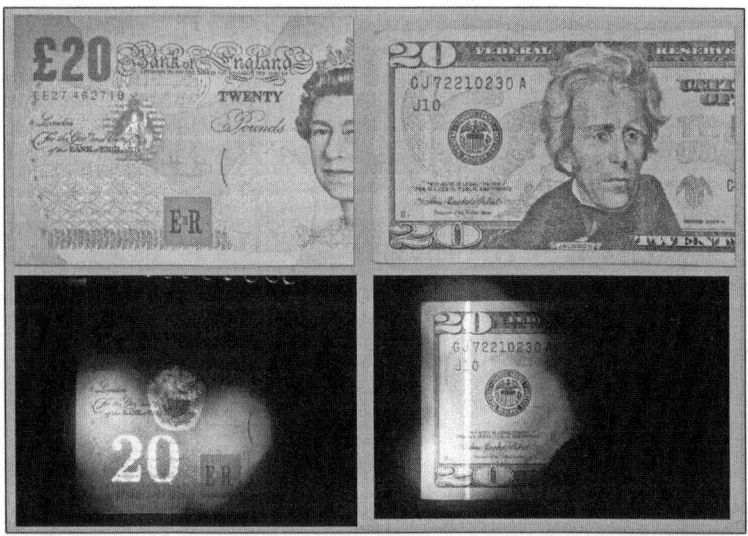

FIGURE 25.16 Currency illumination at 365 nm.

intensity. This enables UV curing to offer a "cure on demand," which is an advantage compared to the working time limitations for a solvent-based processes.

During the 1950s, research started to speed up the curing process by cross-linking polymers using photoinitiators that are sensitive to UV light, using mercury-based lamps. UV curing of adhesives and coatings has been in existence commercially for about 35 years. Medium and high-pressure mercury UV lamps serve the majority of the industrial UV curing equipment market owing to the high output power levels achieved (200 to 300 W/in.). Potential drawbacks are short operating life of the lamp, a 360° emission pattern, susceptibility to equipment vibration, high operating temperature, RoHS materials, and the potential for pressurized lamps to explode upon failure. The operating temperature, as we have already discussed, can be quite high (600°C to 900°C) and potentially harm the substrate material.

The basic factors driving the growth of UV curing methods are as follows:

- Elimination of volatile organic compounds (VOCs) and the associated environmental compliance reporting and permitting costs required for VOCs. Grand fathered manufacturing facilities using VOCs are limited in expanding their operations.
- Worker health and safety concerns—VOC-based products require ventilation and breathing apparatus/health monitoring in addition to higher insurance premiums.
- Manufacturing yield improvements, increased product performance, and lower total costs.
- Lower energy costs to operate and a much smaller UV equipment footprint when compared to traditional drying kilns and ovens.
- Greatly increased working time with UV—"a cure on demand process" with UV light.

The UV curing processes have evolved over the past 35 years depending on how a particular photoinitiator performs with a specific type of UV lamp (medium or high pressure). Some of these lamps have spectral wavelength outputs ranging from UV-C through the visible and into the IR spectrum. Most UV curing applications were developed and based on the specific UV lamp output and the curing performance of a particular coating or adhesive. Chemical photoinitiator manufacturers are now looking at formulating new products that are better suited for longer UV-A (>365 nm) and visible LED wavelengths. The shorter-wavelength UV will be absorbed more at the surface of the material and the longer visible wavelengths will penetrate deeper. Adding the visible LED for curing further enhances the curing depth, and it is easy to build arrays using LED chips with different wavelengths. UV LEDs are a relatively new technology and have not yet been optimized in performance or cost. However, some benefits and attributes of semiconductor UV LED technology are the following:

- LEDs are instant-on compared to warm-up time and re-strike times of 15 min for UV lamps.
- LEDs are more robust than lamps, and are not prone to equipment vibration and breakage.

- LEDs do not explode upon failure; therefore, factory downtime is less.
- LEDs do not heat the substrate material, unlike medium and high-pressure UV, which do.
- UV-A LEDs have a smaller footprint and do not have a 360° emission pattern.
- LEDs have a predictable degradation output, whereas UV lamps are not as predictable.

25.14 AIR–WATER–SURFACE DISINFECTION

UV air–water–surface disinfection are established market applications that have employed UV mercury-based lamps for some time. Chlorine water treatment systems will continue to face growing competition from both ozone and UV lamp disinfection systems, and it is predicted that they will be gradually phased out in the coming years. Chlorine is chemically active and can react with foreign ingredients such as those as found in industrial waste waters to form toxic compounds. Chlorine can combine with ammonia to form chloramines, which are acutely toxic to fish at low concentration. Chlorination is not effective in inactivating *Cryptosporidium*, which has no cure and causes severe stomach cramps and diarrhea. Recovery time from this illness is typically 14 days for people without a weakened immune system. New EPA legislation requires the removal of 99% of *Cryptosporidium* using several methods that have helped increase the usage of UV disinfection.

UV-C water treatment can be used in a variety of applications to disinfect water for drinking, processing wastewater, in pools and spas, beverages, and industrial processing. Industrial processing would include ultrapure water for pharmaceutical, cosmetic, and semiconductor industries and for obscure applications such as maritime ballast water and eliminating sulfate-reducing bacteria in offshore oil drilling. According to 2007 statistics from the American Water Works Association, there are more than 2000 UV drinking water treatment systems operating in Europe and over 1000 UV systems in the United States.

The largest UV disinfecting treatment facility in the world is being implemented for the city of New York. The New York City UV treatment facility is capable of processing up to 2.2 billion gallons per day and serves over 9 million consumers daily. The UV disinfection treatment facility will cost 1/4th of a comparable filtration plant, and it will require approximately 1/10th the space. The UV treatment facility will comprise 56 separate processing units capable of disinfecting 50 mgd (million gallons per day) under worst-case conditions. The city adopted a very conservative (higher) UV dose of 40 mJ/cm^3, which will ensure a 99.9999% UV kill rate for the deadly *Cryptosporidium* protozoa (City of New York, 2006). The contact time to inactivate microorganisms and disinfect the water is approximately 20 to 30 s in a single pass.

All of the UV water treatment facilities discussed are based on UV lamps. UV-C LED power output levels are at present several orders of magnitude lower than needed to inactivate microorganisms. However, in the coming years, improved LED chip design coupled with higher density packaging and improved thermal management will make inroads. Water treatment application for UV LEDs is unique since the high volume of flowing water in the systems could utilize the water to remove a significant portion of the heat generated. As mentioned earlier, low-pressure UV-C lamps operate best at a wall temperature of approximately 40°C (100°F) and lose efficiency at temperatures below or above it. HVAC systems also could utilize the high-velocity cool air to enhance LED performance. UV surface disinfection technology is employed in many areas of the food industry for many of the same reasons UV technology is utilized in water and air disinfection. Surface disinfection is used in the food industry for both packaging and food products to greatly improve the quality of food and extend product shelf life. UV surface disinfection is very effective for food products having a minimum resistance to microbial contamination.

25.15 MEDICAL AND FORENSIC APPLICATIONS

Another major use UV technology is phototherapy to treat psoriasis and other skin conditions. Phototherapy includes a broad range of medical treatments using light. Psoriasis is a persistent and chronic skin disease that has a tendency to be genetically inherited. Psoriasis may affect a small localized area, but it can also cover the entire body. It can be treated with UV-A or UV-B wavelengths. UV-A is done in conjunction with a photosensitizing agent, which allows for a lower UV dose. After several treatments, improvement can be seen in as little as 3 weeks, with maintenance therapy thereafter. UV dental applications include curing (UV-A to Blue VLED) for cavity fillings, brightening, and UV-C for toothbrush and medical instrument sterilization.

The medical analytical instrument market also utilizes UV light sources in fluorescence spectroscopy and ultraviolet-visible spectroscopy. Fluorescence spectroscopy is a type of electromagnetic spectroscopy that analyzes the fluorescence emitted from a sample being irradiated and evaluated. The light source is generally UV to excite the electrons in the specimen to emit light of a lower energy level, usually in the visible spectrum. In fluorescence spectroscopy, the sample is excited, by absorbing the higher-energy UV light, causing the sample to move from its ground electronic state to one of the various vibrational states in the excited electronic state. Analysis of the emission spectrum will permit the identification of the substance (chemical compound, food processing, cancer tumor, etc.). Fluorescence spectroscopy is also used in forensics and chemical research fields. Ultraviolet-visible spectroscopy (UV/VIS) uses multiple wavelengths of light in the visible, ultraviolet, and near-infrared ranges. The absorbance of light in a solution is directly proportional to the solution's concentration (Beer–Lambert law).

UV light sources are fundamental tools for forensic investigative work. The U.S. Department of Justice in the revised *Processing Guide for Developing Latent Prints*, which includes UV light sources for all types of surfaces (porous and nonporous) issued the FBI Laboratory Division in 2000 (FBI). UV light sources have vastly improved collection of human DNA evidence (oils, amino acids, blood) at a crime scene by making the evidence highly visible to investigators. UV light can also be used by police to discover former wounds, bite marks, and bruises not revealed by the visible spectrum for up to 6 to 9 months after the injury was inflicted and to reveal bruise wound details that were not visible under sunlight. Child Services investigative groups use UV illumination to identify injuries up to 6 to 9 months after they have occurred.

Bibliography

Bhattacharya, P., *Semiconductor Optoelectronic Devices (2nd edition)*, New Jersey, 1997.
Brophy, J., *Basic Electronics for Scientists (2nd)*, McGraw-Hill, New York, 1972.
Chappell (editor), *Optoelectronics Theory and Practice*, Texas Instruments, Ltd., Bedford, England, 1976.
Daly, J. C. and Malipeau, D. P., Analog Bi CMOS Design, CRC Press, 2000 ISBN: 0-8493-0247-1.
Davidson, J.N., *The Biochemistry of the Nucleic*, 1965, London: 5th Methuen and Co., Ltd.
The Electronic Engineer, A Course on Optoelectronics, Chilton Company, Bala Cynwyd, Pennsylvania, 1971.
LMV793 Operational Amplifier d*ata sheet,* National Semiconductor (August 2007 revision).
Gage, S., Evans, D., Hodapp, M., Sorensen, H., *Optoelectronics Applications Manual*, McGraw-Hill, New York, 1977.
Gage, S., Evans, D., Hodapp, M., Sorensen, H., Jamison, D., Krause, B., *Optoelectronics/Fiber-Optics Applications Manual (2nd)*, McGraw-Hill, New York, 1981.
Hecht, E., Zajac, A., *Optics*, Addison-Wesley, Reading, MA, 1974.
Horenstein, Mark N., *Microelectronic Circuits and Devices*, Prentice-Hall Professional Technical Reference, Second Edition, June 1995.
JEDEC Standard No, 77, JEDEC Solid State Products Engineering Council, 1981.
Jagger, John, *Introduction to Research in Ultraviolet Photobiology*, 1967, Prentice-Hall, Englewood Cliffs, NJ.
Schubert Fred, E., *Light-Emitting Diodes, 2nd edition*, Cambridge University Press, Cambridge, 2006.
Tsuboi, K. K., Mouser liver nucleic aids. II, Ultraviolet absorption studies. *Biochim. Biophys. Acta* 1950, 6: 202–209.
Wang, Tony, Erhman, Berry, *Application Report SBOA055A Compensate Transimpedance Amplifiers,* Texas Instruments, March 1993, revised March 1995.
United States Environmental Protection Agency (EPA), *Ultraviolet Disinfection Guidance Manual for the Final Long Term 2 Enhanced Surface Water Treatment Rule*, November 2006.
Wilson, J., Hawkes, J., *Optoelectronics: An Introduction.* Prentice-Hall, Englewood Cliffs, NJ, 1983.
The following application bulletins, published by OPTEK/TT electronics, were used with permission in the preparation of this text:
  A.B. 105, Thermal Behavior of GaAs LEDs, Cognard, W. Nunley.
  A.B. 108, Motion Sensing with Optical Interrupters, T. Sward, and W. Nunley.
  A.B. 111, Soldering to Semiconductor Leads, W. Nunley and H. Brown.
  A.B. 112, Two Channel Optical Interrupters, V. Dahlberg and W. Nunley.
  A.B. 113, Reflective Assemblies, T. Sward.
  A.B. 114, Gallium Aluminum Arsenide, D. Wolfe.
  A.B. 116, Linear and Rotary Encoders, J. Davidson and L. Johnson.
  A.B. 118, Understanding Infrared Diode Power Ratings, K. Bailey.
  A.B. 119, A Comparison of Plastic versus Metal Packaging for Infrared Sensors and Emitters, M. McCrorey.
  A.B. 120, Designing a Wide Gap Optical Switch, T. Eichenberger.
  A.B. 121, Understanding the Dissipation Rating of the Optical Semiconductor, M. McCrorey.
  A.B. 229, Designing Encoder Elements for Two Channel Optical Instruments, Mary Cawley.

Glossary

Acceptable quality level (AQL) — The maximum allowable average percentage of defective components that a supplier is permitted to present for acceptance. Also referred to as *acceptance quality level*.

Acceptance angle — The angular limit off the optical axis for which a photosensor will still detect an energy ray to give at least a half-power level signal.

Acceptance cone — A cone defined by the acceptance angle such that a ray within the cone will be above the half-power level.

Alignment — The process of positioning emitter and sensor for maximum infrared transmission (optical alignment) while placing the beam in the desired location. Infrared-sensitive scopes or luminescent materials are often used as aids for aligning complex infrared optical systems.

Ambient temperature — The temperature of the gas or liquid surrounding a component.

AlInGaP — Semiconductor materials used for red, red-orange, orange, and amber LEDs.

Alloy — A composition of two or more materials, usually metals, combined for the purpose of improved performance characteristics.

Angle of emission — The angle at which a ray exits an optical package, measured from the optical axis.

Angstrom — A unit of length equal to 10^{-10} m, sometimes used to quantify the wavelength of electromagnetic radiation.

Anode — The positive side of a diode to which a positive voltage must be applied to facilitate a forward current.

Aperture — An opening used to control or limit the transmission of electromagnetic radiation.

Aperture angle — (Used interchangeably with beam angle.) It is usually better to specify whether a given measurement refers to the angle measured from the optical axis to the half-power point (half-angle) or between the two half-power points (included angle).

Axis of measurement — The direction from which ratiometric measurements are taken.

Alpha — Used in electrical engineering to refer to emitter–collector gain (transistor connected as a common base amplifier). In a junction transistor, alpha is less than one.

Band gap energy	In semiconductor material, electrons are confined to bands of energy, and are barred from entering other regions. However, electrons are able to move from the valence band of an atom to the conduction band. The band gap of a material is the energy difference between the top of the valence band and the bottom of the conduction band. There is a specific amount of energy required to make this transition, and it is different for different materials.
Base	The P portion of an NPN transistor, and in the case of a phototransistor, the photosensitive region. For a PNP transistor, the N region is the control terminal or base.
Beam angle	(Used interchangeably with aperture angle.) It is usually specified as the included angle from one half-power point to the other.
Beam half angle	The angular displacement of emitted electromagnetic radiation from the optical axis to the point of half maximum power.
Cathode	The negative region of a P-N diode to which a negative voltage or ground must be applied for forward current.
Chromaticity	The qualities of visible color associated with hue and saturation but not intensity.
Collector	The terminal of a transistor to which a bias voltage (positive for NPN types) is normally applied.
Collector–base breakdown voltage	The voltage at which current suddenly rises from normal leakage to a higher specified level. Positive voltage would be applied to the collector, the emitter would be open, and the base grounded or negative for this condition.
Collector current	A measure of the current flowing through the collector, expressed in amperes.
Collector–emitter breakdown voltage	The voltage at which current suddenly rises from normal levels to a higher specified level. The transistor is normally biased with no base input.
Collector–emitter saturation voltage	The voltage measured across the collector to emitter with a specified optical or electrical base input and output load.
Color rendering index (CRI)	A subjective method of determining how well a light source renders color to the average observer (scale is 0 to 100).
Color temperature	The physical temperature of a blackbody emitting radiation peaked at a wavelength corresponding to that emitted by a non-blackbody radiator.
Common emitter	The grounded emitter configuration where the emitter terminal is common to both input and output.

Convection	The transfer of heat by currents in a liquid or gas.
Correlated color temperature (K)	Any set of chromaticity coordinates that are the most similar to the blackbody standard is described as having that color temperature.
Critical angle	The angle of incidence at which light is no longer transmitted through an N1–N2 interface but is instead reflected by the interface.
Current sensing	The application of an optoisolator to detect current (through the IRLED) in a circuit, such as a power supply or telephone network.
Current transfer ratio	Output current divided by IRLED current under specified test conditions (a ratio).
Dark current	A measure of the current in a phototransistor or other photosensor with no radiation present.
Dark repair	Enzyme-driven microbial repair of DNA or RNA derived from chemical energy as a result of damaging exposure to UV.
DC current gain	Collector current divided by base current (h_{FE} [DC], h_{FE} [AC]; a ratio).
DC isolation current	The current between input and output at a specified test voltage with all input leads shorted together and all output leads shorted together.
DC isolation voltage	The rated (and tested) input-to-output DC isolation voltage limit with input leads shorted together and output leads shorted together.
Deep UV (DUV)	Ultraviolet electromagnetic radiation that encompasses both the UV-C range (200 to 280 nm) and the UV-B range (280 to 315 nm).
Diode	A two-terminal semiconductor device that conducts freely in one direction only.
Dominant wavelength (nm)	The color the eye actually sees regardless of the peak wavelength emitted.
Duty cycle	Pulse width divided by total cycle time, usually expressed as a percentage of time that the device is on.
Duty factor	Duty cycle expressed as a ratio.
Efficiency	Referring to a photodiode in photovoltaic mode, the ratio of maximum power output to total incident radiation energy.
Emission angle	See Beam angle.
Emitter	(1) The negative voltage terminal of an NPN transistor under normal operating bias. (2) Used to refer to an infrared-emitting diode. (3) Refers to any source of radiation (i.e., infrared emitter).

Emitter–collector breakdown voltage	The voltage at which emitter–collector current rises to a nondestructive specified level, that is, significantly higher than normal leakage current.
Emitter current	(1) A measure of the current flowing through a transistor's emitter terminal. (2) The current through an IRLED.
Epitaxy	A method of depositing a monocrystal film on a monocrystal substrate.
Fall time	A measure of the time required, under specified test conditions, for a waveform to drop from 90% to 10% of its original level.
Fluorescent bulb	A sealed tube containing mercury vapor and lined with a phosphor coating. When current is passed through the vapor, the ultraviolet emission excites the phosphor-emitting visible light. The ultraviolet emissions are absorbed by the glass.
Flux	The total amount of light emitted from a source in all directions.
Focal plane	The plane at which rays converge to form a focused image.
Germicidal	Using ultraviolet electromagnetic radiation to inactivate microorganisms (200 to 300 nm).
Germicidal lamp	Generally refers to a low-pressure mercury-based UV lamp with a primary emission at 253.7 nm.
Goniometer	A spectroradiometer attachment capable of measuring VLED angular light projections.
Hydride vapor-phase epitaxy (HVPE)	An epitaxial growth technique often employed to produce semiconductors such as GaN, GaAs, InP, and their related compounds.
Illuminance (lumens/m^2 = 1 lux = 1 ft-candle)	The flux per unit area falling on a particular surface.
Incandescent bulb	A bulb that emits light when an electric current passes through a resistant metallic filament in a vacuum enclosure.
InGaN	Semiconductor material that transmits in UV, blue, cyan, greenish-yellow, green, and blue source for white phosphor LEDs.
Interrupter	Synonym for transmissive assembly.
Interrupter module	Synonym for transmissive assembly.
Junction temperature	The temperature of the P-N junction of an LED under bias.
Luminance	The visual perception of how bright a source is at some distance and angle.

Glossary

Molecular beam epitaxy (MBE)	A method of depositing single crystals by using a "beam" of evaporated atoms that do not interact with one another or vacuum chamber gases until they reach the wafer, owing to the long mean free paths of the atoms.
Metal organic chemical vapor deposition (MOCVD)	A chemical vapor deposition method of epitaxial growth of materials, especially compound semiconductors from the surface reaction of organic compounds or metal-organics and metal hydrides containing the required chemical elements.
Multimode fiber	A type of optical fiber mostly used for communication over shorter distances, such as within a building or on a campus.
Operating life	The minimum length of time that a device may safely be expected to perform within established device specifications.
Optical axis	The line designated as such and used as a measurement reference; usually, the locus of emitted or received rays, perpendicular to the focal plane.
Optical coupling	Energy transfer from a source to a photosensor.
Optical matching	Refers to similar optical package designs for IRLEDs and photosensors.
Optocoupler	A device designed to provide electrical isolation through conversion to an optically coupled signal and back to an electrical signal.
Optoelectronics	A term used for electronic systems or components that interface with optical systems or components.
Optoisolator	Synonym for optocoupler.
Parabolic reflector	A reflective disk-shaped sheet formed such that a cross section corresponds to a mathematically calculated parabola. A parabolic reflector will have a clearly defined central focal point useful for accurately collimating a point source or focusing collimated light to a single point.
Peak wavelength	The wavelength for which the emitted power is greatest; that is, the wavelength of maximum power output.
Phase	A concept used to measure the relative positions of separate waveforms.
Phonon	A phonon is a mode of vibration occurring in a rigid crystal lattice, such as the atomic lattice of a solid, and is also the primary mechanism by which heat conduction takes place.
Phosphor	A powdered substance that exhibits fluorescence when excited by ultraviolet or blue radiation.

Photoconductive	A term referring to increased conductivity as a function of increasing incident radiation.
Photocoupler	Synonym for optocoupler.
Photocurrent	Current that is generated in a photosensor as a result of incident radiation.
Photodarlington	A photosensor consisting of a Darlington transistor (i.e., two cascaded transistors together on a single piece of semiconductor material) with a photosensitive base region in the first transistor.
Photodetector	A general term usually referring to a photodiode, phototransistor, photodarlington, photo IC, or other device that responds electrically to radiation.
Photodiode	A two-terminal semiconductor device designed to conduct when exposed to incident radiation.
Photoemissive device	Synonym for emitter or IRLED.
Photoemitter	Synonym for emitter or IRLED.
Photometry	The measurement of visible light based on the response of the average human observer.
Photon	A single unit of electromagnetic radiation; equal in energy to the product of frequency and Planck's constant.
Photosensitive device	Synonym for photosensor.
Photosensor	A general term usually referring to a photodiode, phototransistor, photodarlington, photo IC, or other device that responds electrically to radiation.
Phototransistor	A transistor designed with a photo-sensitive base region so that conduction occurs when radiation is incident upon the base.
Phototriac	An integrated circuit designed to pass current in both directions when incident radiation exceeds a specified threshold.
Photopolymerization	A chemical reaction that is initiated with light energy to create polymer chains.
Photorepair	Enzyme-driven microbial repair of DNA or RNA derived from light energy as a result of damaging exposure to UV.
Phototherapy	The use of light, including UV-A and UV-B, for medical purposes such as treating skin conditions such as psoriasis.
Point source	A single point that is the starting place for all emitted rays; however, in actual practice, optoelectronic design may treat an emitting area as a point source as long as specified ratios are not exceeded.
Polar coordinates	A system for location using distance and angle as point coordinates, which are sometimes used to plot emission patterns.

Power	A measure of the rate at which work is done, generally expressed in watts.
Power derating	The system used to correlate the cooling effect possible at a given ambient temperature with the power dissipation capability of a device.
Radiance	The intensity of the energy passing through an area, divided by the measure of the area.
Radiant flux	The measure of radiant power, usually expressed in watts.
Radiant intensity	The measure of radiant flux per unit of solid angle, usually expressed in milliwatts per steradian.
Radiant incidence	The measure of radiant flux incident upon a surface, that is, milliwatts per square centimeter.
Radiation pattern	A description of radiant incidence of radiant intensity as a function of position from the source, usually in a single plane.
Ray path	An imaginary line perpendicular to a wave front, which depicts the path of the wave.
Reflective assembly	Synonym for reflective sensor.
Reflective sensor	An assembly consisting of IRLED and photosensor used to detect objects by reflecting an infrared or light beam off the object.
Responsivity	A measure of the sensitivity of a photodetector equal to output current divided by radiant flux incident upon the sensing surface.
Reverse bias	For a diode junction, interconnection in the nonconductive mode.
Reverse current	The current that flows in the reverse-bias condition.
Rise time	A measure of the time required, under specified test conditions, for a waveform to increase from 10% to 90% of its peak level.
Rotary encoder	A position sensor that signals rotary motion. Optical rotary encoders may detect speed, direction, and absolute position, depending on complexity.
Secondary optics	Any optical lens or diffuser placed over the LED but not a part of the original manufactured device.
Silicon	The base material for photosensors and other semiconductor materials when used in relatively pure crystalline form with specific materials added.
Single-mode fiber (SMF)	An optical fiber designed to carry only a single ray of light (mode).
Slotted switch	Synonym for transmissive assembly.

Snell's law	The principle that predicts the change in ray path as radiation crosses an optical interface; namely, $n_1 \sin \theta_1 = n_2 \sin \theta_2$, where n_1 and n_2 are respective indices of refraction and incident angles are measured from a normal to the interface.
Spectroradiometer	An instrument used for measuring the radiant energy and wavelengths in specific bands of the electromagnetic spectrum. All photometric, radiometric, and colorimetric parameters can be mathematically calculated from these readings.
Spectrum	A term used to refer to the broad range of wavelengths within a given category of radiation; that is, visible spectrum.
Spectral bandwidth	The range of wavelengths between outer half-power points for a given source.
Steradian	A solid angle that intersects a surface area equal to the squared radius of a sphere.
Storage temperature	The maximum and minimum safe temperatures for storage of a device without damage occurring.
Supply voltage	The voltage required to operate a circuit.
Thermal conduction	The transmission of heat across matter.
Thermal radiation	Electromagnetic radiation from an object's surface due to its temperature.
Thermal resistance	A measure of the LED's ability to dissipate internally generated heat.
Transistor	Semiconductor device with two junctions and three terminals known as *collector*, *base*, and *emitter*.
Transmissive assembly	An assembly utilizing an infrared ray path from IRLED to photosensor to detect the presence of an object that breaks the infrared ray path.
Transfer ratio	See Current transfer ratio.
Trigger level	As applied to photosensors, the levels of radiant intensity at which the device turns on or off.
Trigger voltage	The voltage levels that trigger a device to turn on or off.
UV	Abbreviation for ultraviolet electromagnetic radiation with wavelengths ranging from 1 nm to 400 nm.
UV-A	Ultraviolet electromagnetic radiation with wavelengths from 315 to 400 nm.
UV-B	Ultraviolet electromagnetic radiation with wavelengths from 280 to 315 nm.
UV-C	Ultraviolet electromagnetic radiation with wavelengths from 200 to 280 nm.
UV dose	It is the product of the UV intensity (I) expressed as energy per surface area (microwatt-second/cm^2) and dosage residence time (T) in seconds.

UV intensity	Amount of energy per surface area (microwatt-second/cm^2).
Vacuum UV	Ultraviolet electromagnetic radiation with wavelengths from 10 to 200 nm.
VOC	Acronym for volatile organic compound; refers to toxic organic solvents that are traditionally evaporated with heat, leaving behind a solid ink or coating material.
Visible emitter	Synonym for VLED.
Visible LED	Synonym for VLED.
VLED	A diode that emits visible radiation when forward current passes through its junction.
Visible spectrum	That region of the electromagnetic spectrum by which the eye sees. It encompasses that region lying between 400 to 780 nm.
Wavelength	A measure of the length of an electromagnetic wave, calculated by dividing the speed of light by the specific frequency of the light, but usually measured empirically.
Zero voltage crossover	A type of relay output that will turn-on when the line of output is near zero.

Index

A

AC input AC output, solid-state relays, 171–172
Accelerated degradation, 182
Active thermal management, 185–187
Air disinfection, ultraviolet electromagnetic radiation, 296, 312–313
Aircraft lighting, visible-light-emitting diodes, 288
Alignment, 52, 129, 199, 317
Alloys, 21, 28–30, 317
Aluminum arsenide, utilization in light-emitting diodes, 5
Aluminum foil tape, reflective optical switch, 144
Aluminum gallium indium, utilization in light-emitting diodes, 5
Aluminum nitride, utilization in light-emitting diodes, 5
Anodes, 18, 62, 108–109, 306, 317
Apertures, 36, 48, 51–53, 125–134, 139, 147, 317–318
Architectural applications, 287
    visible-light-emitting diodes, 287
Automotive applications
    optical sensors, 241–243
    visible-light-emitting diodes, 287
Automotive mood lighting, visible-light-emitting diodes, 288

B

Backside processing, in processing photointegrated circuits, 106–107
Beam angle optical measurement, 278
Billboards, visible-light-emitting diodes, 287–288
Black carbon-based ink, reflective optical switch, 144
Black dye-based ink, reflective optical switch, 144
Black ionized aluminum, reflective optical switch, 144
Blue smooth plastic surface, reflective optical switch, 144
Boat lighting, visible-light-emitting diodes, 288
Bond paper, reflective optical switch, 144
Britain, certification agency, 157
British Standards, certification, 157
Bubble sensing for medical applications, optical sensors, 257–260
Buried layer, in processing photointegrated circuits, 101–102
Business equipment applications, 227–230
    check/card reader, 229
    copiers, 227
    data interface, 227–229
    keyboards, 227
    mice, 227
    optical couplers/isolators, 229–230
    printers, 227
    touch screens, 227

C

Camera applications, optical sensors, 254
Canadian Standards Association, certification, 157
Capacitors, 110
Capacity, fiber-optic communication, 196
Cathode, 18, 22, 62, 108–109, 291–292, 306, 318
CD discs, optical sensors, 249–250
Certification agencies, 157
Channel letter/contour lighting, visible-light-emitting diodes, 287
Check/card readers, optical sensors, 229
Chip, visible-light-emitting diodes, 269–271
Chip centering
    IRLED packaging, 29
    plastic, metal package comparison, 29
Chromaticity, 275, 277, 318–319
Circuits, photointegrated, 99–113
    backside processing, 106–107
    buried layer, 101–102
    capacitor, 110
    characterization, 110–113
    contact, 105
    deep N+, 102–103

Epi reactor (epitaxial reactor), 102
isolation, 102
lateral pnp, 109
metal, 105–106
N+, 104
other IC circuit devices, 107–110
P+, 103
passivation, 106
photodiodes, 108–109
processing, 101–107
resistors, 104–105
starting wafer, 101
theory, 99–100
vertical pnp, 110
CMOS, interface circuits, photosensor interfacing, 221
Coin changers, optical sensors, 250–251
Color rendering index, 277, 287, 318
optical measurement, 277
Color sensors, 120–121
Color temperature, 181–182, 275–278, 282, 284–287
Computer peripheral applications
check/card reader, 229
copiers, 227
data interface, 227–229
keyboards, 227
mice, 227
optical couplers/isolators, 229–230
optical sensors, 227–230
printers, 227
touch screens, 227
Consumer applications, optical sensors, 249–254
camera applications, 254
CD disc, 249–250
coin changers, 250–251
dollar bill changer, 250
DVD disc, 249–250
game controls, 249
household appliance controls, 254
optical golf game, 254
slot machines, 253–254
smoke detectors, 251–253
TV controls, 249
Consumer portables, visible-light-emitting diodes, 287
Contact, in processing photointegrated circuits, 105
Contour lighting, visible-light-emitting diodes, 287

Copiers, optical sensors, 227
Correlated color temperature, optical measurement, 277
Cost of fiber-optic communication, 195
Coupled emitter (IRLED) photosensor pair, 123–152
Coupler types, 156
Critical angle, 17–20, 28, 192, 274, 319
CSA (Canada), certification agency, 157
Currency validation, ultraviolet electromagnetic radiation, 309–310
Current transfer ratio, 163, 238, 319, 324

D

Data interface, optical sensors, 227–229
DC input AC output, solid-state relays, 170–171
DC input DC output, solid-state relays, 169–170
Deep N+, in processing photointegrated circuits, 102–103
Deep UV, 290, 304, 319
Definition of visible-light-emitting diodes, 269
Degradation
IRLED packaging, 29
plastic, metal package comparison, 29
Degrees, for optical measurement, 278
DEMKO (Denmark), certification agency, 157
Denmark, certification agency, 157
Dermatology, instrumentation, UV application, 296
Diameter, fiber-optic communication, 196
Digital signals, fiber-optic communication, 196
Disinfection, UV application, 296, 312–313
Document validation applications, ultraviolet electromagnetic radiation, 309–310
Dollar bill changer, optical sensors, 250
Dominant wavelength, 275, 291, 296–297, 335
Driving light-emitting device, 211–214
linear operation, 213–214
variables, 211–213
Dual configuration, transmissive optical switch, 138
Duty cycle, 28, 31, 38, 47, 133, 198, 205, 207, 319
DVD discs, optical sensors, 249–250

Index

E

Electrical circuit analogy, 183–184
Electrical isolation systems, optical sensors, 255
Electrical measurements, 275–279
Electronic billboards, visible-light-emitting diodes, 287–288
Emission angle, 31, 199, 319
Emitter assemblies, 179
Encoder function, single-channel slotted switch, 131
Encoder wheel design, transmissive optical switch, 131
End of travel function, single-channel slotted switch, 131
Entertainment lighting, visible-light-emitting diodes, 288
Epi reactor (epitaxial reactor), in processing photointegrated circuits, 102
Epitaxial growth, 7
Epitaxy, 7, 9, 12, 15, 270, 320–321
EPROM erasure, UV application, 296
EPROM ICS reprogramming, UV application, 296

F

Fabrication of light-emitting diodes, 7–16
  epitaxial growth, 7
  gallium aluminum arsenide, 10–13
  gallium arsenide, 9–10
  hydride vapor-phase epitaxy, 14–15
  liquid-phase epitaxy, 7–10
  metal organic chemical vapor deposition, 13–14
  molecular beam epitaxy, 15–16
Fall time, 68, 93, 159–160, 320
Fiber-optic communication, 189–201
  advantages, 195–196
  capacity, 196
  cost, 195
  diameter, 196
  digital signals, 196
  nonconductivity, 196
  power, 196
  security, 196
  signal loss, 196
  weight, 196
  wireless communication, 197–201

Fiber-tip pen, black, reflective optical switch, 144
Finland, certification agency, 157
Flag configuration, transmissive optical switch, 138
Flag switches, 136
Flow configuration, transmissive optical switch, 138
Fluid sensing for medical applications, optical sensors, 257–260
Fluorescent lamps, 291–292
Flux, 70, 259, 275–276, 286, 320
Forensic applications, ultraviolet electromagnetic radiation, 296, 313
Forward voltage, determining junction temperature from, 184
Fountains, visible-light-emitting diodes, 288

G

Gallium aluminum arsenide, utilization in light-emitting diodes, 5
Gallium aluminum nitride, utilization in light-emitting diodes, 5
Gallium arsenide, utilization in light-emitting diodes, 5
Gallium arsenide phosphide, utilization in light-emitting diodes, 5
Gallium nitride, utilization in light-emitting diodes, 5
Gallium phosphide, utilization in light-emitting diodes, 5
Game controls, optical sensors, 249
Gas discharge lamps, UV, failure modes in, ultraviolet electromagnetic radiation, 295
Germany, certification agency, 157
Germicidal effects of UV light, 297–303
Graphite on white bond paper, reflective optical switch, 144

H

Hazardous fluid sensing, 232–235
  optical sensors, 232–235
Heat dissipation
  IRLED packaging, 29
  plastic, metal package comparison, 29
Heat generation, 182
Heat removal, 182–183
Hemodialysis equipment application, optical sensors, 257

H

Hermeticity
  IRLED packaging, 29
  plastic, metal package comparison, 29
High-power light-emitting diodes, junction temperature, 181–187
Household appliance controls, optical sensors, 254
Hydride vapor-phase epitaxy, 320

I

Illuminance, 274–275, 320
  optical measurement, 274–275
Incandescent bulb, 269, 320
Indium gallium arsenide, utilization in light-emitting diodes, 5
Indium phosphide, utilization in light-emitting diodes, 5
Industrial applications, optical sensors, 231–239
Industrial curing, UV application, 296
Industrial UV curing applications, 310–312
Infrared applications, 203–265
  automotive applications, 241–243
  business equipment applications, 227–230
  computer peripheral applications, 227–230
  consumer applications, 249–254
  driving light-emitting device, 211–214
  interfacing to photosensor, 215–225
  medical applications, 255–261
  military applications, 245–248
  optical sensors, 231–239
  pulse operation, 205–210
  telecommunications, 263–265
Infrared-emitting diodes, 1–57
  IRLED packaging, 17–57
  light-emitting diode fabrication, 7–16
  theory, 3–6
Infrared pulse operation, 205–210
Infusion pump application, optical sensors, 255–257
Intensity, optical measurement, 275
Interfacing to photosensor, 215–225
  CMOS, interface circuits for, 221
  photodarlington, 219–225
  photodiode, 215–218
  photointegrated circuit, 225
  phototransistor, 219–225
  TTL, interface circuits for, 222–225

Interrupter, 131, 249, 320
Intertek Group, certification, 157
Intravenous drop monitor, optical sensors, 260
Ionized aluminum, reflective optical switch, 144
IRLED, pulse operation, 205–209
IRLED packaging, 17–57
  characterization of packaged IRLED, 25–39
  chip centering, 29, 31–32
  degradation, 29
  heat dissipation, 29, 32
  hermeticity, 29, 35
  lens quality, 29–31
  measurement of radiant energy, 46–52
  measurement techniques, 51–52
  mechanical shock, 29, 35
  operating temperature, 29, 34
  optical consideration, 35–39
  package cost, 29
  package height, 29–30
  package side emission, 28–29
  parameter definitions, 51–52
  photon emission efficiency, 19–21
  plastic, metal package comparison, 29
  radiant energy measurement, 51–52
  reliability, 32–33, 52–57
  side emission, 29
  solvent resistance, 29, 34–35
  storage temperature, 29
  storage temperature range, 34
  temperature shock, 29
  thermal impedance, 39–46
  thermal impedance calculations, 39–46
  thermal shock, 34
  vibration, 29, 35
Isolation, in processing photointegrated circuits, 102

J

Junction temperature, 39–44, 181–187
  accelerated degradation, 182
  high-power light-emitting diodes, 181–187

K

Keyboards, optical sensors, 227
Kodak diffuse card, reference reflective surface, reflective optical switch, 144

Index

L

L configuration, transmissive optical switch, 138
Laboratory testing, visible-light-emitting diodes, 279–283
Landscaping, visible-light-emitting diodes, 287
Lateral pnp, 109
LCD applications, visible-light-emitting diodes, 287
Leak detection, UV application, 296
LEDs. *See* Light-emitting diode
Lens quality
    IRLED packaging, 29
    plastic, metal package comparison, 29
Lifetime, visible-light-emitting diodes, 283–284
Light-emitting diode fabrication, 7–16
    epitaxial growth, 7
    gallium aluminum arsenide, 10–13
    gallium arsenide, 9–10
    hydride vapor-phase epitaxy, 14–15
    liquid-phase epitaxy, 7–10
    metal organic chemical vapor deposition, 13–14
    molecular beam epitaxy, 15–16
Light generation, visible-light-emitting diodes, 273–274
Linear operation, driving light-emitting device, 213–214
Liquid-phase epitaxy, 7–10
    gallium aluminum arsenide, 10–13
    gallium arsenide, 9–10
Luminance, 121, 274–275, 320
    optical measurement, 275
Luminous efficiency, 278–279

M

Market segments, UV application, 296
Materials utilized for infrared-emitting diodes, 5
Mechanical aids, optical sensors, 236–238
Mechanical shock
    IRLED packaging, 29
    plastic, metal package comparison, 29
Medical applications, 296
    bubble sensing, 257–260
    electrical isolation systems, 255
    fluid sensing, 257–260
    hemodialysis equipment application, 257
    infusion pump application, 255–257
    intravenous drop monitor, 260
    optical sensors, 255–261
    pill-counting systems, 255
    pulse rate detection, 260–261
    ultraviolet electromagnetic radiation, 313
Metal, in processing photointegrated circuits, 105–106
Metal organic chemical vapor deposition, 13, 321
Military applications, optical sensors, 245–248
Molecular beam epitaxy, 15, 321

N

N+, in processing photointegrated circuits, 104
N configuration, transmissive optical switch, 138
NEMKO (Norway), certification agency, 157
Newspaper with ink, reflective optical switch, 144
Nonconductivity, fiber-optic communication, 196
Nordic Certification Service, certification, 157
Norges Elektriske Materiallkntroll, certification, 157
Norway, certification agency, 157

O

Open air, fiber-optic communication, 189–201
    fiber-optic communication, 191–196
    wireless communication, 197–201
Operating life, 33, 47, 57, 292, 308, 311, 321
Operating temperature
    IRLED packaging, 29
    plastic, metal package comparison, 29
Optical axis, 317–318, 321
Optical couplers/isolators, 238–239
    optical sensors, 229–230, 238–239
Optical coupling, 162, 321
Optical golf game, optical sensors, 254
Optical isolator, solid-state relay, 153–187
    electrical considerations, 155–175
    mechanical considerations, 177–187
    thermal considerations, 177–187

Optical measurements, visible-light-emitting diodes, 274–279
Optical sensors, 238, 249–261
  automotive applications, 241–243
  military applications, 245–248
  telecommunications, 263–265
Optical switch, transmissive, 125–139
  configurations, 138
  dual configuration, 138
  encoder function, single-channel slotted switch, 131
  encoder wheel design, 131
  end of travel function, single-channel slotted switch, 131
  flag configuration, 138
  flag switches, 136
  flow configuration, 138
  L configuration, 138
  N configuration, 138
  object recognition function, single-channel slotted switch, 131
  P configuration, 138
  performance characteristics, 131–136
  side configuration, 138
  single-channel slotted switch functions, 131
  slotted optical switch, 125–131
  T configuration, 138
  TS configuration, 138
  wide configuration, 138
Optimization of photodiodes, 64–66
Optocoupler, 161, 163–166, 169, 321–322
Optoisolators, 179–180
  discrete components, 180
Overcurrent surge protection, solid-state relays, 173
Overvoltage surge protection, solid-state relays, 173

P

P+, in processing photointegrated circuits, 103
P configuration, transmissive optical switch, 138
P-N junction injection electroluminescence, 3–5
Package cost
  IRLED packaging, 29
  plastic, metal package comparison, 29

Package height
  IRLED packaging, 29
  plastic, metal package comparison, 29
Package lens effects, phototransistors, 97–98
Packaging, visible-light-emitting diodes, 271–272
Packaging IRLED, 17–57
  characterization, 25–39
  chip centering, 29, 31–32
  degradation, 29
  heat dissipation, 29, 32
  hermeticity, 29, 35
  lens quality, 29–31
  mechanical shock, 29, 35
  operating temperature, 29, 34
  optical consideration, 35–39
  package cost, 29
  package height, 29–30
  package side emission, 28–29
  parameter definitions, 51–52
  photon emission efficiency, techniques for improving, 19–21
  plastic, metal packages compared, 29
  radiant energy measurement, 46–52
  reliability, 32–33, 52–57
  side emission, 29
  solvent resistance, 29, 34–35
  storage temperature, 29
  storage temperature range, 34
  temperature shock, 29
  thermal impedance, 39–46
  thermal impedance calculations, 39–46
  thermal shock, 34
  vibration, 29, 35
Parameter measurement, visible-light-emitting diodes, 274–279
Passivation, in processing photointegrated circuits, 106
Passive thermal management, 184–185
Passport-stock certificate-currency, UV application, 296
Peak wavelength, 39–40, 48–49, 71, 108–109, 181, 275, 319, 321
Pedestrian signals, visible-light-emitting diodes, 288
Phase, 11, 16, 116, 131–132, 294, 321
Phosphide, utilization in light-emitting diodes, 5
Phosphor, 181–182, 269, 272–274, 291, 320–321

Index

Phosphor coatings, visible-light-emitting diodes, 272–273
Photodarlington
 characterization, 80–98
 package lens effects, 97–98
 photosensor interfacing, 219–225
 switching characteristics, 82–97
 theory, 75–80
Photodetector, 52, 322–323
Photodiodes
 characterization, 66–73
 optimization, 64–66
 photoelectric effect, 61–64
 photosensor interfacing, 215–218
 theory, 61–64
Photoelectric effect, photodiodes, 61–64
Photointegrated circuits, 99–113*
 backside processing, 106–107
 buried layer, 101–102
 capacitor, 110
 characterization, 110–113
 contact, 105
 deep N+, 102–103
 Epi reactor (epitaxial reactor), 102
 isolation, 102
 lateral pnp, 109
 metal, 105–106
 N+, 104
 other IC circuit devices, 107–110
 P+, 103
 passivation, 106
 photodiodes, 108–109
 photosensor interfacing, 225
 processing, 101–107
 resistors, 104–105
 starting wafer, 101
 theory, 99–100
 vertical pnp, 110
Photometry, 274, 322
Photopolymerization, 310, 322
Photorepair, 298, 322
Photosensor assemblies, 179
Photosensor interfacing, 215–225
 CMOS, interface circuits for, 221
 photodarlington, 219–225
 photodiode, 215–218
 photointegrated circuit, 225
 phototransistor, 219–225
 TTL, interface circuits for, 222–225

Photosensor pulse operation, 209–210
Phototherapy, 289–290, 296–297, 313, 322
Phototransistors, 75–98
 characterization, 80–98
 package lens effects, 97–98
 photosensor interfacing, 219–225
 R_{BE} phototransistor, 79–80
 switching characteristics, 82–97
 theory, 75–80
Pill-counting systems, optical sensors, 255
Point source, 17–18, 36, 321–322
Pools
 UV application, 296
 visible-light-emitting diodes, 288
Power, fiber-optic communication, 196
Power derating, 45, 213, 323
Printers, optical sensors, 227
Pulse operation, 205–210
 infrared applications, 205–210
 IRLED, 205–209
 photosensor, 209–210
Pulse rate detection, optical sensors, 260–261

R

Radiant efficiency, 278
Radiant energy measurement, IRLED packaging, 51–52
Radiant incidence, 70–71, 323
Radiant intensity, 30–32, 50, 52–53, 323–324
Radiation pattern, 51, 231, 271, 279, 323
R_{BE} phototransistor, 79–80
Red smooth plastic surface, reflective optical switch, 144
Reduced MTTF, 182
Reflective materials, reflective optical switch, 144
Reflective optical switch, 141–152
 electrical considerations, 141–151
 mechanical considerations, 151–152
Reliability, visible-light-emitting diodes, 283–286
 factors affecting, 284–286
 operation, 283–284
Reprogramming EPROM ICS, UV application, 296
Resistors, in processing photointegrated circuits, 104–105
Responsivity, 65–66, 69–72, 323

Restriction of use, hazardous substances, ultraviolet electromagnetic radiation, 295
Retail display, visible-light-emitting diodes, 287
Reverse bias, 64–66, 68, 81, 216, 323
Robotics, optical sensors, 236–238
RoHS. *See* Restriction of use, hazardous substances
Rotary encoder, 131, 254, 323

S

Safety-related optical sensors, 231–232
  optical sensors, 231–232
SDDs. *See* Synchronous driver detector
Secondary optics, 273, 323
  visible-light-emitting diodes, 273
Security
  fiber-optic communication, 196
  UV application, 296
Security systems, optical sensors, 235–236
SEMKO (Sweden), certification agency, 157
SGS FIMKO (Finland), certification agency, 157
Side configuration, transmissive optical switch, 138
Side emission
  IRLED packaging, 29
  plastic, metal package comparison, 29
Signage billboards, visible-light-emitting diodes, 287–288
Signal loss, fiber-optic communication, 196
Silicon photosensors, 59–121
  photodarlington, 75–98
  photodiode, 61–73
  photointegrated circuit, 99–113
  phototransistors, 75–98
  special-function photointegrated circuits, 115–121
Single-mode fiber, 195
Skin, instrumentation, UV application, 296
Slot machines, optical sensors, 253–254
Slotted optical switch, 125–131
Slotted switch, 50, 125, 127–131, 136, 138, 235, 251–252
Smoke detectors, optical sensors, 251–253
Solid-state relays, 167–168
  AC input AC output, 171–172
  applications, 168
  DC input AC output, 170–171
  DC input DC output, 169–170
  optical isolator, 153–187
    electrical considerations, 155–175
    mechanical considerations, 177–187
    thermal considerations, 177–187
  overcurrent surge protection, 173
  overvoltage surge protection, 173
  temperature, 172–173
  theory of operation, 168
  zero voltage crossover, 173–175
Solvent resistance
  IRLED packaging, 29
  plastic, metal package comparison, 29
Spas
  UV application, 296
  visible-light-emitting diodes, 288
Special-function photointegrated circuits, 115–121
  color sensors, 120–121
  synchronous driver detector, 118–119
  theory, 115
  triac driver photosensors, 115–118
Spectroradiometer, 274–275, 283, 320, 324
SSRs. *See* Solid-state relays
Stage lighting, visible-light-emitting diodes, 288
Starting wafer, in processing photointegrated circuits, 101
Stock certificate-currency, UV application, 296
Storage temperature, 29, 34, 324
  IRLED packaging, 29
  plastic, metal package comparison, 29
Supply voltage, 82, 170, 210, 324
Surveillance systems, optical sensors, 235–236
Sweden, certification agency, 157
Switching characteristics, phototransistors, 82–97
Synchronous driver detector, 118–119

T

T configuration, transmissive optical switch, 138
Telecommunications, optical sensors, 263–265

Temperature, solid-state relays, 172–173
Temperature shock
    IRLED packaging, 29
    plastic, metal package comparison, 29
Theory of operation, solid-state
    relays, 168
Thermal conduction, 34, 182, 308, 324
Thermal equilibrium, 183
Thermal measurements, visible-light-emitting
    diodes, 274–279
Thermal radiation, 183, 324
Thermal resistance, 42, 183–185, 187, 271,
    284, 308, 324
Touch screens, optical sensors, 227
Traffic signals, visible-light-emitting
    diodes, 288
Train lighting, visible-light-emitting
    diodes, 288
Transfer ratio, 163, 238, 319, 324
Transmissive assembly, 320, 323–324
Transmissive optical switch, 125–139
    configurations, 138
    dual configuration, 138
    encoder function, single-channel slotted
        switch, 131
    encoder wheel design, 131
    end of travel function, single-channel
        slotted switch, 131
    flag configuration, 138
    flag switches, 136
    flow configuration, 138
    L configuration, 138
    N configuration, 138
    object recognition function,
        single-channel slotted switch, 131
    P configuration, 138
    performance characteristics,
        131–136
    side configuration, 138
    single-channel slotted switch
        functions, 131
    slotted optical switch, 125–131
    T configuration, 138
    TS configuration, 138
    wide configuration, 138
Triac driver photosensors, 115–118
TS configuration, transmissive optical
    switch, 138
TTL, interface circuits, photosensor
    interfacing, 222–225

TV controls, optical sensors, 249
TVs, visible-light-emitting diodes, 287

U

UL (U.S.), certification agency, 157
Ultraviolet, visible technologies,
    267–313
    ultraviolet electromagnetic radiation,
        289–313
    visible-light-emitting diodes, 269–288
Ultraviolet doses, 300–302, 312–313, 324
Ultraviolet electromagnetic radiation,
    289–313
    air disinfection, 312–313
    applications, 296–297
    currency, 309–310
    document validation applications,
        309–310
    failure modes, UV gas discharge
        lamps, 295
    fluorescent lamp, 291–292
    forensic applications, 313
    germicidal effects, UV light, 297–303
    industrial UV curing applications,
        310–312
    medical applications, 313
    RoHS, 295
    ultraviolet LED, 303–305
    UV lamp aging, 294
    UV lamp types, 292–294
    UV packaging technology, 305–309
    UV spectrum, 289–291
    water surface disinfection, 312–313
    WEEE, 295
Ultraviolet LED, ultraviolet
    electromagnetic radiation, 303–305
Underwriters Lab, certification, 157
UV-A, 212, 289–290, 296, 303, 306–307,
    310–313, 322, 324
UV application, 296
UV-B, 212, 289–290, 296, 298, 313, 319,
    322, 324
UV-C, 212, 289–291, 293, 296–298, 309,
    311–313, 319, 324
UV gas discharge lamps, failure
    modes in, 295
UV intensity, 301, 324–325
UV lamp aging, ultraviolet electromagnetic
    radiation, 294

UV lamp types, 292–294
UV packaging technology, ultraviolet electromagnetic radiation, 305–309
UV spectrum, ultraviolet electromagnetic radiation, 289–291

V

Vacuum UV, 289–290, 325
Validation/security, UV application, 296
VDE (Germany), certification agency, 157
Verband Deutscher Elektrotechniker, certification, 157
Vertical pnp, 110
Vibration
   IRLED packaging, 29
   plastic, metal package comparison, 29
Visible, ultraviolet technologies, 267–313
   ultraviolet electromagnetic radiation, 289–313
   visible-light-emitting diodes, 269–288
Visible-light-emitting diodes, 179, 212, 227, 247, 253–272, 311, 325
   applications, 287–288
   chip, 269–271
   defined, 269
   generating light, 273–274
   laboratory testing, 279–283
   lifetime, 283–286
   operation, 283–284
   optical measurement, 274–279
   packaging, 271–272
   parameter measurement, 274–279
   phosphor coatings, 272–273
   reliability, 283–286
   secondary optics, 273
   thermal measurement, 274–279
Visible spectrum, 313, 324–325
VLEDS. *See* Visible-light-emitting diodes

W

Wafer fabrication, UV application, 296
Waste electrical and electronic equipment, ultraviolet electromagnetic radiation, 295
Water surface disinfection, ultraviolet electromagnetic radiation, 296, 312–313
WEEE. *See* Waste electrical and electronic equipment
Weight, fiber-optic communication, 196
Wells, UV application, 296
White bond paper, reflective optical switch, 144
White smooth plastic surface, reflective optical switch, 144
Wide configuration, transmissive optical switch, 138
Wireless communication, 197–201

Z

Zero voltage crossover, 170, 173, 325
   solid-state relays, 173–175